INTERDISCIPLINARY CONTRIBUTIONS TO ARCHAEOLOGY

Series Editor: Jelmer Eerkens, *University of California, Davis, CA, USA*
Founding Editor: Roy S. Dickens, Jr., *Late of University of North Carolina, Chapel Hill, NC, USA*

For further volumes:
http://www.springer.com/series/6090

Nicholas Tripcevich • Kevin J. Vaughn
Editors

Mining and Quarrying in the Ancient Andes

Sociopolitical, Economic, and Symbolic Dimensions

Springer

Editors
Nicholas Tripcevich
Archaeological Research Facility
University of California
Berkeley, CA, USA

Kevin J. Vaughn
Department of Anthropology
Purdue University
West Lafayette, IN, USA

ISSN 1568-2722
ISBN 978-1-4614-5199-0 ISBN 978-1-4614-5200-3 (eBook)
DOI 10.1007/978-1-4614-5200-3
Springer New York Heidelberg Dordrecht London

Library of Congress Control Number: 2012951615

Printed on acid-free paper

Springer is part of Springer Science+Business Media (www.springer.com)

Acknowledgment

As researchers, examining ancient mines and quarries in the Andes, we follow in the path of a number of scholars, especially geologists, who noted the traces of Prehispanic mining, and to Georg Petersen who combed the Peruvian landscape and the colonial literature for insights into ancient mining practices. We thank our contributors who produced excellent papers in a timely fashion and were a pleasure to work with. We would like to thank Teresa Krauss and Morgan Ryan of Springer for their direction through the publishing process. This volume benefited from the detailed comments of one anonymous reviewer and the insightful commentary from our two discussants Drs. Izumi Shimada and Richard Burger, and we thank them for their efforts.

Contents

Part III Metals

Part IV Discussion

Contributors

César Borie Departamento de Antropología, Universidad de Chile, Ñuñoa, Santiago, Chile

William E. Brooks Geologist, Reston, VA, USA

Richard L. Burger Department of Anthropology, Yale University, New Haven, CT, USA

Gabriel E. Cantarutti Department of Anthropology, University of Illinois at Chicago, Chicago, IL, USA

Luis Enrique Castillo Museo Nacional de Arqueología, Antropología e Historia, Lima, Peru

Daniel A. Contreras Institute for Ecosystem Research, Kiel University, Kiel, Germany

Jelmer W. Eerkens University of California-Davis, Davis, CA, USA

Valentina Figueroa Université de Paris 1 Panthéon-Sorbonne, Paris, France

Mark Golitko Field Museum of Natural History, Chicago, IL, USA

Benedikt Gräfingholt Institut für Archäologische Wissenschaften, Fach Ur- und Frühgeschichte, Bochum, Germany

Jean Louis Guendon Université de Provence, UMR 6636 du CNRS (Centre National de la Recherche Scientifique), Aix-en-Provence cedex 2, France

Donald Jackson Departamento de Antropología, Facultad de Ciencias Sociales, Universidad de Chile, Casilla, Santiago, Chile

John Wayne Janusek Department of Anthropology, Vanderbilt University, Nashville, TN, USA

Justin Jennings Royal Ontario Museum, Toronto, ON, Canada

Elizabeth Klarich Department of Anthropology, Smith College, Northampton, MA, USA

Carlos Lémuz Aguirre Sociedad de Arqueología de La Paz, Universidad Mayor de San Andrés, La Paz, Bolivia

Moises Linares Grados Arqueocare, Lima, Peru

Dennis Ogburn University of North Carolina at Charlotte, Charlotte, NC, USA

Jorge Olano Canales Nasca Mining Project, Universidad Nacional Federico Villareal, Ventanilla, Callao, Peru

Camila Oñate Departamento de Antropología, Universidad de Chile, Ñuñoa, Santiago, Chile

Félix Palacios Universidad Nacional San Andres, Arequipa, Peru

Markus Reindel German Archaeological Institute, Commission for Archaeology of Non-European Cultures, Bonn, Germany

Andrew Roddick Department of Anthropology, McMaster University, Hamilton, ON, Canada

Diego Salazar Departamento de Antropología, Facultad de Ciencias Sociales, Universidad de Chile, Casilla, Ñuño, Santiago, Chile

Hernán Salinas Departamento de Antropología, Facultad de Ciencias Sociales, Universidad de Chile, Casilla, Santiago, Chile

Carol A. Schultze Historical Research Associates, Seattle, Washington

Gabriela Schwörbel (deceased) Museo Nacional de Arqueología, Antropología e Historia, Lima, Peru

Izumi Shimada Southern Illinois University, Carbondale, IL, USA

Thomas R. Stöllner Deutsches Bergbau-Museum Bochum, Forschungsstelle Archäologie und Materialwissenschaften, Fachbereich Montanarchäologie, Bochum, Germany

Institut für Archäologische Wissenschaften, Lehrstuhl Ur- und Frühgeschichte, Bochum, Germany

Nicholas Tripcevich University of California – Berkeley, Berkeley, CA, USA

Hendrik Van Gijseghem Université de Montreal, Montreal, Quebec, Canada

Kevin J. Vaughn Department of Anthropology, Purdue University, West Lafayette, IN, USA

Verity H. Whalen Department of Anthropology, Purdue University, West Lafayette, IN, USA

Patrick Ryan Williams Field Museum of Natural History, Chicago, IL, USA

Willy Yépez Álvarez Proyecto Arqueológico La Real, Arequipa, Perú

Part I
Introduction

Chapter 1
An Introduction to Mining and Quarrying in the Ancient Andes: Sociopolitical, Economic and Symbolic Dimensions

Kevin J. Vaughn and Nicholas Tripcevich

Introduction

Geological resources have long contributed in significant ways to economic, social, political, and ritual life in Andean communities. From the first mobile forager groups to the vast Inca empire, raw materials from mineral resources such as architectural, lapidary, ornamental, and knappable stone; clay for pottery, mineral ores for prized metals such as gold, silver, and copper; minerals for pigments such as hematite, cinnabar, and manganese; and salt have all had a profound—if sometimes unacknowledged—role in the Andean world. While archaeologists have used a number of analytical techniques on the materials that people have procured from the earth, these materials all have one thing in common: they were extracted from a mine or quarry, and despite their importance, comparative analyses of mines and quarries have been exceptionally rare in the New World, especially in the Andes. The papers in the current volume focus on archaeological research at primary deposits of raw materials extracted through mining or quarrying in the Andean region (Fig. 1.1).

In recent decades, the study of ancient mining and quarrying has seen increased interest by archaeologists worldwide, and a number of synthetic treatises on mining have appeared, although many have an emphasis on European or Mediterranean prehistory (Abu-Jaber et al. 2009; Boivin and Owoc 2004; Brewer-La Porta et al. 2010; Craddock 1995; Craddock and Lang 2003; Ericson and Purdy 1984; La Niece and Craddock 1993; La Niece et al. 2007; Stöllner 2008b; Topping and Lynott 2005a).

K.J. Vaughn (✉)
Department of Anthropology, Purdue University,
700 West State Street, West Lafayette, IN, USA
e-mail: kjvaughn@purdue.edu

N. Tripcevich
University of California-Berkeley, 2251 College Bldg, UC Berkeley,
Berkeley, CA 94720-1076, USA
e-mail: nico@mapaspects.org

N. Tripcevich and K.J. Vaughn (eds.), *Mining and Quarrying in the Ancient Andes*,
Interdisciplinary Contributions to Archaeology, DOI 10.1007/978-1-4614-5200-3_1,

Fig. 1.1 Map showing locations by chapter

In comparison, the study of mining and quarrying has not been a topic of significant archaeological discourse in the New World. Indeed, detailed study of ancient mining has been lacking in the Andes, and there are few synthetic treatments of the topic. One exception is Cruz and Vacher's (2008) volume presenting recent research on colonial and precolonial mining and metallurgy in the southern Andes, with a particular emphasis on the rich evidence from Potosí. Beyond this significant work, there have been few attempts to evaluate mines and quarries in the Andes from a synthetic and comparative perspective. In the 1990's, Craig and West (1994) published a valuable treatise on mining and metallurgy in the New World in their volume "In Quest of Mineral Wealth: Aboriginal and Colonial Mining and Metallurgy in Spanish America." While much of the focus in this volume was on post-conquest mining and metallurgy, several important papers including those of Shimada and Ramirez evaluated the topic from archaeological and ethnohistoric perspectives, respectively. Shimada's paper has proven to be especially valuable to scholars studying Prehispanic mining and metallurgy in the Andes, as it provided a comparatively early effort to evaluate actual mines in the archaeological record. Another exception has been the work of Núñez (2006) who has recorded and worked at many mining sites in northern Chile and is one of the few researchers in the Andes to have a long-term research program considering these archaeological sites.

Earlier treatises on mining and metallurgy in the Andes include Lechtman's (1976) "A metallurgical site survey in the Peruvian Andes." In this seminal work, Lechtman surveyed on the North Coast of Peru and the Titicaca Basin focusing on sites (both mining and smelting sites) related strictly to metallurgy. The work of Georg Petersen (1970), recently translated into English by Brooks (Petersen 2010), was a comprehensive effort to describe known metallurgical and mining sites in the Andes; however, the sites were not archaeologically evaluated by the author.

These efforts laid the foundation for the research that is presented in this volume. Lechtman paved the way for future scholars by asking fundamental questions about mining such as "who were the miners?" and "how was mining organized?," while Petersen demonstrated the ubiquity of mining and metallurgy sites in the Andes. Focusing on North Coast data and sites in northern Chile, Shimada and Nuñez, respectively, provided actual evidence from ancient mines and attempted to place these sites into a broader archaeological and Andean context. The papers presented in this volume attempt to build on these original studies to bring the topic of mining and quarrying in the ancient Andes to the forefront by evaluating primary evidence for these activities.

On Studying Mining and Quarrying in the Archaeological Record

Using these preceding efforts as a springboard, our goal in preparing this volume was to bring together a number of archaeologists investigating the primary evidence for mining and quarrying in the Prehispanic Andes. Beyond this, however, previous efforts to explore this topic (including our own research) had revealed that mining was deeply embedded in Andean life and had significant dimensions beyond the strictly economic. We therefore felt it was imperative that the scholars who contributed to our volume made an attempt to go beyond simply documenting primary evidence for raw material extraction. Indeed, we asked them to consider how this activity was embedded in other aspects of Prehispanic Andean life. Thus, our volume's principal goal was to place the activity of mining and quarrying into an archaeological and Andean context, rather than aiming to be geographically and chronologically comprehensive. Even so, the papers in this volume cover Early Holocene hunter gatherers through the Inca Empire and geographically span from Central Ecuador through Chile including coastal and highland Peru and the Titicaca Basin. The geographic and temporal scope of these papers is consistent with the pervasiveness of mining-associated activities for many people in the Andean past.

While in theory evidence for ancient mines should be exceptionally rare because of contemporary (and even Prehispanic) mining activities (see Eerkens et al. 2009; Stöllner 2009), the papers in this volume demonstrate that ancient mines, while certainly not common, are nevertheless present in the archaeological record. Perhaps not surprisingly, one exception is that we still lack direct evidence for ancient gold mines. Gold as a raw material was clearly important for Prehispanic cultures including the Inca Empire, but nothing matched the Spanish appetite for the metal. The acceleration of gold mining in colonial and recent times has surely obliterated most traces of ancient gold mining from the archaeological record, and we still lack direct evidence for this past activity in many regions. Petersen (2010: 23–24) does list several locations in the Andes where there have been reported evidence for pre-Colonial mining of gold including Cotabamba, Cajamarca, the south coast, Pataz, and Paucartambo. Despite this, the direct evidence of mining is mostly unsubstantiated.

Dimensions of Mining and Quarrying

Of course, mining is only the first stage in what is a long and complex *chaîne opératoire*. As important as evaluating how materials were mined is determining how they were processed and then eventually used in production. This is best exemplified by Shimada's work on mining and metallurgy, work that he calls the "holistic approach" to craft production (Shimada 2007). While we recognize that it is difficult to separate any of these "stages" of craft production (from extraction to production), our focus in this volume is on a comparative view of the extraction of raw materials that we believe has been largely overlooked in Andean archaeology. We therefore structure the remainder of our discussion around several themes that are explored by the papers in this volume (1) the raw materials mined and the problems miners faced when extracting those materials, (2) the processing and transport of raw materials, (3) the symbolic dimensions of mining and quarrying, and (4) the organization of Prehispanic mining and how mining was embedded into other realms of Prehispanic life.

The Raw Materials: Technological Considerations

Many previous treatises of our topic have focused almost exclusively on the materials necessary for metallurgy. However, in its broadest sense mining includes the various processes for extracting solid resources from the earth's crust and it comprises a variety of procurement methods including quarrying, tunneling, trenching, and placer mining. Thus, excluding a consideration of non-metallurgical materials is rather arbitrary. Indeed, virtually all crafts in the Prehispanic Andes (with the notable exception of textiles[1]) required mining or quarrying of raw materials. These raw materials included stone of all kinds including lapidary, architectural, sculptural, knappable, and ornamental stone; minerals for pigments such as cinnabar (HgS), hematite or red ochre (Fe_2O_3), yellow ochre or limonite ($FeO(OH)\cdot nH_2O$), manganese; clays and tempers; salt; and the by-now-familiar metals of copper, silver, and gold. Each of the papers in this volume considers at least one of these raw materials in their analysis.

Regardless of the materials that originate in mines and quarries, many of the technological considerations that an ancient miner must face are shared. One must decide where best to dig and which minerals are worth recovering and processing. There are broad engineering issues such as preventing mines from collapsing, contending with water seepage, reducing poisoning effects on miners, and finding tools suitable for digging in the surrounding matrix. Are these minerals combined with other materials in order to be useful? How are the target minerals to be transported away from the source area, and how does one conduct some form of processing adjacent to the source? Mining requires a sophisticated knowledge of the properties of raw materials as Roddick and Klarich's paper in this volume (Chap. 5) clearly shows, but as their paper and other chapters demonstrate, considerations beyond geological or technical properties were oftentimes critical in the selection of raw material. For Janusek and colleagues (Chap. 4), a change in raw material from sandstone to andesite at Tiwanaku between AD 500 and AD 700 underscores the importance of the materiality of stone for articulating a broad suite of changes as Tiwanaku grew into an expansive urban center. Their paper provides original data identifying the primary sources of sandstone and andesite at Tiwanaku and notes that the shift in material type corresponds with a larger change in color and durability from sandstone to andesite. Further, the spatial provenance of the andesite, originating at a more distant volcanic peak, corresponds with the expanding sphere of Tiwanaku influence.

Another technological consideration was the accessibility and composition of an ore deposit (see Stöllner 2008a). In the quarrying of architectural stone and material for tool production, size and material properties of the stone are limiting factors. The tools involved in extraction are sometimes found in archaeological contexts, with one of the most well-known cases being the Chuquicamata miner's tool set

---

[1] While mining was not required for the extraction of the raw materials used in weaving (e.g., cotton and wool), textile production often needed colorants and tools (e.g., for cutting and shearing) requiring initial mining of raw materials.

(Bird 1979: Figs. 3 through 10; Craddock et al. 2003). Indeed, as the papers here by Salazar and colleagues, Vaughn and colleagues, and Cantarutti (Chaps. 7–9) demonstrate, one of the most common artifacts found at well-preserved ancient mines are the stone, and sometimes wooden, tools used to extract raw material from mines themselves (see also Salazar et al. 2011; Vaughn et al. i.p.).

One theme apparent in many of the papers in this volume is that mining technology appears to have remained fairly simple (and conservative) through time in the Prehispanic Andes. Hard minerals in the Andes were extracted using a remarkably similar technology. Stone tools from the Peruvian north coast (Shimada 1994) to the south coast (Chaps. 8 and 14), to Chile (Chaps. 7 and 12) all bear a resemblance to each other, and all bear resemblance to the Chuquicamata "Copper Man's" tool kit referenced above (Bird 1979: Figs. 3 through 10; Craddock et al. 2003). Given that Salazar et al. (2011) have recorded very similar tools at an early Holocene mine in North-Central Chile (see also Chap. 7), the geographic and temporal scope of this toolkit is quite impressive. Today, much of the toolkit of the itinerant miner consists of iron tools (that replicate the stone tools used in the past) and sometimes dynamite (see Eerkens et al. 2009; Chap. 8).

While the toolkit and mining technology appeared to be similar across a broad swath of the Andes, a range of techniques were employed to allow access to deposits, prevent erosion or collapse, and reduce the risk associated with mining. In their paper, Jennings and colleagues (Chap. 6) describe structural columns in a salt mine that are today painted with a skull and crossbones. There are examples of mines for building stone that were highly dangerous due to roof collapses and only by using modern machinery they are safely exploited as open-cast mines (Stöllner 2008a). Among the earliest mining evidence in the new world, it was found that lateral buttressing allowed for the exploitation of trench mines (Salazar et al. 2011; see also Chap. 7).

One final consideration is that mines often require regular maintenance and with neglect shafts and quarries may become blocked by collapsed or colluviated material, or by material that has simply been moved by previous miners. Such concerns may bracket the viability of exploiting specific deposits, particularly surface works subject to erosion, because when a mine is abandoned the effort involved in restoring the location to productivity may be prohibitive.

From Source to Region

Once materials are mined, they are either processed at the site as shown in the papers by Vaughn and colleagues (Chap. 8), Cantarutti (Chap. 9), Salazar and colleagues (Chap. 12), and Van Gijseghem and colleagues (Chap. 13), or the materials need to be moved to other sites for processing. This is especially the case when the raw material is cumbersome or difficult to manage (an obvious case being architectural stone). Evidence for the porting of mined material from the quarry site to processing areas or onwards to the sites where the material was utilized is sometimes apparent archaeologically, as shown in the papers by Tripcevich and Contreras

(Chap. 2), Ogburn (Chap. 3), Janusek and colleagues (Chap. 4), and Salazar and colleagues (Chap. 12). Additionally, the determination of where the material was utilized may need to come from further geochemical studies as Tripcevich and Contreras (Chap. 2) discuss in the use of obsidian throughout the Andes, Janusek and colleagues (Chap. 4) show in the use of andesite and sandstone in Tiwanaku, Vaughn and colleagues (Chap. 8) show in the use of hematite in Nasca, and Reindel and Stöllner (Chap. 14) show in the use of obsidian and copper in the Palpa region.

While movement of extraneous and heavy raw material was generally avoided by initial processing near the geological source area, there are exceptions to this pattern. Architectural blocks may be fashioned into rough approximations, although in anticipation of some damage during transport, the fine dressing occurred in the immediate vicinity of the final placement of the stone (Protzen 1983: 185). Obsidian nodules of predictably high knapping quality appear to have often been transported with the cortex intact, as edges of stone tools may be fractured during travel and the cortical flakes may be in themselves useful, sharp cutting implements (Beck et al. 2002; Tripcevich and Contreras 2011). Large and heavy materials like architectural stone particularly require road construction, stream crossings, and other facilities (Protzen 1983), though depending on the level of infrastructure in place (and state apparatus), even relatively portable materials such as copper ores and chrysocolla may have required road building as Salazar and colleagues (Chap. 12) and Shimada (1994) demonstrate.

On an economic level, the characteristics of the material being extracted largely determine the geography of mine siting. The specific mine locations and adjacent processing areas may be preferentially located close to facilities like grazing for herds and sheltered locations for residence, or along routes used regularly by pastoralists (Stöllner 2008a: 71; Tripcevich and Contreras 2011; Tripcevich and Mackay 2011). Salazar and colleagues (Chap. 12) offer compelling evidence for mine siting to be determined, on the contrary, by ritually determined geographical concerns. Indeed, one of the commonalities that we found in many of the papers is that authors are beginning to acknowledge that mines were sacred in the Andes as Shimada (1994) and Nuñez (1999) have pointed out before. We return to this element of mining in further detail below.

Mineral Essence and the Symbolic Dimensions of Mines in the Andes

The power inherent in material from the earth to simultaneously contain physical presence, social linkages to places where mining occurs, and sacred power derived from an animated landscape brings the materiality of mined substances to foreground in many of these studies (Topping and Lynott 2005b). Underlying this is a perspective that humans inhabit a landscape with a moral obligation to relationships not just to other people, but towards the lands within which they dwell, the surrounding mountains, and the entities that permeate the environment (Ingold 2006, 2011). The presence of shrines and offerings at mining sites serves as empirical evidence for their ritual significance, but at a basic level the discovery and exploitation of the mine may have been intertwined with a conception of geological formation processes that diverges from current scientific understanding. Is gold or copper found in a particular seam because it lies in a cosmologically auspicious location (despite our tendency to infer the reverse)? In Chap. 3, Dennis Ogburn discusses the evidence for quarry areas being revered as *huacas,* but further suggests that stones were quarried in some contexts because they were adjacent to pre-existing shrines at sacred landforms. In Chap. 12, Salazar and colleagues find similar links to regional cosmological patterns in the siting of Inca mine features in the Atacama region. While empirical evidence in the form of offerings or ritual structures are important to support claims of sacred power at mine sites, the essential power of the material may derive from its location or other characteristics. Citing ethnohistoric sources, Dean (2010) explores the transubstantial nature of sacredness in stone in the Inca empire. "The challenge here is to understand rocks not as a mineral matter of variable composition that the Inca and other Andeans mistakenly (or even charmingly) endowed with life force, but as ancient Andeans saw them—potentially animate, transmutable, powerful, and sentient." (Dean 2010: 5). While not *all* rocks were considered sacred, Dean argues that the Inca recognized that some rocks were simultaneously standard mineral composites and potentially imbued with power, making them much more than simple rocks. While Dean's book draws principally on ethnohistoric accounts of the contact period and focuses principally on architectural stone among the Inca, it provides an insight into the richness of Prehispanic relationships with the mineral world.

Indeed, ethnohistoric and ethnographic sources demonstrate that mines and quarries were often considered sacred in the Andes. Cobo and Jiménez de la Espada (1890 [1653]) described mines as *huacas* and ethnographic work by sociocultural anthropologists (discussed below) have shown that this conception is still pervasive. Many studies contend that the positioning of mines on the landscape and the act of mining often has ritual and symbolic dimensions that may not be a direct reflection on the geological deposit or the most economical way of extracting the material.

For the Inca, mined materials including metals and ores were considered a product of the earth, analogous to harvested crops (Nuñez 1999). People could "intervene in their emergence to the surface of the earth by mining, thus influencing the sacred power that had engendered them" (Berthelot 1986: 82). Today mines are still considered sacred in the Andes (Nash 1979). The most well-known ethnographic example of the symbolic importance of mines is in Oruro, Bolivia where, despite conquest by the Spanish and centuries of forced labor with institutional attempts to destroy belief systems (e.g., see Nash: 136 and Taussig 1980: 144), miners believe they penetrate the world of *Supay* (sometimes referred to as *Hahuari*, *Huari*, or in Peru, *Muqui* (Salazar Soler 2006), often translated as the "spirit of the hills," and called the "Devil" or "Tío" by the Spanish) when entering mines. Supay (or Hahuari in this telling of the legend) was a

> powerful ogre…believed to live in the hills and was identified with the "Devil" or "Uncle" of the mines. It was he who persuaded the people to leave their work in the fields and enter the caves to find the riches he had in store. They abandoned the virtuous life of tilling the

> soil and turned to drinking and midnight revels paid for by their ill-gained wealth from the mines...Hahuari lives on in the hills where the mines are located, and is venerated in the form of the Tio, or Devil, as the owner of the wealth of the mines. Llama pastoralists say they have seen him at night carrying the mineral on teams of llamas and vicunas into the mines where the animals deposit it and where it is found by the miners, who give their thanks in offerings of liquor, cigarettes and coca. The Tio controls the rich veins of ore and reveals them only to those who give him offerings
>
> (Nash 1972: 223–224).

Supay is propitiated with several kinds of rituals. This was most famously expressed by a Bolivian miner who said to June Nash in her ethnographic fieldwork: "We eat the mines and the mines eat us. For that reason, we have to give these rituals to the spirit of the hills so that he will continue to reveal the veins of metal to us and so that we can live" (Nash 1979: ix). In one ritual called *K'araku*, offerings of a live llama are made to *Supay* for luck and to gain his good will (Nash 1979: 123). These rituals take place inside large gallery mines; in smaller mines, the sacrifice is offered at the mine's entrance. Other rituals called *Ch'alla* involve offerings of liquor, coca, cigarettes, the playing of music on a *chiranga* or guitar, and dancing within the mines (Nash 1979: 137). Clearly, these rituals incorporate elements of Catholicism and are in part due to the legacy of oppressive colonial mining (Nash 1979). We are certainly not suggesting that these provide direct analogies for possible rituals that took place in the Prehispanic Andes; however, the practice is at least consistent with Andean beliefs toward the sacred with roots in the Prehispanic period (Gil García and Fernández Juárez 2008: 106; Harris 2000; MacCormack 1984; MacCormack 1991).

Despite these associations, the remains of ancient rituals and offerings are found, though infrequently, in connection with Prehispanic Andean mines. One example was reported by Shimada (1994: 54) who found a cache of *Spondylus princeps* shells in structures associated with the Cerro Mellizo copper mines. Shimada argues that this cache along with other artifacts strongly suggest evidence for the symbolic importance of this particular mine. Others are reported here for the first time. Vaughn and colleagues as well as Salazar and colleagues (Chaps. 8 and 12) report finding *Spondylus princeps* shells in excavations of mines and their associated infrastructure. For example, Salazar and colleagues recovered spondylus remains in platform complexes at San Jose del Abra that they interpret to indicate that El Abra was a *huaca* in Prehispanic times. For their part, Vaughn and colleagues find the remains of finished and unfinished Spondylus within Mina Primavera and suggest that the mine itself had ritual importance and that ritual practice likely occurred within the mine.

Beyond evidence for ritual taking place within mines, several of the papers presented here suggest that archaeological evidence supports the notion that the essence of the mined materials had symbolic importance. For example, in Chap. 4, Janusek and colleagues find evidence for the primacy of stone in shaping native identities in the contemporary Tiwanaku region and argue that the shift from sandstone to andesite resonated with the broad political and religious transformation of Tiwanaku into an expansive polity. The ceremonial importance of minerals excavated from the earth is evident from the earliest times. As described by Salazar and colleagues (2011), and expanded upon here in Chap. 7, the hematite mine of San Ramón 15 near Taltal in the Antofagasta region of coastal Chile contains the oldest evidence of mining in the New World with hammerstones found in layers dated to over 12,000 cal. BP. While hematite is known to have practical utility (e.g., in hide processing), the clearest local use is in ritual practices where hematite was used in abundance in an Archaic period mortuary tradition that even had resonance 10 millennia later in the Colonial period where ochre continued to be used in burials.

In Chap. 6, Jennings and colleagues report several chunks of salt possibly from the Huarhua salt mine found in tombs from sites 20 and 30 km away from the mine. While the evidence is not direct, the authors believe that their placement in tombs at least raise the possibility of the symbolic significance of the salt. Cantarutti in his paper (Chap. 9) suggests evidence for Inca state ritual on a ceremonial platform constructed at the summit of Mt. Los Puntiudos in Chile (including a possible *capacocha*) offered to ensure success of mining operations there. Salazar and colleagues (Chap. 12) describe three "administrative/ceremonial" sites in Atacama with common elements (Inca imperial pottery, part of sacred geography, and all associated with administration of mining centers).

Furthermore, when archaeologists find evidence for preferential use of particular materials, especially at some distance from the geological source, despite the availability of reasonable alternatives, these may be understood to represent social or essential ties to that source area. Regarding the Quispisisa obsidian quarry Tripcevich and Contreras (Chap. 2) reference the concept of "pieces of places" (Bradley 2000: 88) in recognition that non-local artifacts are not simply objects with their own history; these exotic goods may also impart references to people and places and to the power of the landforms where such materials are originally found. This notion is consistent with the Andean *pacarisca* considered by Ogburn in Chap. 3. He discusses quarries as sacred spaces [first identified as such by Cobo and Jiménez de la Espada (1890 [1653])] and that they were most likely worshipped as huacas. Ogburn also makes the important point echoed by many authors in this volume that if the quarries were *huacas*, their "essence" was incorporated into the major temples of the Inca Empire.

While mined materials are linked to the power of the landscape from which they derived, they may simultaneously communicate social links across space with regions where those items originate (Boivin 2004). Andean archaeologists have long used stylistic features to detect cultural affiliation across long distances, and stylistic motifs and exotic raw materials may both serve as vehicles for communicating social information across distance. However, while style communicates affiliation, motifs can be replicated using local materials. An item made from visually distinctive, exotic raw material may contain both the power inherent in the landscape, and be presented in a way that is palpable and communicative, and it cannot be reproduced using local materials if the use of distinctive exotics was crucial. Thus, access to particular materials and the circulation of goods may be tied to social identity, territorial rights, and an established relationship between social group and the material. In Chap. 5, Roddick and Klarich focus on the "social history of exploitation" of a material and to the persistence of particular pastes and tempers across centuries despite changes in environment that may cause clay sources to disappear. The social context and technological choice involved in ancient mining were likely primary reasons for procuring those materials, and it is also a central puzzle for archaeologists seeking to parse the episodes of use of a particular source. These issues are incorporated into a holistic approach to quarries advocated by Bloxam (2011). Her approach evaluates mines, mining facilities, and the distribution system (to the extent that they are visible and preserved) in seeking socially constructed landscapes of particular use episodes of ancient source areas.

The Organization of Prehispanic Mining in the Andes

Colonial mines such as the well-known Potosí silver mines of southern Bolivia (Bakewell 1984; Platt and Quisbert 2008) and the cinnabar mines of Huancavelica (Burger and Matos Mendieta 2002) were large, extensive operations controlled by the Spanish Crown. The character of mining and concomitant metallurgy obviously changed drastically when the Spanish crown took control of the production process (see Van Buren and Mills 2005: 4), and unfortunately chroniclers are not liberal in their descriptions of the actual organization of Prehispanic mining.

Certainly, the ubiquity of a particular raw material, and the ways in which it was distributed across a landscape, would have had an effect on how efforts to control the resource were mediated. For example, in Chap. 5, Roddick and Klarich demonstrate that clays and tempers—both ubiquitous and widely available in the Andes—may have been difficult if not impossible to control. Other resources that had restricted distribution, especially those related to metallurgy, were much more amenable to state control. Indeed, ethnohistoric documents suggest that the Inca Empire had a fairly high level of authority and administration over mines in Tawantinsuyu.

For example, it is clear that the Sapa Inca's personal property included mines and mining operations (D'Altroy 2002: 301). Additionally, the Inca monopolized the richest mines of the Empire such as the silver mines of Porco, Bolivia (Bakewell 1984; Barba 1923) and Tarapacá (in Iquique, Chile; see Brown and Craig 1994; Cobo 1979 [1653]; Zori and Tropper 2010) and also (Berthelot 1986), and that the mines were exploited through mit'a labor. But, the Inca did not have a claim to all mines as some were exploited under the auspices of local *caciques* so that they would have appropriate gifts to give to the Inca:

> ...Some of these mines were worked at the expense and under the auspices of the Inca himself, and others, constituting the majority, were worked at the expense of the caciques [lords] of the districts in which the mines were located. This was so that they would have things to give as presents to the Inca
>
> (Cobo 1979 [1653]: 249).

Because of this, the largest mines and the most productive sources were reserved for the Inca, while smaller "community" mines were scattered around the empire and had various levels of control (Berthelot 1986: 72). Even with this ostensible autonomy given to local lords, the Inca installed supervisors in the provinces who were responsible for monitoring mining and to collect and weigh ore (D'Altroy

2002: 301) and ultimately, the Inca had absolute authority when the size of the work force and the collection of the minerals were concerned (Berthelot 1986: 74).

Mining seems to have been a seasonal activity (Van Buren and Presta 2010). Rowe (1946: 246) states "(gold mines) were worked only 4 months a year…All the gold was taken by the government, which kept inspectors at the entrance to the mining area to see that none was stolen." In high altitudes, mining was done in the summer, while Berthelot (1986: 74) states that mining was done year round where the weather permitted (e.g., in Huánuco and probably at Carabaya). Polo [(1940 [1561]): 165; cited in Berthelot 1986: 75)] also states:

> When the Indians went to the mines, there were persons who accompanied them in order to collect the gold that they found, no matter how large or small the quantity, since they were solely obliged to supply their labor, and the Indians therefore did not even know how much gold had been amassed, and no one dared to take the smallest piece for himself.

While chroniclers offer some tantalizing clues as to the importance of mining as well as to the extent that mining was controlled by the Inca, direct evidence for how mining was organized in the Andes has been limited. Some argue that one reason for this is that mining was a small-scale itinerant activity (Lechtman 1976: 41). Indeed, small-scale itinerant mining is a pattern seen in many indigenous cultures throughout the world (e.g., Ballard and Banks 2003; Knapp et al. 1998). However, recent indirect evidence for mining in the Andes has suggested that in some contexts it (and related activities such as metallurgy) may have been large scale and possibly state controlled.

For example, in Chap. 11, Carol Schultze (also discussed in Schultze et al. 2009) argues based on excavations at a metallurgical center in the Titicaca Basin that metallurgy was complex and large scale prior to the Middle Horizon Tiwanaku and well before the Inca Empire. Abbot, Cooke and colleagues (Abbott and Wolfe 2003; Cooke et al. 2008, 2009) have argued from indirect evidence (i.e., levels of lead and other metals in lake sediments) for increased metallurgical activities that metallurgy—and by inference mining—increased after the Middle Horizon with the advent of the Tiwanaku state and the Wari Empire. Missing from these arguments, however, is actual evidence for ancient mining. Many of the papers presented here present direct evidence for the organization of mining in the archaeological record.

For one, many papers in this volume demonstrate that mines were clearly an important part of the political economy of states from Tiwanaku to the Inca. Salazar and colleagues (Chap. 12), for example, show that the Inca radically changed the organization of copper mining in the Atacama desert by adding administrative facilities and by controlling the distribution of food and water in the region. Furthermore, these drastic changes to the region's landscape were probably managed by large-scale, state-sponsored feasting as the authors present compelling evidence for this activity in the El Abra complex. In Chap. 9, Cantarutti also presents persuasive evidence for state organization of mining in Chile. While not providing direct evidence for state organized labor, in Chap. 3, Ogburn makes the point that cut stone—a critical imperial commodity—demanded organization of labor for extraction, reduction, and transportation that greatly exceeds the requirements of other materials. Van Gijseghem and colleagues (Chap. 13) suggest that the organization and control of mining in the Ica Valley peaked during the Late Intermediate Period when not only were populations at their highest, but polities were also at their most socio-politically complex. That these increases in social, political, and economic complexity correlated with an increasing concern for the control of the extraction of raw materials through mining is not surprising.

In contrast, Jennings and colleagues (Chap. 6) argue for a *lack* of state control in the organization of mining at the Huarhua salt mines. The disparity here may in fact be the differences in the raw material being mined. The demand for salt, while certainly highly valued and a necessity, created a different circulation pattern than that of copper and architectural stone. There may also be an element of "materialization" here that is inherent in the raw material being exploited. Copper, gold, pigment, and stone (to name a few) are raw materials used in crafts that can be materialized to convey a particular political or religious ideology (whether it is used to paint on ceramics, made into monuments, or crafted into luxury and ceremonial objects). Rather than being primarily a display item or emphasizing status differences, salt is a widely available—though irregularly distributed—necessity, like an "ordinary good" reflecting ethnic ties (Smith 1999). It was therefore likely to have circulated within networks moving staples and other basic goods.

It is important to note, however, that many of these minerals elude archaeologists' attempts at clear classification. While Huarhua salt extraction does not appear to have been organized by state institutions, Jennings and colleagues observe that salt seems to have been ascribed symbolic importance in certain times and places as it was found in Middle Horizon (ca. AD 600–1000) burial contexts in Cotahuasi. Similarly, Tripcevich and Contreras (Chap. 2) discuss the varied associations of obsidian, where fl discarded in middens and yet it also is found in ritual contexts and burials as early as 2000 BC. In sum, the materiality of a substance entangles many factors. We may generalize about the potential for control of a raw material source, and about the processing steps required, but strong evidence for the social and ceremonial importance of a given substance often derives from close attention to variability in the archaeological contexts in which it is found.

Furthermore, with smaller scale societies, not surprisingly, there is far less evidence for administrative control of activities in mines; one possible exception may be at Mina Primavera reported in Chap. 8 by Vaughn and colleagues. They link an intensification of mining and processing of hematite contemporaneous with the emergence of the Nasca polity that correlates with activities taking place at the region's ceremonial center to a demand for the mine's material. Another possible exception is with obsidian where Tripcevich and Contreras in Chap. 2 report on a number of sizable quarry pits at Quispisisa. These may be evidence of large-scale mining or of low levels of excavation sustained over millennia. Quantities of obsidian are exposed in gullies through natural erosion processes, which raises the question: *why dig quarry pits when it is abundantly available in gullies*? This leads the investigators to speculate that intensive procurement that may have occurred during particular times depleted the material exposed naturally in gullies, and hence the quarries are probably due to large-scale or state-sponsored mining when naturally available stone was unavailable.

Imperiled Resources

Ultimately, the papers presented in this volume demonstrate the importance of evaluating ancient mines and quarries in the Andes. These sites do not provide evidence merely of the economic and pragmatic aspects of Prehispanic life, they were important loci in the ancient Andes where sociopolitical, economic, and symbolic aspects of Prehispanic life in the Andes intersected.

As we stated earlier, ancient mines and quarries are exceptionally rare. With rapid development in the Andean countries coupled with an increase in transnational mining industries targeting raw material sources that have been exploited for millennia, we expect to see an increase in the destruction of ancient mines and quarries. One example of immediate concern is that a large portion of the river valley containing the Quispisisa obsidian source is slated to be flooded under a 2,500 million m^3 reservoir as part of the "Pampas Verdes" hydroelectric project (Electropampas 2003). It is this destruction of sites and the rarity of Prehispanic mines and quarries that make their study all the more urgent. Of course, we are not the first to declare urgency in the study of these sites (see, for example, Bloxam 2011 who also makes a broad appeal), but we hope that this volume is the first of many efforts to evaluate the importance of mines and mining in our study of the past.

References

Abbott, M. B., & Wolfe, A. P. (2003). Intensive pre-Incan metallurgy recorded by lake sediments from the Bolivian Andes. *Science, 301*(5641), 1893–1895.

Abu-Jaber, N., Bloxam, E. G., Degryse, P., & Heldal, T. (Eds.). (2009). *QuarryScapes: Ancient stone quarry landscapes in the Eastern Mediterranean*. Oslo: Geological Survey of Norway.

Bakewell, P. J. (1984). *Miners of the Red Mountain: Indian labor in Potosí 1545–1650.* Albuquerque: University of New Mexico Press.

Ballard, C., & Banks, G. (2003). Resource wars: the anthropology of mining. *Annual Review of Anthropology, 32*, 287–313.

Barba, A. A. (1923). *El Arte de los Metales*. New York: Wiley.

Beck, C., Taylor, A. K., Jones, G. T., Fadem, C. M., Cook, C. R., & Millward, S. A. (2002). Rocks are heavy: transport costs and Paleoarchaic quarry behavior in the Great Basin. *Journal of Anthropological Archaeology, 21*, 481–507.

Berthelot, J. (1986). The extraction of precious metals at the time of the Inka. In J. V. Murra, N. Wachtel, & J. Revel (Eds.), *Anthropological history of Andean polities* (pp. 68–88). Cambridge: Cambridge University Press.

Bird, J. B. (1979). The "Copper Man": A prehistoric miner and his tools from northern Chile. In E. P. Benson (Ed.), *Pre-Columbian metallurgy of South America* (pp. 105–132). Washington, D.C.: Dumbarton Oaks Research Library and Collections Trustees for Harvard University.

Bloxam, E. (2011). Ancient quarries in the mind: pathways to a more accessible significance. *World Archaeology, 43*(2), 149–166.

Boivin, N. (2004). From veneration to exploitation: Human engagement with the mineral world. In N. Boivin & M. A. Owoc (Eds.), *Soils, stones and symbols: Cultural perceptions of the mineral world* (pp. 1–30). London: UCL.

Boivin, N., & Owoc, M. A. (Eds.). (2004). *Soils, stones and symbols: Cultural perceptions of the mineral world*. London: UCL.

Bradley, R. (2000). *An archaeology of natural places*. New York: Routledge.
Brewer-La Porta, M., Burke, A., & Field, D. (Eds.). (2010). *Ancient mines and quarries: A trans-Atlantic perspective*. Oxford: Oxbow Books.
Brown, K. W., & Craig, A. K. (1994). Silver mining at Huantajaya, viceroyalty of Peru. In A. K. Craig & R. C. West (Eds.), *Quest of mineral wealth: Aboriginal and colonial mining and metallurgy in Spanish America* (pp. 303–327). Department of Geography and Anthropology, Louisiana State University, Baton Rouge, LA: Geoscience Publications.
Burger, R. L., & Matos Mendieta, R. (2002). Atalla: a center on the periphery of the Chavín Horizon. *Latin American Antiquity, 13*(2), 153–177.
Cobo, B. (1979 [1653]). *History of the Inca empire (translated and edited by Roland Hamilton)*. Austin: University of Texas Press.
Cobo, B., & Jiménez de la Espada, M. (1890 [1653]). *Historia del Nuevo mundo. Sociedad de bibliófilos andaluces*. Sevilla: Imp. de E. Rasco.
Cooke, C. A., Abbot, M. B., & Wolfe, A. P. (2008). Late-Holocene atmospheric lead deposition in the Peruvian and Bolivian Andes. *Holocene, 18*(2), 353–359.
Cooke, C. A., Balcom, P. H., Biester, H., & Wolfe, A. P. (2009). Over three millennia of mercury pollution in the Peruvian Andes. *Proceedings of the National Academy of Sciences of the United Sates of America, 106*(22), 8830–8834.
Craddock, P. T. (1995). *Early Metal Mining and Production*. Washington, D.C.: Smithsonian Institution Press.
Craddock, B. R., Cartwright, C. R., Craddock, P. T., & Wray, W. B. (2003). Hafted stone mining hammer from Chuquicamata, Chile. In P. T. Craddock & J. Lang (Eds.), *Mining and metal production through the ages* (pp. 52–68). London: British Museum Press.
Craddock, P. T., & Lang, J. (2003). *Mining and metal production through the ages*. London: British Museum Press.
Craig, A. K., & West, R. C. (1994). *Quest of mineral wealth: Aboriginal and colonial mining and metallurgy in Spanish America* (Geoscience and man (33)). Baton Rouge, LA: Louisiana State University.
Cruz, P. J., & Vacher, J. (Eds.). (2008). *Mina y metalurgia en los Andes del sur desde la época prehispánica hasta el siglo XVII*. Sucre, Bolivia: Instituto Francés de Estudios Andinos.
D'Altroy, T. N. (2002). *The Incas*. Oxford: Blackwell Publishers.
Dean, C. (2010). *A culture of stone: Inka perspectives on rock*. Durham, NC: Duke University Press.
Eerkens, J. W., Vaughn, K. J., & Linares Grados, M. (2009). Prehispanic mining in the Southern Nasca Region, Peru. *Antiquity, 83*, 738–750.
Electropampas. (2003). Resumen Ejecutivo Proyecto "Pampas Verdes".
Ericson, J. E., & Purdy, B. A. (Eds.). (1984). *Prehistoric quarries and lithic production*. New York: Cambridge University Press, Cambridge Cambridgeshire.
Gil García, F. M., & Fernández Juárez, G. (2008). El culto a los cerros en el mundo andino: estudios de caso. *Revista Española de Antropología Americana, 38*(1), 105–113.
Harris, O. (2000). *To make the earth bear fruit: Essays on fertility*. Institute of Latin American Studies, London: Work and Gender in Highland Bolivia.
Ingold, T. (2006). Rethinking the animate, re-animating thought. *Ethnos: Journal of Anthropology, 71*(1), 9–20.
Ingold, T. (2011). *Being alive: Essays on movement, knowledge and description*. New York: Routledge.
Knapp, A. B., Pigott, V. C., & Herbert, E. W. (Eds.). (1998). *Social approaches to an industrial past: The archaeology and anthropology of mining*. London: Routledge.
La Niece, S., & Craddock, P. T. (1993). *Metal plating and patination: Cultural, technical, and historical developments*. Boston: Butterworth-Heinemann, Oxford.
La Niece, S., Hook, D. R., & Craddock, P. T. (Eds.). (2007). *Metals and mines: Studies in archaeometallurgy*. London: Archetype.
Lechtman, H. (1976). Metallurgical site survey in the Peruvian Andes. *Journal of Field Archaeology, 3*(1), 1–42.

MacCormack, S. (1984). From the sun of the Incas to the virgin of Copacabana. *Representations, 8*, 30–60.
MacCormack, S. (1991). *Religion in the Andes*. Princeton: Princeton University Press.
Nash, J. (1972). Devil in Bolivia's nationalized tin mines. *Science and Society, 36*(2), 221–233.
Nash, J. C. (1979). *We eat the mines and the mines eat us: Dependency and exploitation in Bolivian tin mines*. New York: Columbia University Press.
Nuñez, A. L. (1999). Archaic adaptation on the South-Central Andean Coast. In T. Michael Blake (Ed.), *Pacific Latin America in prehistory: The evolution of archaic and formative cultures* (pp. 199–211). Pullman: Washington State University Press.
Núñez, A. L. (2006). La orientación minero-metalúrgica de la producción atacameña y sus relaciones fronterizas. In H. Lechtman (Ed.), *Esferas de interacción prehistóricas y fronteras nacionales modernas: losAndes sur centrales/Lechtman, Heather* (pp. 205–254). New York: Institute of Andean Research.
Petersen, G. (1970). *Minería y metalurgía en el antiguo Perú*. Instituto de Investigaciones Antropológicas, Lima: Arqueológicas.
Petersen, G. (2010). *Mining and metallurgy in ancient Perú (trans: W.E. Brooks. Special paper)*. Boulder, Colo: Geological Society of America.
Platt, T., & Quisbert, P. (2008). Tras las huellas del silencio: Potosí, los Inkas y el virrey Toledo. In P. José Cruz & J. Vacher (Eds.), *Mina y metalurgia en los Andes del sur desde la época prehispánica hasta el siglo XVII* (pp. 231–277). Sucre, Bolivia: Instituto Francés de Estudios Andinos.
Polo de Ondegardo, J. (1940 [1561]). Informe al licenciado Briviesca de Munatones. *Revista Historica, 13*, 125–196.
Protzen, J.-P. (1983). Inca quarrying and stonecutting. *Ñawpa Pacha, 21*, 183–214.
Rowe, J. H. (1946). Inca culture at the time of the Spanish Conquest. In J. Steward (Ed.), *Handbook of South American Indians, vol. 2* (The Andean civilizations, Vol. 143, pp. 183–330). Washington DC: Bulletin of the Bureau of American Ethnology.
Salazar Soler, C. (2006). *Supay Muqui dios del socavón*. Congreso del Perú, Lima, Peru: Vida y mentalidades mineras.
Salazar, D., Jackson, D., Guendon, J. L., Salinas, H., Morata, D., Figueroa, V., Manríquez, G., & Castro, V. (2011). Early evidence (ca. 12,000 BP) for iron oxide mining on the Pacific coast of South America. *Current Anthropology, 52*(3), 463–475.
Schultze, C. A., Stanish, C., Scott, D. A., Rehren, T., Kuehner, S., & Feathers, J. K. (2009). Direct evidence of 1,900 years of indigenous silver production in the Lake Titicaca Basin of Southern Peru. *Proceedings of the National Academy of Sciences of the United Sates of America, 106*(41), 17280–17283.
Shimada, I. (1994). Pre-hispanic metallurgy and mining in the Andes: Recent advances and future tasks. In A. K. Craig & R. C. West (Eds.), *Quest of mineral wealth: Aboriginal and colonial mining and metallurgy in Spanish America* (Vol. 33, pp. 37–73). Baton Rouge, LA: Louisiana State University.
Shimada, I. (Ed.). (2007). *Craft production in complex societies: Multicraft and producer perspectives*. Salt Lake City: University of Utah Press.
Smith, M. L. (1999). The role of ordinary goods in premodern exchange. *Journal of Archaeological Method and Theory, 6*(2), 109–135.
Stöllner, T. (2008a). Mining landscapes in early societies—Imprinting processes in pre- and protohistoric economies? In C. Bartels & C. Küpper-Eichas (Eds.), *Cultural heritage and landscapes in Europe—Landschaften: Kulturelles Erbe in Europa* (Proceedings of the International Conference Bochum, 8.-10. Juni 2007, pp. 65–92). Bochum: Selbstverlag des deutschen Bergbaumuseums Bochum 161.
Stöllner, T. (2008b). Montan-archaeology and research on old mining: just a contribution to economic history? *Anatolian Metal, 4*, 149.
Stöllner, T. (2009). Gold in Southern Peru? Perspectives of research into mining archaeology. In M. Reindel & G. A. Wagner (Eds.), *New technologies for archaeology: Multidisciplinary investigations in Palpa and Nasca, Peru* (pp. 393–407). Berlin: Springer.

Taussig, M. T. (1980). *The devil and commodity fetishism in South America: xiii, 264 p.* Chapel Hill: University of North Carolina Press.

Topping, P., & Lynott, M. J. (Eds.). (2005a). *The cultural landscape of prehistoric mines*. Oxford: Oxbow Books.

Topping, P., & Lynott, M. J. (2005b). Miners and mines. In P. Topping & M. J. Lynott (Eds.), *The cultural landscape of prehistoric mines* (pp. 181–191). Oxford: Oxbow Books.

Tripcevich, N., & Contreras, D. A. (2011). Quarrying evidence at the Quispisisa Obsidian source, Ayacucho, Peru. *Latin American Antiquity, 22*(1), 121–136.

Tripcevich, N., & Mackay, A. (2011). Procurement at the Chivay obsidian source, Arequipa, Peru. *World Archaeology, 43*(2), 271–297.

Van Buren, M., & Mills, B. H. (2005). Huayrachinas and tocochimbos: Traditional smelting technology of the southern Andes. *Latin American Antiquity, 16*(1), 3–25.

Van Buren, M., & Presta, A. M. (2010). The organization of Inka silver production in Porco, Bolivia. In M. A. Malpass & S. Alconini Mujica (Eds.), *Distant provinces in the Inka empire: Toward a deeper understanding of Inka imperialism* (pp. 173–192). Iowa City: University of Iowa Press.

Vaughn, K. J., Van Gijseghem, H., Linares Grados, M., & Eerkens, J. W. (still in press). Minería hematita en la costa sur del Perú: Investigaciones arqueológicas en Mina Primavera. Chungará: Revista de Antropología Chilena.

Zori, C., & Tropper, P. (2010). Late prehispanic and early colonial silver production in the Quebrada de Tarapacá, Northern Chile. *Boletín del Museo Chileno de Arte Precolombino, 15*, 65–87.

Part II
Pigment, Clay, Salt and Stone

Chapter 2
Archaeological Approaches to Obsidian Quarries: Investigations at the Quispisisa Source

Nicholas Tripcevich and Daniel A. Contreras

Introduction

At a global scale, and spanning human history from the Paleolithic through recent times, the material remnants of mining and quarrying have sustained the interest of a segment of the archaeological community. This volume provides us with an opportunity to reflect on how and why archaeologists have studied the vestiges of mining and quarrying, and to consider a specific Andean context: the extraction of Quispisisa-type obsidian from the Jichja Parco obsidian quarries over a period of 10,000 years of Central Andean prehistory. Preliminary research at the source (Contreras et al. in press; Tripcevich and Contreras 2011) has documented large-scale extraction of obsidian, while regional consumption patterns (Burger and Glascock 2002) demonstrate that the material was used and widely distributed not long after humans arrived in the Central Andes. We use the example of ongoing research at the Quispisisa source to examine what study of mining and quarrying in the Prehispanic Andes can contribute to perspectives on the Andean past. More generally, we reflect on the actual questions that archaeologists hope to address by examining mines and quarries, and consider how we can approach mining and quarrying evidence in such a way as to be able to answer such questions.

N. Tripcevich (✉)
University of California-Berkeley, 2251 College Bldg, UC Berkeley,
Berkeley, CA 94720-1076, USA
e-mail: nico@mapaspects.org

D.A. Contreras
Institute for Ecosystem Research, Kiel University,
Olshausenstr. 75, 24118 Kiel, Germany
e-mail: dcontreras@ecology.uni-kiel.de

N. Tripcevich and K.J. Vaughn (eds.), *Mining and Quarrying in the Ancient Andes*, Interdisciplinary Contributions to Archaeology, DOI 10.1007/978-1-4614-5200-3_2,

Archaeology of Lithic Procurement

Source areas present particular difficulties to archaeologists but may also provide opportunities for research. As single locations that are linked to sites dispersed throughout a larger region, source areas enable holistic archaeological approaches to regional lithic economies and technical reduction processes. Moreover, in the case of obsidian, the links provided by geochemical source assignment offer definitive connections that are relatively rare in archaeology. At the same time, the challenges of working at sources are many. Theoretically and methodologically, research at quarries is complicated by a material record that is predominantly “shattered, overlapping, sometimes shallow, nondiagnostic, undatable, unattractive, redundant, and at times voluminous” (Ericson 1984: 2). Difficulties in temporal control, as well as the effort involved in differentiating an abundance of naturally fractured raw material from cultural products, present obstacles to the study of quarry use over time.

The priorities of quarry studies, viewed globally, have long included (1) linking production at quarry sites and the transport of material to evidence from lithic consumption at sites in the region, (2) inferring the rates of production, and (3) considering the regional contexts of lithic access and distribution through time. Evidence for technology and manufacturing changes is abundant at quarries and workshops, as these locations are typically rich in primary reduction material. However, a deficiency in later stage lithics (often the exported product) limits the usefulness of typological approaches and methods that focus on the characteristics of finished tools. Opportunities for research in source areas are particularly strong for approaches that take a technical and sequence-based approach to understanding the use of stone material, but as sequences or operational chains explicitly link early and larger stages of production, using a sequence-based approach at a source area forces the analyst to carefully study regional assemblages as well as source area materials. Clearly it is easier to build such links with geochemically or petrographically sourceable stone (among these the “chain” can be better demonstrated), and regional consumption patterns are of course best assessed in regions with published bodies of lithics research.

Studies at lithic sources often seek to address questions concerning the material type, appearance, and morphology of source material, and the degree of reduction performed at the source area. This characterization and quantification enables the investigation of social considerations of broader interest to archaeologists:

- Who procured the raw material and produced the evidence of quarrying that archaeologists may document?
- Were the knappers the same individuals who procured material?
- Were either of these groups specialized or supported? What sort of infrastructure facilitated the quarrying for material (and does architectural or depositional evidence remain)?
- What kind of sociopolitical organization underpinned raw material procurement? Was access to the resource limited to particular communities due to ethnic or political restrictions?

- Who consumed the material (i.e., were miners procuring for their own use, for trade or exchange, or at the behest of others)? Was material consumed locally or widely distributed?
- Were source area visits embedded in other activities or were these special purpose journeys? Were particular social or ceremonial practices associated with access to the source area or procurement, and use of the material?
- Is the source area and, by association, distinctive material from that source, prominent in the ritual or cosmological landscape in the region? This may be evident from activities at the source or in special treatment of the material.

We explore means of addressing such questions in this chapter focusing on obsidian procurement at the Quispisisa source.

Building from a Production System Approach

Procurement at lithic sources represents the first step in a progression conceptualized by frameworks such as the lithic reduction sequence and *chaîne opératoire* (Edmonds 1990; Schiffer 1975; Sellet 1993; Shott 2003; Torrence 1986). These sequences aid researchers in positioning geological source areas within the larger context of lithic tool use life, maintenance, and discard. The prevalence of early-stage reduction material in source areas, however, means that a complete operational chain will probably depend upon incorporating evidence from lithic materials recovered in other contexts elsewhere in the region.

In recent decades, archaeologists have sought to place procurement and lithic production into its regional context by identifying principal indicators of changes in procurement through time. Ericson's (1984: 4) approach to the study of "lithic production systems" is shown in Table 2.1.

These indices depend, upon general artifact-type categories and provide a basis for comparing empirical data from reduction activities between workshops, local sites, and the more distant consumption zone. A discussion of these measures is beyond the scope of this chapter, but Ericson's approach collapses variability, both in time and space, in the interest of comparability between archaeological datasets. This was intended to reflect a production and distribution system with "feedback mechanisms" in the form of regional demand (Ericson 1984: 2). Because it principally relies on metrics that are commonly gathered in laboratory analysis, Ericson's approach provides a means of generating a composite view of particular production zones where consistent data are available (Fig. 2.1).

Ericson presents the spatial distribution of lithic production in terms of stages of production and zones of geographic proximity to the source area. In principle, this regional approach provides a clear set of expectations about reduction patterns against which to examine actual archaeological data. In practice, however, the intermingling of artifacts from different episodes of quarrying and variable quarrying strategies often undermines the value of such generalizations. Furthermore, developing indices at a regional scale depends upon consistently implemented and

Table 2.1 Measurement indices for lithic production analysis (after Ericson 1984: 5)

Name	Variable (numerator)	Normalizer (denominator)	Unit(s)	Relevance
Exchange index	Single source	Total lithic material	Count, weight, %	Regional exchange
Debitage index	Debitage	Total tools and debitage	Count, weight, size, %	General production index[a]
Cortex index	Primary and secondary reduction flakes	Total debitage	Count, %	Indicative of the import of raw materials on site[a]
Core index	Spent cores	Total cores and tools	Count, %	Use if cores were transported or a medium of exchange
Biface index	Bifacial thinning flakes	Total debitage	Count, %	Biface production

[a]Excluding retouch/sharpening flakes

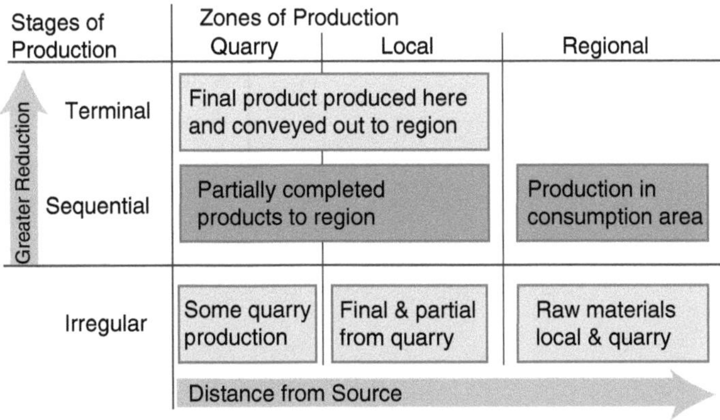

Fig. 2.1 Production system diagram based on table by Ericson (1984: 5)

comparable methods in lithics research, thus broadly applying such an approach to published data in the Central Andes is still difficult. Moreover, Ericson's approach provides a composite view of particular production zones by documenting the predominant production strategy at a given source area, but at the cost of characterizing variability within a production context.

It may take many years of work to decipher the complex record of activities at a large quarry zone over past millennia. For example, the Tosawihi opalite quarries in the US Great Basin (Elston and Raven 1992) have been examined in detail over some years using an energetics and ecology approach. In recent decades, a greater number of archaeologists are considering the ritual significance and meaning of quarrying in the past, often using empirical evidence that includes ceremonial structures at quarries and links to stone objects from that source found in ritual contexts (Bradley 2000: 81–96; Bradley and Edmonds 1993; Cooney 1998; Edmonds 1995; O'Connor et al. 2009; Skeates 1995; Topping 2010). In the Andes, the evidence for the symbolic and ceremonial importance of quarries is strongest at architectural stone sources used during later periods of the Prehispanic period (Chaps. 3 and 4). The systems-based approaches to production provide comparative information about broader patterns of regional interaction and prompt investigators to make explicit many of their assumptions. However, a focus on broad systems may lead to reduced attention to detail and less documentation of variability, as well as eliding any changes in contexts of consumption.

Obsidian Quarrying in the Central Andes

The breadth of chapters in this volume suggests that the evidence of mining and quarrying behavior in the Central Andes reveals a range of ways to manage (and perhaps even conceptualize) resources. Compare, for example, the patterns of access

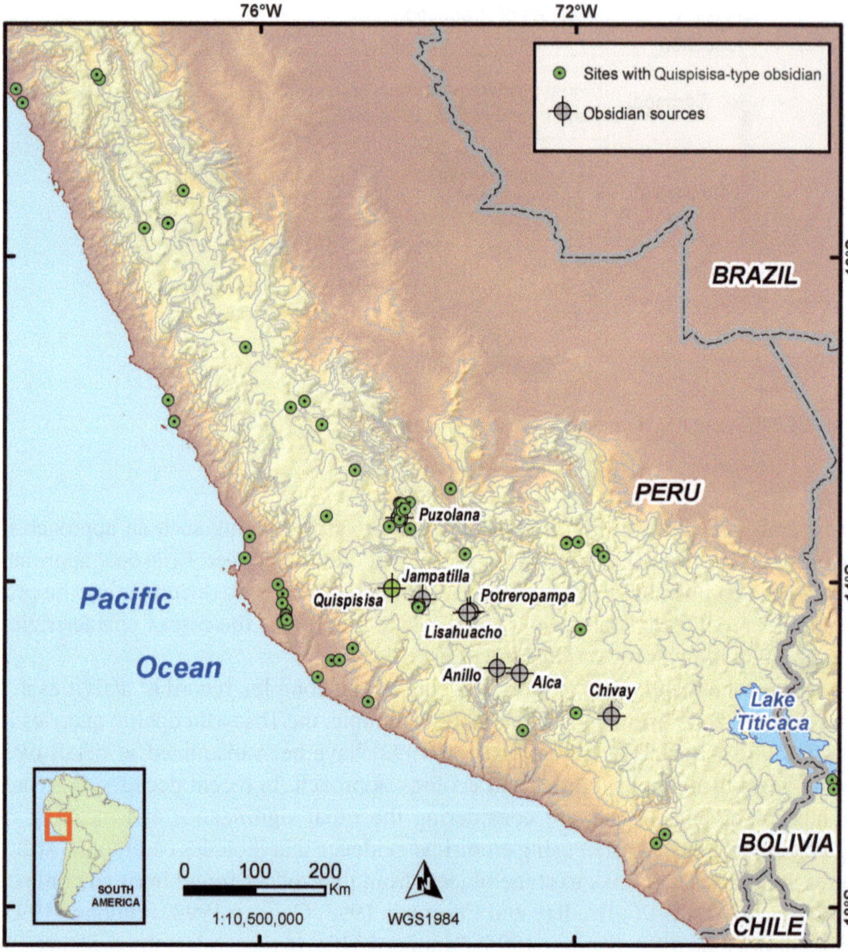

Fig. 2.2 Map of central Andes showing major obsidian sources (labeled) and the location of archaeological sites containing artifacts made from Quispisisa-type obsidian

described by Jennings et al. at Cotahuasi and Inca period mining by Salazar et al. at Atacama. What can obsidian contribute to this discussion? That is, what do we know about the way(s) in which obsidian was procured, distributed, and circulated in the Prehispanic past of the Central Andes?

We examine these issues of lithic procurement and production using the case study of the source of Quispisisa-type obsidian in southern Ayacucho, Peru (Fig. 2.2). As a reflective natural glass with conchoidal fracture producing extremely sharp edges, obsidian has been employed by humans since the earliest tool-making periods in world prehistory. Obsidian is of great utility to present day archaeologists as well due to its high visibility, distinctive material characteristics, and analytical potential (Shackley 2005). Even prior to the advent of geochemical analysis methods, obsidian procurement was emphasized by archaeologists studying ancient mining (Holmes 1900, 1919: 214–227). Over the past 50 years, research interest in obsidian has increased around the world, largely as a consequence of the discovery that chemical composition of obsidian artifacts and source areas could be used to link artifacts to geological sources (Burger and Asaro 1977; Cann and Renfrew 1964; Glascock et al. 2007; Shackley 2011). Obsidian also provides a means of direct chronological control through estimates derived from the rate of absorption of water on culturally modified materials (Eerkens et al. 2008; Liritzis and Laskarisa 2011; Tripcevich et al. 2012). While there are limitations to the obsidian hydration dating method in some circumstances, it has proven to be of broad utility for improving chronological sequences. Hydration dating can be particularly useful at quarry sites where supporting evidence from culturally diagnostic artifacts or datable organic material is frequently unavailable (Tripcevich et al. 2012).

Early archaeological attention to obsidian in the Central Andes focused not on procurement but rather on its appearance in archaeological contexts. The material has been the subject of archaeological attention for at least a century—Max Uhle described dart foreshafts with obsidian points from the Nazca cemetery at Chaviña as early as 1909 (Uhle 1909) and also collected obsidian from sites such as Marcahuamachuco (Fig. 2.3) in the early years of the twentieth century (Burger and Glascock 2009; McCown 1945). Uhle and his successors were primarily interested in obsidian *cum* artifact, and in using those artifacts to infer the behavior of their makers and users. By the 1970s, when it had become analytically possible to separate Central Andean obsidians into geochemical groups (Burger and Asaro 1977; Burger et al. 2000: 271–272), the obsidian sources became foci of interest as a first step to permit subsequent research into regional procurement. The immediate goal was to identify geological source areas in order to tie geochemical groupings of obsidian to specific origin points.

Geochemical links between artifacts and obsidian types enabled discussion not just of tool use, but also of the circulation of specific obsidians. In the Andes, systematic research into obsidian sourcing that had begun in the 1970s was delayed by the remoteness of many of the sources and by dangerous political conditions during the 1980s (Burger and Glascock 2002; Burger et al. 2000; Glascock et al. 2007). Obsidian sources in the Central Andes are confined to the South-Central Andes; the next sources to the north are in the highlands of Ecuador (Burger et al. 1984; Burger and Glascock 2009), and while material from Ecuadorian sources has been found transported 450 km south in Tumbes, Peru (Moore 2010: 406), the region forms a sphere of circulation separate from the sources of southern Peru (Burger 1984). Similarly, there are regionally significant obsidian sources in southern Bolivia, Argentina, and Chile (Barberena et al. 2011; Yacobaccio et al. 2004), but material from those sources circulated in a distinctive sphere from that of the central Andean obsidian sources.

Geochemical sourcing thus added a vital dimension to our understanding of the procurement and circulation of obsidian in the Central Andes (summarized in Glascock et al. 2007). However, geochemical sourcing is not in itself sufficient to approach questions about the organization of procurement and manufacture, its

Fig. 2.3 Obsidian biface measuring 53.4 mm with grey banding and an opaque red tip. Photo courtesy of the Phoebe A. Hearst Museum of Anthropology and the Regents of the University of California. Photography by Nicholas Tripcevich (Catalogue No. 4-3531)

links to consumption, and to further interpretive work concerning, for example, the conceptualization of resource ownership and access through prehistory. Andean obsidian source research thus parallels the history of work in Mesoamerica where there was a "contagious enthusiasm for obsidian sourcing" (see Clark 2003: 19) in the late 1960s and 1970s. In the case of the Andes, however, it was somewhat less contagious being largely the result of efforts of one person: Richard L. Burger. In the Andes, we thus saw an initial focus on distribution patterns as a means of reconstructing networks of trade and exchange, and presently, with all the major sources identified, a shift to emphasizing obsidian sources themselves as resources, to be exploited or conserved, controlled, or communally maintained.

Richard Burger's collaborative studies (Burger et al. 1994, 1998a, 1998b, 1998c, 2000, 2006; Burger and Glascock 2000, 2001, 2002; Glascock et al. 2007) have now located and sampled at seven of the principal obsidian sources in the Central Andes (Fig. 2.2): Alca, Chivay, Jampatilla, Lisahuacho, Potreropampa Puzolana, and Quispisisa, while the Acarí type has recently been linked to the Anillo source in northern Arequipa. However, survey and excavation characterizing Prehispanic procurement at these sources remain scarce, represented in print only by Tripcevich's work at Chivay (Tripcevich 2007; Tripcevich and Mackay 2011). Ongoing geoarchaeological survey and geochemical analysis at the Alca obsidian source (Burger et al. 1998b; Jennings and Glascock 2002; Rademaker

2006, 2012) have documented a few quarry pits and limited tunneling into tuff for obsidian procurement, as well as identifying distinct geochemical signatures for particular flows at Alca that may provide analytical possibilities (Eerkens and Rosenthal 2004). For example, Rademaker (2012) is able to explore patterns in the use of particular sectors of the Alca source, such as an apparent shift during the later Holocene towards greater use of one particular subsource that lies along a travel corridor.

This research notwithstanding, the long-term widespread use of obsidian in the Central Andes suggests that quarries have been underused as research foci. In particular, Quispisisa-type obsidian has a remarkably long history of use, and eventually was transported great distances, reaching nearly 1,000 km from the source to the site of Pacopampa by the first millennium B.C.E. (Burger and Glascock 2009: 25). While Peru's other two major sources—Alca and Chivay—display relatively little evidence of quarrying, the project that we have begun at the source of Quispisisa-type obsidian in southern Ayacucho, which we describe here and in (Contreras et al. in press; Tripcevich and Contreras 2011) focuses on an obsidian source that contains the most large-scale evidence of obsidian quarrying found in the Central Andes to date.

Quarrying of Quispisisa-type Obsidian

Our preliminary work at the Quispisisa source (Tripcevich and Contreras 2011: 125) has demonstrated that the area features an array of quarrying evidence unique in the Central Andes. In our initial visit to the source area, we used Burger and Glascock's (2000, 2002) description of the two-hour hike to the source area, and we were guided by a local resident Jesus Vilchez who described large pits across the Urabamba river from the obsidian exposure encountered by Burger's team. Over the ensuing 4 years, we conducted numerous reconnaissance visits to the source area, and have thus far documented 34 quarry pits on a hill known as Jichja Parco. The pits themselves are mainly ellipsoidal, ranging in size from about 10 m on their long axes and 1 m deep to 45 m across and 3 m deep. The pits documented thus far are spread over an area of 90 ha, comprising in total a mined surface of at least 13,000 m^2 and an estimated excavated volume of at least 32,000 m^3. We have also observed but not yet documented other pits, both comparable in size and shallower, meaning that these figures are minimal counts.

The pits occur in clusters across the hillslope, often adjacent to one another. They are virtually carpeted with obsidian, primarily small discarded nodules, and surface scatters also include flake debris from the initial stages of reduction. Spoils piles were apparently routinely heaped downslope, forming a berm following the circumference of at least part of each pit; these berms are similar in composition (judging by surface inspection) to the bottoms of the pits in most cases (Fig. 2.4). They give the pits the appearance of the "Doughnut quarries" described at the Ucareo-Zinapécuaro obsidian source in West Mexico (Healan 1997).

Fig. 2.4 Photograph showing quarry pits aligned along a contour (pits 7005 and 7006 are visible) with a downslope berm on the *left*

Approaching the quarry pit area on foot takes several hours by trail from the road, and one passes a few locations where obsidian is exposed through erosion on the sides of the Urabamba drainage (Fig. 2.5). Crossing the river, one encounters a considerable density of obsidian nodules up to 30 cm across eroding out in the headcuts of quebradas, alluvially transported in quebrada channels, or colluvially transported on slopes. The availability of large nodules in these contexts brings up a question: why excavate large pits to acquire obsidian when it is readily available in these erosion contexts? The extensive evidence of quarrying also prompts a more general question about the exploitation of Quispisisa-type obsidian: do big pits imply organized labor? More generally, such evidence focuses our attention on an issue particularly germane to this volume: how are we to interpret quarrying evidence?

The quarrying evidence at the Quispisisa source may be the product of a long history of exploitation: regional archaeological evidence demonstrates that the source was exploited as early as the Archaic Period by mobile foragers. Later, in the first millennium BCE, Quispisisa-type obsidian was widely distributed in the interaction network associated with the Chavín phenomenon, and during the Middle Horizon the Wari Empire made extensive use of obsidian from this source. In other words, we may presume that the source area has been subject to extraction under divergent sociopolitical formations. Whether the extraction activity itself changed as the consuming populations changed remains an open question; this drives the issue of how distinct Central Andean sociopolitical forms organized the production of lithics and the exploitation of raw materials more generally.

We suggest three (overlapping rather than mutually exclusive) factors at play in quarrying behavior, and elaborate below (Table 2.2) specific models of differing

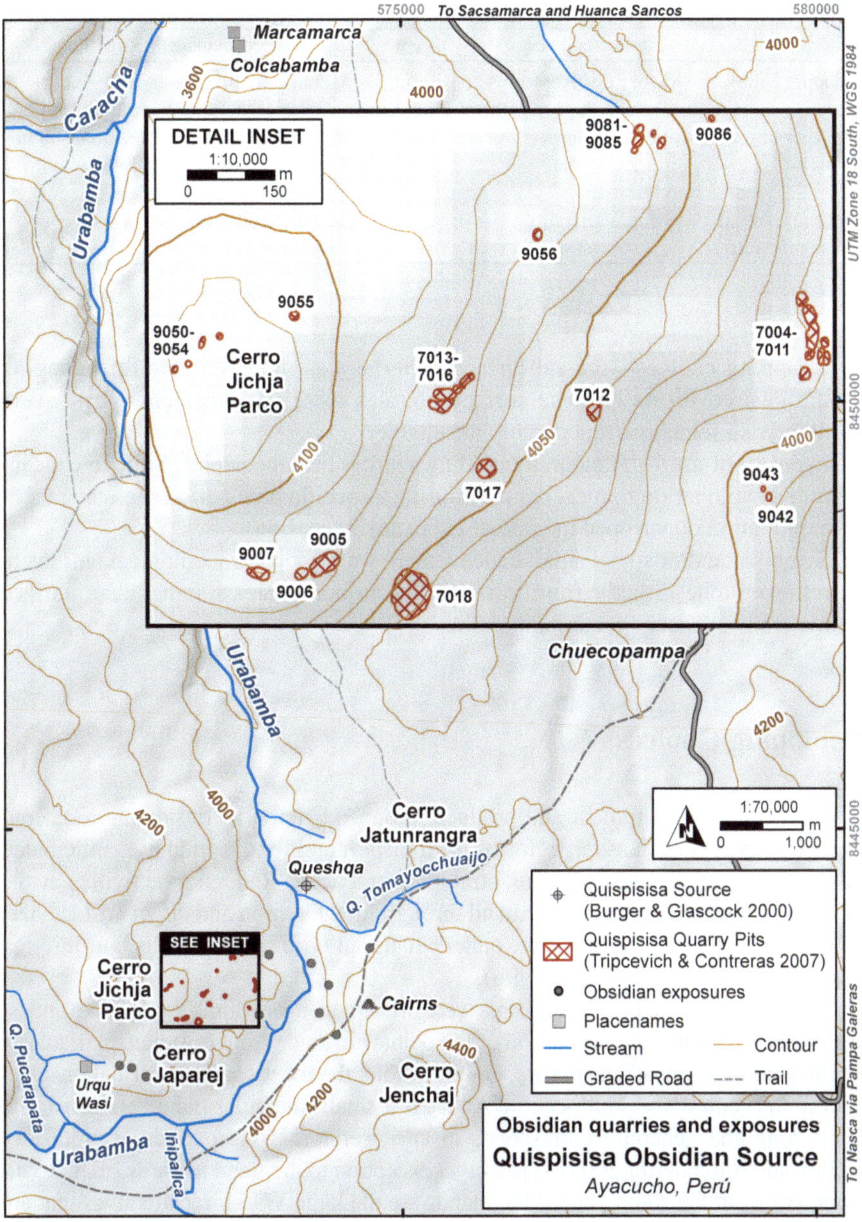

Fig. 2.5 Map of Jichja Parco quarry area at the Quispisisa Source

modes of exploitation and their material correlates for two extremes: unmanaged and low-intensity extraction by foragers (or pastoralists, or even nearby agriculturalists) on one end of the spectrum, and coordinated, perhaps state-run, access/extraction/production on the other.

Table 2.2 Expected material correlates in different scenarios of obsidian exploitation

	Unmanaged access	Coordinated/managed extraction
Knapping choices	More diverse; forms consistent with hunter-gatherer toolkits	More standardized knapping but greater variability of target forms
Intensity of use	Lower intensity; use spread out over time	Higher intensity, use more concentrated in time and potentially involving more coordinated labor
Symbolic/social significance	More variability	Less variability; higher investment

1. Knapping choices—the subsurface material was more suitable for knapping the desired forms because surface nodules were not large enough or were otherwise somehow functionally unsuitable.
2. Intensity of use—the naturally eroding sources became depleted due to concentrated exploitation to meet demand during a specific time period (i.e., the rate of exploitation outstripped the rate of exposure by erosion).
3. Symbolic and/or social significance—there were social or ideological reasons to retrieve stone directly from quarried subsurface contexts rather than surface materials or eroded material in gullies.

Knapping Choices

Understanding procurement and production depends upon a consideration of consumption factors such as target forms for transport and for eventual use, anticipated degree of curation, and reduction strategies practiced. Reduction activities at the obsidian source area reflected demand throughout the region and target artifact size, as well as the ability to transport bulkier material with pack animals and through established routes (e.g., Close 1996).

Preceramic foliate projectile points were often made from durable stone like andesite or quartzite but some proportion of the points are made from obsidian, particularly in the vicinity of obsidian sources. On the whole, the most common formal obsidian tool in the central and south-central Andes is a small triangular point between 1 and 2 cm long and generally corresponds to later, ceramic-using periods of prehistory (Klink and Aldenderfer 2005). A principal exception to the trend towards small, triangular points in the south-central Highlands are the large Wari bifacial knives that are up to 5 cm long, found sometimes far from the geological obsidian source (Bencic 2000; Burger and Glascock 2009; McCown 1945; Nash 2002; Owen and Goldstein 2001; Williams et al. in press). These are commonly made from Quispisisa-type obsidian, the obsidian source used heavily by Wari during the Middle Horizon.

Examples of advanced reduction obsidian artifacts from sites throughout the region thus inform our approach to analysis of materials at the source, as later stage and discarded obsidian artifacts provide insights into the trajectory for artifacts transported from the source area. In upcoming work at the Quispisisa source, our analysis of flaked stone from the source and nearby workshops will focus on changes in blank production and core reduction over time. For example, variation in core to flake ratios may indicate whether producers sought cores or blanks and patterns in the size and shape of late flake removals may indicate preferred blank form. At workshops further from the source, changes in core size at discard may reflect changes in the relative cost of material acquisition. The significance of linking these changes to larger patterns in regional production/consumption highlights the importance of locating datable stratified deposits at production areas and dumps, or in their absence the value of being able to chronologically relate distinct quarrying areas through radiocarbon dating of associated organic material or through examining the evidence of locally-calibrated obsidian hydration rates.

Documenting intra-quarry variability—e.g., color or knapping properties—may be examined in tandem with spatial variability in extraction, assessment, and reduction strategies, offering the possibility of examining chronological changes. Combined with the ongoing study of lithic collections at a regional scale, this approach can draw out patterns in standardization and continuity in lithic production practices that may aid in addressing questions such as whether procurement was practiced to fill local needs or to create products for exchange, as well as addressing questions of potential control of access and/or coordination of production by local or regional authority. As other chapters in this volume demonstrate (Chaps. 6 and 12), there are Prehispanic Andean examples of mineral resources being treated as communal, open-access resources (e.g., rock salt) *and* state-owned and controlled resources (metal ores).

Intensity of Exploitation

We suggest two models to account for the formation of "doughnut quarries" as observed at Jichja Parco. The first posits the gradual formation of pits over the long-term due to continual low-intensity quarrying, while in the second model quarrying is the result of coordinated, intensive exploitation during a specific time period that exceeded the supply of materials available on the surface. Both models may have been in effect simultaneously at certain times during the site's history.

Ad Hoc Quarry Activity

This model suggests two possibilities (1) a preference for subsurface material, and (2) excavation of quarry pits as a process incidental to material acquisition. In the first case, subsurface material might be superior, perhaps better insulated from erosion and thermal fluctuations, or might be preferred for other reasons. In the second case, target nodules might be recovered from the surface resulting in the gradual formation of a pit in that location. As such pits become deeper and labor required to remove material from them increases, new quarries might be started nearby. A small amount of maintenance may have been required for such pits, but on the whole the process need not imply large-scale organization. Exploitation might be by individuals or small groups, with use-rights either unrestricted or perhaps structured by kinship. Quarrying that results from such a system would produce pits spanning a wide period of time and with general heterogeneity in factors such as nodule selection and reduction strategies.

Coordinated Extraction with Intensified Use

The second model posits that the effects of growing regional demand might exceed the obsidian available naturally due to pronounced need during a particular time period. While it is evident that in modern circumstances where erosion on slope surfaces, along riverbanks, and in incised gullies exposes large nodules, it is possible that during specific times extraction outstripped this supply. If obsidian available in surface and erosional contexts was not sufficient to meet demand, excavation for additional material would have become necessary. This might occur in case of moderately intensified extraction in a relatively short (multi-decadal or centennial) span of time, even without extra-local coordination, or in case of significantly intensified momentary (decadal or annual) demand, for instance imposed by a regional polity like Wari. In either case coordination of labor and/or use-rights would lead to many contemporary pits and more homogeneity in material selection and reduction strategies.

At a regional scale, the density and spatial extent of the consumption zone provide some clues about the demand during particular time periods, although consistent quantification of obsidian at many sites is lacking and the sampling of obsidian for sourcing from archaeological contexts in the Central Andes, while improving steadily, remains inconsistent. There is little correlation between the extent of obsidian distribution and size or number of quarry pits: Alca material was as widely geographically distributed as Quispisisa-type obsidian, but the primary source area has only a few modest pits (Rademaker 2012), while Chivay only has one pit (Tripcevich 2007). Such comparisons are further complicated by differences in the accessibility at obsidian at the sources and in the irregular distribution of consumer sites sampled for obsidian sourcing, largely a reflection of the history of archaeological research in the region (Burger 2000). The intensity of extraction is presumably more directly correlated with the intensity of use, but due to the difficulties of sampling we are not yet able to compare the scale of consumption of Quispisisa obsidian to other types.

These two models may be considered as poles on a spectrum of mining behavior. Distinguishing where on this spectrum given evidence of mining activity may fall requires assessing the chronology of extraction. At the Quispisisa source, we approach this from two perspectives. First, cultural evidence from archaeological sites and lithic workshops at both local and regional scales may provide datable samples with which to address the intensity of use over time. Second, directly dating quarry activity through stratigraphic evidence in mining debris and/or through obsidian hydration analysis of flaked obsidian from a broad sample of pit features will provide, at a minimum, a relative chronological evidence for the exploitation of obsidian from these pits.

Symbolic and Social Aspects of Obsidian

> "Of all the things that the Spanish showed him [Atahualpa], there was none he liked more than glass, and he said to Pizarro that he was very surprised that, having things of such beauty in Spain, he would travel to distant and foreign lands looking for metals as common as gold and silver" (Benzoni and Smyth 1857 [1565]).

Ethnohistoric sources and contemporary ethnography in the Andes are rich with accounts of ritual practices associated with mines and mining, and these accounts inform current studies of mining in the Andes, include many of the chapters in this volume. As observed by Bernabé Cobo (1652) and other chroniclers, those who worked the mines also worshiped the ore-rich hills and the mines as shrines. Cobo specifically mentions rituals surrounding silver and gold sources, as well as rituals involved with the procurement of pyrite, sulfur, and cinnabar. We have few details about obsidian mining, however, as it was of little interest to the Spanish chroniclers, and the Inca seem to have made less use of obsidian, apparently focusing on other materials (Burger et al. 2000: 344–346).

These sources suggest that ancient Andean peoples mined and quarried in an animated landscape (sensu Ingold 2006, 2011). Here social and symbolic relationships between human communities and the entities or essence that reside in certain natural features, including particular mountains and the minerals that lay within them, were as significant as economic and political imperatives. Moreover, ethnohistoric and ethnographic evidence of connections between communities and geological landscape features is widespread in the Andes. Thus, while in some contexts it seems that an obsidian source is simply a source of sharp, functional stone, in others obsidian may be charged with social, political, and ideological associations, and the source may be correspondingly prominent in the landscape. The challenge for archaeological approaches to Andean mining is to consider what kind of material evidence would support these assertions of emotive or symbolic attachment to particular places and materials derived from those places. Spatial association with ritually important locations, for example, or a pattern of interment together with other unusual materials believed to have been high value may imply that a source location had ritual significance. Andean ethnohistoric sources do not specify whether obsidian was considered a ritually significant material or if its mining was associated with ritual activity. There are examples of ritual structures at obsidian sources, such as a cluster of Inca period chullpa mortuary structures at the Chivay obsidian source (Tripcevich and Mackay 2011) and a relatively high density of saylluas (cairns) marking the location where a major trail overlooks the Quispisisa obsidian source (Tripcevich and Contreras 2011: 125). However, chullpa and cairn structures are not uncommon in the central Andes, underscoring the difficulty in finding straightforward material indicators of the ritual significance at mining sites.

Archaeological evidence from the Central Andes suggests that obsidian may have been valued as a functionally important and/or exotic material in some times and places, while valued for symbolic and ritual reasons in others. Obsidian generally is found in mundane contexts such as household middens, but it also is sometimes found in ritual contexts (discussed below). Obsidian sometimes appears to be a valued exotic material—for example it has been found included in burials where the material is scarce—but it also is sometimes found in contexts that span intrasite status differences, such as at Pukara (Klarich 2005: 255–256) or at Chavín de Huántar (Contreras and Nado in press).

In addition to focusing on commoner versus elite distinctions in use of this material, it is useful to consider social practices and further details about the use of obsidian at a distance from the geological source area. Obsidian may have been both commonplace and symbolically rich as a marker of group identity or as an "ordinary good" (Smith 1999) incorporated into household practices but communicating social meaning. For example, where it is familiar and known to come from adjacent volcanic regions obsidian might be interpreted as a marker of ethnic affiliation or exchange with communities associated with volcanic, obsidian producing areas. This may have been the case with Chivay obsidian in the Titicaca region where it is available only from the lands well to the west of Titicaca (Tripcevich 2010); it is commonly found at rockshelters and herder sites, and yet it is also found in burials and ritual mounds (Couture 2003; Giesso 2003).

Bradley (2000: 81–90) describes attachment to the character of flint sources in Neolithic Britain where material from unusual or sometimes dangerous locations on mountain sides appears to be found at greater distances and collected in ceremonial contexts. He links these materials to an affinity for qualities of the source areas that may have led to artifacts made from these materials being placed in the ground with some formality in graves and hoards as "Pieces of place." Stone artifacts from these sources circulated despite the presence of accessible and suitable alternative materials, suggesting to Bradley that these artifacts served perhaps as a reminder or a "piece of" the culturally significant source area in the Neolithic landscape. In some cases, flints from particular sources are visually distinguishable, in other cases it has been argued that a small portion of cortex may be left to aid in identifying sources (Rudebeck 1998). A key feature of these interpretations, then, is that knappers were concerned with more than knapping characteristics and easy availability.

In the Andes, ethnohistoric accounts describe social links, in particular ethnic affiliation and identity, incorporated in the movement of stone from the ethnic place of origin or *pacarisca* as discussed by Ogburn (Chap. 3). While no such accounts have come to light for obsidian, its characteristic appearance and limited number of sources suggest that research on obsidian in the Central Andes should consider such models. Among ethnographies from stone tool-using groups, such as in Australia (Brumm 2010; Gould 1980; McBryde 1997; Taçon 2004), people would endeavor to obtain stone from outcrops associated with their totemic ancestors or otherwise acquire stone over great distances despite many functional alternatives. Objects may accrue meaning by virtue of cultural biographies (Gosden and Marshall 1999), and possession of particular objects could have been bound to social ties resulting from such objects being traded and transported across the landscape (Lechtman 1984; Saunders, 2004). Thus, while it is difficult to determine if ancient peoples retained knowledge about geological origin of particular materials, we may look for patterns in the distribution and/or consumption of stone artifacts from sources that are visually indistinguishable but assignable to sources today (e.g., using obsidian geochemistry). We may investigate whether provenance information may have been communicated in association with artifacts as they were traded, or whether materials from particular sources were directly procured or used and treated in distinctive ways. Phenomena such as direct transport from the source area by displaced populations during the Inca period may explain, for instance, a collection of 29 small, unmodified obsidian nodules from the Chivay source found by Bingham in a ritual context over 300 km distant at the gateway to Machu Picchu (Burger et al. 2000: 347; Burger and Salazar 2004: 103, 161). This possible evidence of ritual use of obsidian at Machu Picchu is complemented back at the Chivay source by an Inca-style cutstone masonry structure (possibly a square *chullpa*) and pottery adjacent the principal obsidian workshop (Tripcevich and Mackay 2011).

While ethnohistoric, ethnographic, and archaeological evidence suggests that geological source areas of visually significant materials were prominent in the landscapes of communities throughout the Central Andean region, Andean peoples used these materials in eminently practical activities for which sharp, workable materials are desirable, and also engaged with them also through ritual practices and ceremonial display, potentially signaling social identity, and/or status. Artifacts may have been imbued with essential power derived from associations with the procurement zone, and may have communicated obligations and relationships between individuals, communities, and sacred entities manifested in various ways in an animated landscape—but in the end, these relationships must be demonstrated in particular cases using specific evidence. Compelling arguments along these lines have been made for other sourceable stones in the Central Andes for Inca stones (Ogburn 2004 and Chap. 3), Tiwanaku stones (Chap. 4), and possibly exotic granite and limestone at Chavín (Turner et al. 1999). This raises an intriguing comparative possibility: did Andean peoples consider obsidian, basalt, and andesite equivalent materials, or was structural stone perhaps distinct from tool stone? Again, considerations of both procurement and consumption evidence may be mobilized to consider this question.

Conclusion

The scale and relatively undisturbed state of the Quispisisa source provide an opportunity to examine Prehispanic organization of production and extraction. The absence of large workshop areas in the immediate area of the source, while initially puzzling, may be resolved with further survey in the region and could ultimately benefit archaeological analyses because workshops from different time periods may have more spatial separation. Even when reduction areas are identified, temporal control over these deposits will be difficult, but in lieu of temporally diagnostic materials like pottery, obsidian hydration dates guided by ^{14}C dates on organic materials may be the best method of parsing the changing use of the source area thorough time. Various aspects of quarrying behavior discussed here—including intensity and character of use—will require chronological control to address.

As source-area research is integrated with evidence of regional production and consumption, a systematic approach to these research topics becomes possible. The varied importance of obsidian from the mundane to the socially significant suggests that source-area research can shed light not just on the contribution of obsidian to subsistence activities, but also on highland people's relationship with this stone source, and its meaning in their broader social and ceremonial milieu.

Ultimately, quarries are of interest for what they can tell us about the ways in which mineral resources were accessed, exploited, controlled, and understood in the Prehispanic Central Andes. Obsidian quarries are particularly well suited for examining diachronic changes in resource use in that they offer the possibility of links to consumption zones via geochemical sourcing and direct chronological control via hydration dating. In the case of Quispisisa-type obsidian, the timespan in question includes the development of agropastoral economies and the florescence and demise of regional polities, offering the possibility of a long-term study of Central Andean approaches to mineral resources.

Acknowledgments The authors would like to thank Jesus Vilchez, Steve Shackley, Katharina Schreiber, Richard Burger, Michael Glascock, Yuichi Matsumoto, Yuri Cavero, Cirilo Vivanco, and Roger Murillo, and acknowledge support from the Council of American Overseas Research Centers, the Stanford University Branner Library of Earth Sciences, and the UC Berkeley Archaeological Research Facility.

References

Barberena, R., Hajduk, A., Gil, A. F., Neme, G. A., Durán, V., Glascock, M. D., et al. (2011). Obsidian in the South-Central Andes: Geological, geochemical, and archaeological assessment of north Patagonian sources (Argentina). *Quaternary International, 245*(1), 25–36.

Bencic, C. M. (2000). Industrias líticas de Huari y Tiwanaku. In P. Kaulicke, & W. H. Isbell (Eds.), *Huari y Tiwanaku: Modelos vs. evidencias. Boletín de Arqueología PUCP*, (Vol. 4, pp. 89–118). Lima: Pontificia Universidad Católica del Perú Fondo Editorial.

Benzoni, G., & Smyth, W. H. (1857 [1565]). *History of the new world.* London: Hakluyt Society.

Bradley, R. (2000). *An archaeology of natural places*. London/New York: Routledge.

Bradley, R., & Edmonds, M. R. (1993). *Interpreting the axe trade: Production and exchange in Neolithic Britain. New Studies in Archaeology*. Cambridge: Cambridge University Press.

Brumm, A. (2010). Falling Sky: Symbolic and cosmological associations of the Mt. William Greenstone Axe Quarry, Central Victoria, Australia. *Cambridge Archaeological Journal, 20*(2), 179–196.

Burger, R. L. (1984). Archaeological Areas and Prehistoric Frontiers: The case of Formative Peru and Ecuador. In D. L. Browman, R. L. Burger, & M. A. Rivera (Eds.), *Social and economic organization in the Prehispanic Andes* (Vol. 194, pp. 33–71). Oxford, England: B.A.R. International Series.

Burger, R. L., & Asaro, F. (1977). Trace element analysis of obsidian artifacts from the Andes: New perspectives on Pre-Hispanic economic interaction in Peru and Bolivia—LBL6343. Lawrence Berkeley Laboratory.

Burger, R. L., Asaro, F., Michael, H. V., & Stross, F. H. (1984). The Source of obsidian artifacts at Chavín de Huántar. Appendix E. In R. L. Burger (Ed.), *The prehistoric occupation of Chavín de Huántar, Peru* (Vol. 14, pp. 263–270). Berkeley: University of California Press.

Burger, R. L., Asaro, F., Michael, H. V., Stross, F. H., & Salazar, E. (1994). An initial consideration of obsidian procurement and exchange in prehispanic Ecuador. *Latin American Antiquity, 5*(3), 228–255.

Burger, R. L., Asaro F., Salas G., & Stross F. (1998a) The Chivay obsidian source and the geological origin of Titicaca Basin type obsidian artifacts. *Andean Past* 5, 203–223.

Burger, R. L., Asaro, F., Trawick, P., & Stross, F. (1998b). The Alca obsidian Source: The origin of raw material for Cuzco Type obsidian artifacts. *Andean Past, 5*, 185–202.

Burger, R. L., Fajardo Rios, F. A., & Glascock, M. D. (2006). Potreropampa and Lisahuacho obsidian sources: Geological origins of Andahuaylas A and B Type obsidians in the province of Aymaraes, Department of Apurimac, Peru. *Ñawpa Pacha, 28*, 109–127.

Burger, R. L., & Glascock, M. D. (2000). Locating the Quispisisa obsidian source in the department of Ayacucho, Peru. *Latin American Antiquity, 11*(3), 258–268.

Burger, R. L., & Glascock, M. D. (2001). The Puzolana obsidian source: Locating the geologic source of ayacucho type obsidian. *Andean Past, 6*, 289–307.

Burger, R. L., & Glascock, M. D. (2002). Tracking the source of Quispisisa Type obsidian from Huancavelica to Ayacucho. In W. H. Isbell & H. Silverman (Eds.), *Andean archaeology I: Variations in sociopolitical organization* (pp. 341–368). New York: Kluwer Academic/Plenum Publishers.

Burger, R. L., & Glascock, M. D. (2009). Intercambio prehistórico de obsidiana a larga distancia en el norte Peruano. *Revista del Museo de Arqueología, Antropología e Historia, 11*, 17–50.

Burger, R. L., Mohr Chávez, K. L., & Chávez, S. J. (2000). Through the Glass Darkly: Prehispanic obsidian procurement and exchange in southern Peru and northern Bolivia. *Journal of World Prehistory, 14*(3), 267–362.

Burger, R. L., & Salazar, L. C. (2004). *Machu Picchu: Unveiling the mystery of the Incas*. New Haven, CT: Yale University Press.

Burger, R. L., Schreiber, K. J., Glascock, M. D., & Ccencho, J. (1998c). The Jampatilla obsidian source: Identifying the geological source of pampas type obsidian artifacts from Southern Peru. *Andean Past, 5*, 225–239.

Cann, J. R., & Renfrew, C. (1964). The characterization of obsidian and its application to the Mediterranean region. *Proceedings of the Prehistoric Society, 30*, 111–133.

Clark, J. E. (2003). A review of twentieth-century mesoamerican obsidian studies. In K. G. Hirth (Ed.), *Mesoamerican lithic technology: Experimentation and interpretation* (pp. 15–54). Salt Lake City: University of Utah Press.

Close, A. E. (1996). Carry that weight: The use and transportation of stone tools. *Current Anthropology, 37*(3), 545.

Cobo, B. (1979 [1653]). History of the Inca Empire, translated and edited by Roland Hamilton Retrieved from http://hdl.handle.net/2027/heb.01916

Contreras, D. A. and Nado, K. (in press). Interpreting geochemically characterized obsidian from Chavín de Huántar, Peru. In D. Peterson and J. Dudgeon (Eds.), *The archaeology of circulation, exchange and human migration: Techniques, cases, evidence*. Equinox Publishing, Ltd.

Contreras, D. A., Tripcevich, N., & Palomino, Y. C. (2013). *Investigaciones en la fuente de la obsidiana tipo Quispisisa, Huancasancos Ayacucho*. In Proceedings of the XVI Congreso Peruano del Hombre y la Cultura Andina y Amazonica, UNMSM, Lima, Peru, October 2009, Lima, Peru.

Cooney, G. (1998). Breaking stones, making places: The social landscape of axe production sites. In A. M. Gibson, & D. D. A. Simpson (Eds.), *Prehistoric ritual and religion* (pp. xiii, 242 p.). Thrupp, Stroud, Gloucestershire: Sutton.

Couture, N. C. (2003). Ritual, monumentalism, and residence at Mollo Kontu. In A. L. Kolata (Ed.), *Tiwanaku and its hinterland: Archaeology and paleoecology of an Andean civilization,* (Vol. 2, pp. 202–225). Washington, DC: Smithsonian Institution Press.

Edmonds, M. M. (1990). Description, understanding and the chaîne opératoire. *Archaeological Reviews from Cambridge, 9*(1), 55–70.

Edmonds, M. R. (1995). *Stone tools and society: Working stone in Neolithic and Bronze Age Britain*. London: Batsford.

Eerkens, J. W., & Rosenthal, J. S. (2004). Are obsidian subsources meaningful units of analysis? Temporal and spatial patterning of subsources in the Coso Volcanic Field, southeastern California. *Journal of Archaeological Science, 31*(1), 21–29.

Eerkens, J. W., Vaughn, K. J., Carpenter, T. R., Conlee, C. A., Grados, M. L., & Schreiber, K. (2008). Obsidian hydration dating on the South Coast of Peru. *Journal of Archaeological Science, 35*, 2231–2239.

Elston, R. G., & Raven, C. (Eds.). (1992). *Archaeological investigations at Tosawihi. A Great Basin Quarry, Part 1: The periphery, Vol. 1. Report prepared for the Bureau of Land Management, Elko Resource Area, Nevada.* Silver City, NV: Inter-mountain Research and Bureau of Land Management.

Ericson, J. E. (1984). Towards the analysis of Lithic production systems. In J. E. Ericson & B. A. Purdy (Eds.), *Prehistoric quarries and Lithic production. New directions in archaeology* (pp. 1–10). Cambridge: Cambridge University Press.

Giesso, M. (2003). Stone tool production in the Tiwanaku heartland. In A. L. Kolata (Ed.), *Tiwanaku and its hinterland: Archaeology and paleoecology of an Andean civilization,* (Vol. 2, pp. 363–383). Washington, DC: Smithsonian Institution Press.

Glascock, M., Speakman, R. J., & Burger, R. L. (2007). Sources of archaeological obsidian in Peru: Descriptions and geochemistry. In M. Glascock, R. J. Speakman, & R. S. Popelka-Filcoff (Eds.), *Archaeological chemistry: Analytical techniques and archaeological interpretation* (pp. 522–552). Washington, DC: Oxford University Press.

Gosden, C., & Marshall, Y. (1999). The cultural biography of objects. *World Archaeology, 31*(2), 169–178.

Gould, R. A. (1980). *Living archaeology. New studies in archaeology*. New York: Cambridge University Press.

Healan, D. M. (1997). Pre-hispanic quarrying in the Ucareo-Zinapécuaro Obsidian Source Area. *Ancient Mesoamerica, 8*, 77–100.

Holmes, W. H. (1900). The Obsidian Mines of Hidalgo, Mexico. *American Anthropologist, 2*, 405–416.

Holmes, W. H. (1919). *Handbook of Aboriginal American Antiquities, Part I, Introductory, The Lithic Industries*. Bulletin of American Ethnology 60. US Government Printing Office, Washington, DC.

Ingold, T. (2006). Rethinking the animate, re-animating thought. *Ethnos: Journal of Anthropology, 71*(1), 9–20.

Ingold, T. (2011). *Being alive: Essays on movement. Knowledge and description*. New York: Routledge.

Jennings, J., & Glascock, M. D. (2002). Description and method of exploitation of the Alca obsidian source, Peru. *Latin American Antiquity, 13*(1), 107–118.

Klarich, E. A. (2005). *From the monumental to the mundane: Defining early leadership strategies at late formative Pukara, Peru* (Unpublished Ph.D. Dissertation). University of California, Santa Barbara.

Klink, C., & Aldenderfer, M. (2005). A projectile point chronology for the South-Central Andean Highlands. In C. Stanish, A. Cohen, & M. Aldenderfer (Eds.), *Advances in Titicaca basin archaeology,* (Vol. 1, pp. 25–54). Los Angeles, CA: Cotsen Institute of Archaeology.

Lechtman, H. (1984). Andean value systems and the development of prehistoric metallurgy. *Technology and Culture, 25*(1), 1–36.

Liritzis, I., & Laskarisa, N. (2011). Fifty years of obsidian hydration dating in archaeology. *Journal of Non-Crystalline Solids, 357*(10), 2011–2023.

McBryde, I. (1997). 'The landscape is a series of Stories'. Grindstones, quarries and exchange in aboriginal Australia: A lake eyre case study. In A. Ramos-Millán & M. A. Bustillo (Eds.), *Siliceous Rocks and Culture* (pp. 587–607). Granada: Universidad de Granada.

McCown, T. (1945). Pre-Incaic Huamachuco: Survey and excavations in the region of Huamachuco and Cajabamba. *University of California Publications in American Archaeology and Ethnology, 39*, 223–399.

Moore, J. D. (2010). Making a Huaca: Memory and Praxis in Prehispanic far northern Peru. *Journal of Social Archaeology, 10*(3), 398–422.

Nash, D. J. (2002). *The archaeology of space: Places of power in the Wari empire* (Unpublished Ph.D. dissertation). University of Florida.

O'Connor, B., Gabriel, C., & John, C. (2009). *Materialitas: Working stone, carving identity*. Prehistoric Society research papers. Oxford: Oxbow Books.

Ogburn, D. E. (2004). Evidence for long-distance transportation of building stones in the Inka empire, from Cuzco, Peru to Saraguro, Ecuador. *Latin American Antiquity*, 15(4), 419–439.

Owen, B. D., & Goldstein, P. S. (2001). Tiwanaku en Moquegua: Interacciones regionales y colapso. In P. Kaulicke (Ed.), *Horizonte Medio* (Boletin de arqueología PUCP, Vol. 2, pp. 169–188). Lima: Fondo Editorial de la Pontificia Universidad Católica del Perú.

Rademaker, K. (2006). Geoarchaeological investigations of the Waynuna Site and the Alca Obsidian Source. M.S., University of Maine.

Rademaker, K. (2012). *Early human settlement of the high-altitude Pucuncho Basin, Peruvian Andes* (Unpublishe Ph.D. dissertation). Department of Anthropology and Quaternary Studies, University of Maine, Orono.

Rudebeck, E. (1998). Flint extraction, axe offering, and the value of cortex. In M. R. Edmonds & C. Richards (Eds.), *Understanding the Neolithic of north-western Europe* (pp. 312–327). Glasgow: Cruithne Press.

Saunders, N. J. (2004). The cosmic earth: Materiality and mineralogy in the Americas. In N. Boivin & M. A. Owoc (Eds.), *Soils, stones and symbols: Cultural perceptions of the mineral world* (pp. 123–141). London: UCL.

Schiffer, M. B. (1975). Behavioral chain analysis: Activities, organization, and the use of space. *Fieldiana, 65*(103–174).

Sellet, F. (1993). Chaîne opératoire: The concept and its applications. *Lithic Technology, 18*, 106–112.

Shackley, M. S. (2005). *Obsidian: Geology and archaeology in the North American southwest*. Tucson: University of Arizona Press.

Shackley, M. S. (2011). *X-ray fluorescence spectrometry (XRF) in geoarchaeology*. New York: Springer.

Shott, M. J. (2003). Chaîne opératoire and reduction sequence. *Lithic Technology, 28*(2), 95–105.

Skeates, R. (1995). *Animate objects: A biography of prehistoric 'axe-amulets' in the central Mediterranean region 61*. London: Prehistoric Society.

Smith, M. L. (1999). The role of ordinary goods in premodern exchange. *Journal of Archaeological Method and Theory, 6*(2), 109–135.

Taçon, P. S. C. (2004). Ochre, clay, stone and art. In N. Boivin & M. A. Owoc (Eds.), *Soils, stones and symbols: Cultural perceptions of the mineral world* (pp. 31–42). London: UCL.

Topping, P. (2010). Neolithic axe quarries and flint mines: Towards an ethnography of prehistoric extraction. In M. B. La Porta, A. Burke, & D. Field (Eds.), *Ancient mines and quarries: A trans-atlantic perpective* (pp. 23–32). Oxford: Oxbow Books.

Torrence, R. (1986). *Production and exchange of stone tools prehistoric obsidian in the Aegean. New studies in archaeology*. UK: Cambridge University Press.

Tripcevich, N. (2007). *Quarries, caravans, and routes to complexity: Prehispanic obsidian in the South-Central Andes* (Unpublished Ph.D. dissertation). University of California, Santa Barbara.

Tripcevich, N. (2010). Exotic goods and socio-political change in the south-central Andes. In C. Dillian & C. White (Eds.), *Trade and exchange: Archaeological studies from history and prehistory*. New York: Springer.

Tripcevich, N., & Contreras, D. A. (2011). Quarrying evidence at the Quispisisa obsidian source, Ayacucho, Peru. *Latin American Antiquity, 22*(1), 121–136.

Tripcevich, N., Eerkens, J. W., & Carpenter, T. R. (2012) Obsidian hydration at high elevation: Archaic quarrying at the Chivay source, Southern Peru. *Journal of Archaeological Science, 39*(5), 1360–1367.

Tripcevich, N., & Mackay, A. (2011). Procurement at the Chivay obsidian source, Arequipa, Peru. *World Archaeology, 43*(2), 271–297.

Turner, R. J. W., Knight, R. J., & Rick, J. W. (1999). Geological landscape of the pre-Inca archaeological site at Chavin de Huantar, Peru. Current Research 1999-D. *Geological Survey of Canada*, 47–56.

Uhle, M. (1909). Peruvian throwing-sticks. *American Anthropologist, 11*(4), 624–627.

Williams, P. R., Dussubieux, L., & Nash, D. J. (2012). Provenance of Peruvian Wari obsidian: Comparing INAA, LA-ICP-MS, and portable XRF. In I. Liritzis & C. Stevenson (Eds.), *The Dating and Provenance of Obsidian and Ancient Manufactured Glasses* (pp. 75–85). Albuquerque, NM: University of New Mexico Press.

Yacobaccio, H. D., Escola, P. S., Pereyra, F. X., Lazzari, M., & Glascock, M. D. (2004). Quest for Ancient Routes: Obsidian sourcing research in Northwestern Argentina. *Journal of Archaeological Science, 31*, 193–204.

Chapter 3
Variation in Inca Building Stone Quarry Operations in Peru and Ecuador

Dennis Ogburn

High quality cut stone architecture was one of the hallmarks of Inca culture, employed in the construction of palaces, temples, and other important buildings (Fig. 3.1). However, being a symbol of very high status, this type of construction was rather limited in distribution. Naturally, dressed stone architecture is most common in Cusco and in other royal sites in the greater Cusco region. But outside the imperial heartland, the Incas limited its use to selected, significant sites, which were primarily concentrated in the highlands of Chinchaysuyu (the northwest quarter of the empire); well-known examples include Huánuco Pampa, Vilcashuamán, Cajamarca, Tomebamba, and Ingapirca. That quarter contains the most distant sites with cut stone architecture: Callo to the south of Quito near the volcano Cotopaxi, and Caranqui to the north of Quito. This type of stonework is much rarer in the other three quarters of the empire. Along the coast, adobe architecture predominates in the more important Inca sites, but cut stone architecture is found at a few sites, such as the major shrine of Pachacamac and the administrative center of Paredones in the Nasca Valley.

Despite the limited distribution of cut stone architecture in the Inca Empire as compared to other types of construction, it was still utilized in a significant number of buildings, terraces, and other structures, and thus required the production of a tremendous number of blocks. The process of locating and extracting suitable material not only involved a mammoth amount of labor, but must have also been considered a very important enterprise given the high status of cut stone architecture in the empire. Thus it is worth exploring Inca cut stone quarrying operations as a high status, labor intensive operation. Here I address the variability in those endeavors, including variations in the type, size and locations of quarries, the properties of the raw materials extracted, and the range of infrastructure associated with the quarries.

D. Ogburn (✉)
University of North Carolina at Charlotte, Charlotte, NC, USA
e-mail: dogburn@uncc.edu

N. Tripcevich and K.J. Vaughn (eds.), *Mining and Quarrying in the Ancient Andes*, Interdisciplinary Contributions to Archaeology, DOI 10.1007/978-1-4614-5200-3_3,

Fig. 3.1 Classic Inca imperial cut stone architecture at Pisaq, a royal estate of the Emperor Pachacuti in the Urubamba Valley

I will also discuss the cultural and religious connotations of both the stone and the quarries themselves and how those may have affected the choice of material to be exploited for building stones.

Of course, we would expect a level of variability in many aspects of Inca dressed stone quarrying given that Inca imperial strategies in the provinces were hardly uniform. In particular, it is well-understood that the Incas implemented political and economic policies that ranged greatly from low-impact, locally supervised arrangements to those with high investment and direct imperial supervision (e.g., D'Altroy 1992; Menzel 1959; Schreiber 1992). Thus, in the regions of the empire where the Incas constructed buildings of cut stone, their approach to extracting stone would have been guided to an extent by broad-based political, economic, and cultural factors in conjunction with imperial goals and needs specific to the region in question. Additionally, the variability in quarrying was to a large extent determined by nature. While various types of igneous, metamorphic, and sedimentary rock were available throughout the Andes, only a fraction of those deposits contained material that had properties suitable for producing dimension stone. Moreover, those deposits varied widely in terms of volume of accessible material and their location in relation to intended building sites, among other factors that could directly affect their potential for exploitation.

Furthermore, the nature of extracting, working, and transporting material for dressed stone masonry differed from the exploitation of other rock or mineral resources. As opposed to obsidian, gems, and other mined materials, cut building stone could not be collected in or reduced to easily transportable sizes, nor could it be processed and refined at the source in the same manner as metal ores. Instead, the rock had to be extracted in relatively large blocks that then had to be transported to

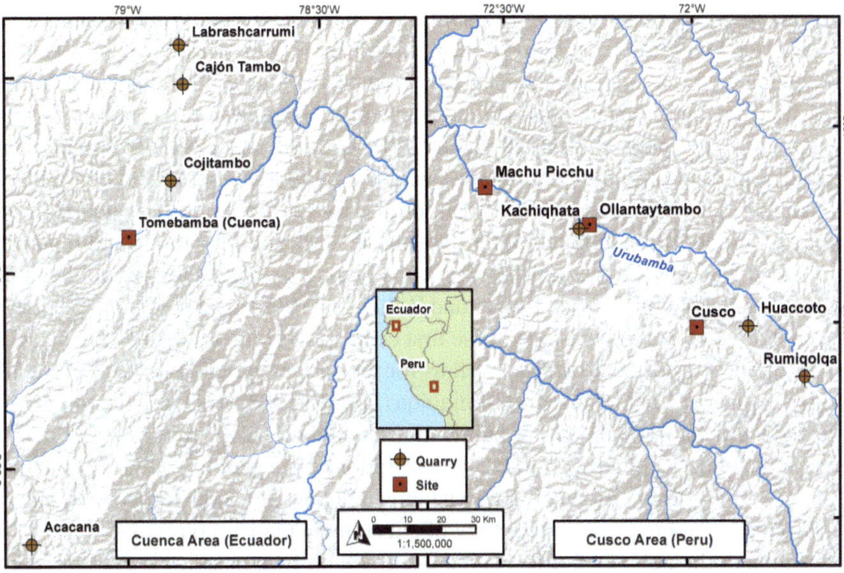

Fig. 3.2 Locations of cut stone quarries and major Inca sites discussed in the text

the building site. As a result, quarry operations for cut building stone required large investments in labor in all stages, from extraction, reduction to rough blocks, to transportation of blocks within and out of the quarries. Only the final stages of finishing and fitting the stones occurred on the building sites.

For this analysis, I draw on observations I have made between 2001 and 2011 at a number of quarries in Peru and Ecuador (Fig. 3.2), including Rumiqolqa, Huaccoto, and Machu Picchu in the Cusco region, and Cojitambo, Cajón Tambo, Labrashcarrumi, and Acacana in the southern highlands of Ecuador. The primary focus of my exploration of these quarries has been to obtain rock samples for use in geochemical sourcing studies; primarily this has entailed locating the quarries, examining the extent of quarrying debris and related infrastructure, and collecting a number of samples from dispersed locations throughout the exploited geological deposits. To date, my most extensive explorations have focused on Rumiqolqa, Cojitambo, and Huaccoto for my original geochemical provenance project (Ogburn 2004a) and an on-going project to extend geochemical sourcing to include more quarries and sites in the Cusco area, conducted in collaboration with Bill Sillar of University College, London. Although my quarry-related fieldwork has not entailed detailed survey or mapping of all features of the quarries, the information I have recorded has covered the axes of variability discussed here.

To supplement my first-hand observations, I also draw on published information regarding Inca building stone quarries, although the literature is rather sparse despite the fact that these places provided the material for the most iconic of Inca achievements. Early documents from the Spanish Colonial period, such as the accounts of Inca history and culture, sometimes mentioned stone quarries but gave few details—if we are lucky, a name would be given. Likewise, few details are given in the Relaciones Geográfics de Indias, even though they were created in part to give the Spanish crown an assessment of the resources available throughout the Andes. The first detailed accounts of Inca quarries come from explorer accounts of later centuries, most notably that of Squier (1877), who described the quarries of Rumiqolqa and Kachiqhata. Some geological works (e.g., Gregory 1916; Kalafatovich 1970) have provided data on the rock from some Inca quarries, but typically do not provide much other relevant detail beyond locations and reporting that there was evidence or knowledge of quarrying. Among the few publications with an archaeological focus, Heizer and Williams' (1968) brief study of stone quarries in Peru and Bolivia was one of the first concerted efforts to identify the quarries used for cut stone by the Incas as well as for the earlier site of Tiwanaku. Protzen's works regarding Kachiqhata and Rumiqolqa (Protzen 1983, 1993) are the most notable and the most in-depth. Other publications include Harth-Terré's (1965) study of Inca stonework, Kalafatovich's (1970) description of the geology of Sacsayhuaman, Gomis' (2003) observations on the Ecuadorean quarry of Cajón Tambo, and Hunt's (1990) petrographic provenance study of cut stone in the Cusco region.

Although the published information dealing with Inca quarries is far from comprehensive, in combination with my field observations, I believe it is more than sufficient for examining the variability of Inca cut stone building quarries. Here, rather than attempting to catalog or describe specific quarries in detail, my aim is to examine specific axes of variability in quarrying, covering the types of stones extracted, the physical types of quarries, the size range of quarry operations, methods of extraction, locations of quarries in relation to building sites, the range of infrastructure found in quarries, and the cultural aspects of quarrying and the meanings attached to quarries and the stone that came from them. I also endeavor to explore how all of these natural and cultural elements influence how the Incas ultimately decided where and what to quarry.

Types of Stone

Cut stone architecture requires massive stone, i.e., rock that can be quarried as blocks (Heldal and Bloxam 2008). Many Inca cut stone quarries in the highlands consist of hard igneous rock, which is no surprise given the volcanic origin of much of the Andes. Types of igneous stone quarried include granite (Machu Picchu), andesite (Rumiqolqa, Huaccoto, Cojitambo), diorite (San Blas in Cusco), and rhyolite (Kachiqhata). The Incas also used some softer rock for cut stone, as with the limestone employed in the megalithic walls of Sacsayhuaman, much of the construction at Chinchero (Heizer and Williams 1968), a number of buildings within and around Cusco, and some structures in the city of Tomebamba. While the Incas used various types of stone, it is challenging to identify all of the types and how frequently they were used because the literature contains many conflicting names applied to the rock coming from some of these quarries. In general, I suggest archaeologists take care in employing specific rock names unless they are derived from recent geological analyses—there are even discrepancies in names given by geologists over time.

In terms of selecting the type of rock to exploit, the Incas had many potential quarries available throughout the Andes. However, they did not always select the source closest to the building site, and it is notable that in many sites with cut stone architecture, they did not rely on just one quarry, but brought in stone from two or more sources. Thus convenience was not always a major factor in determining which source to exploit. As Protzen (1983: 184) noted "the choice of a particular rock type must have been of utmost importance, or they would not have quarried sites so difficult of access or so far away." Various physical properties came into play in choosing the stone to be used in different projects. For one, the Incas needed to select for a level of structural homogeneity that permitted the rock to be worked into large blocks. They needed deposits that were sufficient in volume to extract enough blocks to complete entire buildings: although the Incas sometimes used several different types of stone for constructing a site, they seldom employed more than one type of stone in the construction of individual high-status buildings of dressed stone masonry. This preference for making buildings with a single stone type can be observed at numerous sites that contain material from multiple quarries, including Ollantaytambo (Protzen 1993: 157), Cusco (as seen in the interior structures of the Coricancha), and Ingapirca in Cañar. In contrast, the Incas may have utilized smaller deposits primarily to quarry blocks for lower-status buildings that contained a mixture of stones from different sources. Because they were exploiting quarries that produced stone of varying physical appearance, it is clear that the Incas did not attempt to have a uniform look for all high status buildings throughout the empire by favoring one certain color or texture (although they may have aimed for such in a few cases: see "Cultural Meanings of Stone and Quarries: Choosing Sources of Cut Stone" below). Of course, another important property for any building stone was durability, and in particular the Incas seemed to pay attention to resistance to flaking and weathering.

Igneous deposits often hold the potential to provide good dimension stone because both extrusive and intrusive rock can be uniform, durable, and plentiful. For example, the granite quarried at Machu Picchu and the andesite from Rumiqolqa easily met these requirements. In contrast, Huaccoto, which was closer to Cusco than Rumiqolqa, produced stone that was more prone to flaking and weathering (Hunt 1990). But despite evidence of extensive quarrying activity there, it appears the Incas used comparatively little material from that quarry for construction in Cusco. The Incas faced the most constraints on material choice in their megalithic construction projects. Megaliths could be created from large boulders or outcrops, but to keep them from cracking or splitting when worked into blocks or moved, the rock had to have an internal structure that was not easily fractured and also had to be free from structural issues such as existing fractures or veins of softer material. But a total absence of natural fractures could have inhibited quarrying megaliths from bedrock because the Incas often took advantage of such cracks to separate blocks from the parent rock (Protzen 1983). These natural constraints may in part explain why significant Inca cyclopean construction is limited to the sites of Sacsayhuaman, Ollantaytambo, and Machu Picchu.

Fig. 3.3 Andesite outcrop in the Rumiqolqa quarry that was split with wedges; in the *lower right* along the crack, a roughly rectangular worked hole for inserting a wedge is visible. This presumably occurred in Inca times, but could be of more recent date

Types of Quarries and forms of Extraction

The Incas utilized two different strategies for extracting building stone: opencast and surface quarrying. I am not aware of any gallery quarrying for cut stone. Opencast mining entailed extraction of stone from hillsides or other accessible outcrops of bedrock, creating pits or cuts in the quarry landscape. Rumiqolqa and Cojitambo, two of the quarries most heavily exploited by the Incas, were of this type, providing abundant andesite blocks for Cusco and Tomebamba, respectively. These operations required devoting significant labor to separating material from the parent rock, a difficult step that has been the subject of much wild speculation. However, the Incas did not have to rely on any secret techniques. In many cases, they simply took advantage of natural fractures in the rock, wedging pieces apart using either wooden poles or bronze pry bars (Protzen 1983). In the absence of existing fissures, they would create a line of holes for inserting wedges to break the rock apart; in some cases, those holes may have been created by pounding with rock hammers while in others they may have used bronze chisels (Protzen 1983). Figure 3.3 shows an andesite outcrop at Rumiqolqa that was broken in this manner. For rock with regular fracture planes, they would pound a line along the stone until it split apart, as with the rhyolite blocks from Kachiqhata (Protzen 1993).

The second strategy was to exploit superficial deposits of boulders. Some of these surface operations focused on working boulders near their point of geological origin. Among these is Kachiqhata, which provided the rose-colored megaliths of Ollantaytambo; there boulders were selected from a huge mountainside rock fall (Protzen 1983, 1993). Closer to Cusco, Inca quarrying activity at the andesite quarry of Huaccoto may have focused on boulders available on the surface within the lava

Fig. 3.4 The Labrashcarrumi quarry, a superficial quarry focused on working glacially deposited boulders; the sacred lake of Culebrillas is visible in the background to the *left*

flow, perhaps in combination with extraction from bedrock. In other surface quarries, the Incas exploited boulders transported and deposited by glacial activity. For example, quarrying at Labrashcarrumi in the southern highlands of Ecuador entailed shaping large glacial erratics that had been deposited near the lake of Culebrillas (Fig. 3.4). Likewise, at the nearby quarry of Cajón Tambo, which provided the green stones for the site of Ingapirca in the province of Cañar, the Incas exploited boulders deposited in a glacial cirque (Gomis 2003). Within these surface quarries, most boulders may have been shaped into single ashlars, though the larger boulders may have been split to create multiple building blocks.

After separation from bedrock or boulders, the blocks were further reduced. For some blocks, this took place close to the point of extraction, as seen in the Llama Pit at Rumiqolqa, where around 250 cut stones were found (Protzen 1983). In other cases, blocks were moved to other locations to be further shaped into roughouts. The Rumiqolqa quarry contains a number of such reduction areas throughout the site (Fig. 3.5), and it appears that even some of the megalithic blocks extracted from Kachiqhata were moved to a separate section of the quarry to be shaped (Heizer and Williams 1968; Squier 1877). Final finishing was done at the construction site, where stones were modified to fit into walls and exterior faces and edges were pecked to match the style of surrounding blocks. The separate quarry work areas suggest specialization among workers, with some assigned to prying out rough blocks or trimming boulders and others dressing the stones to a near-finished state before being shipped out. Such division of labor and spatial segmentation of stages of production may have characterized a number of Inca state-directed extractive operations, as noted in the silver mining and refining in Porco (Van Buren and Presta 2010) and copper mining in Chile (see Chap. 12).

Fig. 3.5 Worker activity area in the Rumiqolqa quarry; rounded hammer stone is visible in *center* surrounded by smaller chips from the later stage of reduction, while large pieces broken off in the earlier stage were tossed farther from the work zone

Location in Relation to Building Sites

The location of quarries in relation to intended building sites was important because of the high labor demands for moving blocks to the scene of construction. In some cases, such as Machu Picchu, rock was quarried on site. Other quarries were within 1–4 km of the site of use, as with the quarries of Acacana and Cajón Tambo in Ecuador and Kachiqhata in Peru. Remarkably, the greatest distance between quarry and construction site is seen in largest quarries with which I am familiar. Rumiqolqa, which supplied the majority of cut stone for the buildings of Cusco, is about 35 km from the heart of the city, while Cojitambo, the main source of cut stone for Tomebamba, is about 20 km distant from that site.

However, it is important to note that close proximity to the construction site did not correlate with ease of transport, as stones often had to be transported up or down steep hills and across rivers. For example, the rhyolite megaliths quarried at Kachiqhata had to be moved 700–900 m down a mountainside, across the Urubamba River, and up another hillside to the construction site at Ollantaytambo. Because of the size of the blocks, they could not be brought across a bridge and had to be dragged directly across the river bed, likely during the dry season when the river was at its lowest (Protzen 1993). From other quarries, the Incas could construct bridges to facilitate the transport of smaller sized blocks, as with Cajon Tambo from where the stones had to be carried across the Río Silante. Steep slopes could also add to the difficulty of transport. The andesite quarry of Huaccoto is a notable example, being located up a mountain at an elevation around 4,100 m, roughly 700–800 m higher than the Inca capital. Huaccoto is much closer to the city of Cusco than Rumiqolqa, and Hunt (1990: 28) states "a gravity gradient would help in transport of quarry material to Cuzco," whereas workers would have had to exert more effort to bring Rumiqolqa stones up the valley. However, the simple statistics of elevation and distance are misleading, because the descent from Huaccoto is long and sinuous, as well as steep. Bringing stones down from that quarry would have been dangerous and in places would have required fighting against gravity to keep movement of stones downhill from getting out of control. In contrast, the 100 m rise in elevation over the 35 km distance from Rumiqolqa would have been negligible, and the path to Cusco was rather straight and level. Although there is evidence of Inca activity at Huaccoto, it is not clear how much stone they quarried there during imperial times and where they may have used it; at any rate, it appears the great majority of fancy cut stone in Cusco was brought from Rumiqolqa. Regardless of the location of specific quarries, the fact that the Incas moved 500 or more blocks weighing as much as 700 kg from Rumiqolqa to southern Ecuador (Ogburn 2004a) demonstrates that the Incas were capable of overcoming almost any obstacle when transporting stone from quarry to building site and that saving labor was not always a primary consideration.

While distance to building sites varied, it is interesting that many of these quarries lay along or within 1 km of a major Inca road. The quarries of Acacana, Cojitambo, Cajón Tambo, and Labrashcarrumi in southern Ecuador were all located along or very near the main Chinchaysuyu road. Rumiqolqa was located within 1 km of the Collasuyu road that headed southeast from Cusco toward Lake Titicaca. A notable exception is Huaccoto, which would have required the construction of a spur road up the mountain to link the quarry with the Collasuyu road, if not a direct route to Cusco. In part, the frequent association between quarries and major roads may be tied to the fact that these quarries supplied stone for major sites along those main roads.

Size of Quarries

The size of each quarry operation, including locations for extraction, reduction, storage, and other functions, can be hard to accurately assess. Ideally, the extent of pits, worked outcrops, tailings and abandoned blocks can allow for an estimation of the quarried area. But in cases where exploitation of the quarries continued into Spanish Colonial times or up to the present, as with Rumiqolqa, Huaccoto, and Cojitambo, it is difficult to gauge what resulted from Inca use versus more recent activity. In those cases, the presence of abandoned blocks of Inca style, evidence of being worked using Inca tools such as river cobble hammerstones, and elements of Inca infrastructure would be the best indicators of which areas were worked during the Inca imperial phase. However, because the Spanish employed native workers as quarrymen for their building projects, most of the techniques from Prehispanic times would have been carried over for some time, and rough rectangular blocks created for Spanish structures may have been similar to abandoned Inca blocks. It may be even more difficult to distinguish Inca from pre-Inca quarrying activity, as the techniques employed would have been similar if not identical.

Thus, those caveats should be kept in mind when characterizing the area covered by Inca cut stone quarry operations. In general, where bedrock is being extracted, the quarrying operation may be small in terms of surface area, whereas quarries exploiting glacial or other surface deposits may be more extensive because of the dispersed nature of the raw material. Nonetheless, the largest operation, Rumiqolqa, was a bedrock type quarry, and it undoubtedly produced the greatest number of building stones for the Incas. As a rough estimate that operation may have spanned between 100 and 200 ha. The Cojitambo quarry, which was a similar type of subsurface operation, was probably in the range of 30–40 ha. Kachiqhata, a surface quarry, was probably not exploited to any significant extent after Inca imperial times (Protzen 1993), so most if not all surface evidence should be the result of Prehispanic quarrying; that quarry operation may have covered up to about 100 ha. At the smaller end of the scale, limited extraction for a small site or a single building may only have resulted in small quarries. For example, the quarry pit at Acacana in southern Ecuador is only about 10 m in diameter, and it may have only been quarried, perhaps with other pits, for a few modest sized structures at the nearby sites of Inkapirka and Tambo Blanco. At times it may be difficult to determine quarry size even when a quarry was exclusively used by the Incas, as built-up soil or vegetation could cover significant portions of the spoils and other surface evidence. This appears to be the case at Labrashcarrumi in southern Ecuador (Fig. 3.4), where judging by the partially worked blocks on the surface, the quarry operation there may have covered up 10 ha. Quarries like Labrashcarrumi that focused on working surface material can also be less visible archaeologically because they comprise dispersed, small concentrations of spoils (Heldal and Bloxam 2008) rather than more obvious quarry faces or pits that result from extracting blocks from bedrock.

Quarry Infrastructure

A certain level of added infrastructure was necessary to support the workers and meet the physical needs of extracting stone. The associated facilities and infrastructure at Inca quarries shows a very wide range of variation, indicating that there was not a standard approach to formal organization of the quarries. Instead, the added infrastructure was related to the distance from the construction site and from other Inca settlements, the size of the quarry itself, and its geology and topography.

In terms of human needs, because these operations entailed extended stints of labor on-site, workers needed to be provided with social infrastructure, i.e., food, water, and housing. How those needs were met would have depended both on how the labor was mobilized and the distance from other Inca imperial sites as well as from local settlements. If the workforce comprised people from the local area who were fulfilling their *mita* labor obligations, they could have traveled daily between their homes and the quarry, obviating the need for housing. In contrast, housing would have been necessary in cases where *mita* laborers had to travel far from their homes, when workers were *mitmaqkuna* resettled permanently to work at the quarries, or when quarries were distant from inhabited zones. Nearby Inca state settlements could have fulfilled the need for housing as well as storing food and providing

Fig. 3.6 Road surface in the Inca section of the Huaccoto quarry, with surfacing of small pebbles

water for workers; in lieu of such sites, facilities may have been constructed at or adjacent to the quarry.

As for physical requirements, quarrying operations frequently required constructed roads to move blocks within the quarry itself and to the nearest road. Subsurface operations and hillside quarries sometimes required extensive construction of ramps and roadbeds to move stone out of pits and around the quarry; roads on steep slopes required significant shoring up, and ramps and routes through pits had to be frequently changed or augmented as pits were enlarged or the locus of extraction shifted. Steep slopes often required the construction of level spaces to serve as activity areas, such as for shaping rough stone into finished blocks. Construction of roads, ramps, etc. was often done with the spoils of quarrying activity, which were abundant on-site.

The specific infrastructure created and methods used to facilitate movement of stones within and out from these quarries have been the subject of much speculation aimed at explaining how the Incas managed to move large blocks weighing up to 100 t. Accounts from the Spanish Colonial era all agree that the Incas used ropes and sheer manpower to drag stones within quarries and to construction sites (e.g., Betanzos 1996: 157 [1557]; Cieza de León 1984: 145 [1553], Cieza de León 1985: 190 [1553]; Garcilaso de la Vega 1966: 464 [1609]). It is often assumed that the Incas reduced the needed labor by employing roller logs or rounded stones akin to large ball bearings. But contrary to popular belief, there is no solid supporting evidence for such techniques, though it is likely that the Incas employed other means for reducing friction when transporting stones. Protzen (1993) suggests that the Incas may have put clay on the roadbed leading to Ollantaytambo, which would have been wetted to provide a slick surface to ease dragging large blocks. The best evidence for efforts at easing transport may be seen at Huaccoto, where the surface of the main roadbed running through an Inca section of the quarry is littered with rounded pebbles about 2–5 cm in diameter (Fig. 3.6). As stream-rounded rocks, these were not local to the area and were clearly brought to the quarry for a specific purpose. The pebbles would have eased the dragging of blocks within and out from the quarry. However, it probably would not have helped with moving cyclopean blocks, which were not produced at Huaccoto. It should also be noted that because Huaccoto may have been exploited more heavily in the Spanish Colonial era than in Inca times, an Inca origin for the pebble surface is not totally certain.

Within the other parameters of human and physical requirements, there was considerable variation in associated infrastructure. On-site quarries, as at Machu Picchu, may have had no facilities specific to the quarry, as workers could have been housed at the settlement. Roads and ramps for moving stones would have been the same routes utilized for other purposes at the site, or would have been dismantled or built over as construction progressed. Thus evidence of quarrying operations is comparatively minimal at Machu Picchu; not even the extensive spoils from quarrying and stone working are visible because they were used in part to serve as fill and drainage for the sacred plaza, agricultural terraces, and other sectors of the site (Wright et al. 1999). In comparison, Labrashcarrumi was not an on-site quarry (and at present it is unclear where the cut stone of that quarry was used), but it needed no dedicated support facilities because it was situated within 1 km of the *tambo* of Paredones de Culebrillas, which was located along the main highland road. There are no indications of roads or causeways for moving finished blocks in the area of that quarry, but that may be due to the covering of *páramo* vegetation that has also obscured most of the debris from the quarrying operation.

The greatest extent of infrastructure is found at the large quarries that were located far from building sites. At Cojitambo, it is difficult to identify Inca features because of the heavy continued use of the quarry and the encroachment of modern roads and habitation, but there are remains of some rough structures that could date to that time period. There are also some structures of definite Inca origin at the north end of the hill of Cojitambo; those may have combined quarry support with *tambo* functions. In contrast, Rumiqolqa contains remains of extensive associated infrastructure that clearly date to Inca use of the quarry. These include a network of roads, ramps, and causeways to move stones around the site and to haul them out of the pits and the quarry. There are remains of a number of buildings on the west side of the quarry, in the zone McEwan (1984) referred to as Qharanqayniyoq, where a wide stairway leads up to structures including a long hall facing a rectangular plaza and a number of other rectangular buildings (Fig. 3.7). These presumably housed workers, supervisors and support personnel. However, some of the buildings in that zone may have been of Wari construction, perhaps supporting quarrying activities during the Middle Horizon (McEwan 1984; Fig. 3.8). Across a small ravine, there is a series of storehouses situated near the crest of the hill and facing toward Qharanqayniyoq, undoubtedly holding the supplies to support and feed the workforce. Also within the quarry settlement area I observed several small beehive-shaped structures that may have been *chullpas* (burial chambers), and Protzen (1983) noted Inca structures in other parts of the quarry. All of this indicates a large-scale, full time work force was employed at the site, provisioned by the state. I am not aware of any records that mention the labor force of Rumiqolqa, so I cannot say whether it comprised men serving their rotational *mita* obligations or a more permanent

Fig. 3.7 Elements of Inca infrastructure at the Rumiqolqa quarry, including a stairway leading up to a number of rectangular structures that presumably housed workers

Fig. 3.8 Remains of possible Wari structures at the Rumiqolqa quarry; note *rounded corners*, which contrast with squared corners of Inca structures elsewhere in the quarry

population of local workers or *mitmaqkuna*. But people were certainly living, working, and dying at the site. Kachiqhata, another large quarry, also had a great amount of infrastructure (Protzen 1983, 1993; Squier 1877) even though it was not located too far from the building site at the high status sector of Ollantaytambo. The series of retaining walls, ramps, roads, and slipways that cross the quarry are directly related to its location on a steep mountainside. There are also sets of structures that probably housed workers and supervisors, and *chullpas* are found around the quarry area. It is interesting that workers and supervisors may have been housed on-site, as opposed to within the settlement of Ollantaytambo itself, which was only 4 km distant and contained a substantial residential sector.

Cultural Meanings of Stone and Quarries: Choosing Sources of Cut Stone

Finally, and perhaps most importantly, these quarries and the stone extracted from them had significant meanings to the Incas (see Dean 2010 for a discussion of the varied meanings applied to stones), and certain cultural preferences may have influenced which source of rock the Incas ultimately chose to build their most important buildings (see Chap. 4 for a discussion of how cultural meanings may have also factored into choice of building stone for the earlier Tiwanaku culture). While geological processes determined the location of potential quarries and affected the suitability of rock for different applications, cultural concerns played a great role in determining which of those potential sources were quarried and how much material was extracted. At times, these decisions ran counter to the notions of efficiency and ease of access and exploitation that predominate in contemporary capitalist economic operations.

To begin with, the act of building with cut stone was in itself a choice heavily laden with cultural meanings. To meet their needs for constructed spaces to serve myriad functions, the Incas could have much more easily built everything with adobe or fieldstone and mortar. While some very important structures were built of those materials, it is clear that cut stone buildings were the epitome of the Inca architecture of power (Gasparini and Margolies 1980). By using cut stone, the Incas of the ruling class were not just building nice, durable structures, but sending multiple messages to various audiences. They set themselves apart from their subjects, provincial elites, Incas of lower status, and other groups through different physical spaces and materials. By using a very labor-intensive form of construction to create what was essentially monumental architecture (in terms of labor consumed, if not always in size), the Incas were signaling their ability to command resources, skilled craftsmen, and immense amounts of labor (Trigger 1990: 122) and emphasizing their authority and control (Rapoport 1993: 36). This conspicuous consumption of labor to create these buildings served to not only advertise Inca power but also to reinforce it (Ogburn 2004b).

Of course, the decision of the ruling Incas to utilize dressed stone for imperial architecture did not occur in a vacuum. While cut stone architecture was relatively rare in the pre-Inca highlands, it was actually employed around Cusco during the Late Intermediate Period, as seen at Cruz Moqo (Fig. 3.9), contiguous with Sacsayhuaman. Beyond local precedents, much has been made of the historical claims that the Incas

Fig. 3.9 Pre-imperial cut stonework at Cruz Moqo above Cusco

remodeled Cusco and created their imperial architecture in imitation of the Middle Horizon city of Tiwanaku (see Chap. 4 for illustrations), which was seen in ruins by the Emperor Pachacuti (e.g., Cieza de León 1984 [1553]; Cobo 1990 [1653]). However, it is clear from studies of the details of Inca architecture and stone working that the Incas were not precisely replicating Tiwanaku masonry nor were they using the exact same techniques: it is more accurate to say they were inspired by rather than imitating what they had seen there (Gasparini and Margolies 1980; Protzen and Nair 1997). Nonetheless, high status Inca architecture gained significant cultural capital by referring to a powerful and significant polity of the past.

Once the Incas had decided to build with cut stone (first in Cusco, then in other locales), this then led to the task of finding suitable material and deciding which sources to exploit. As noted (see "Types of Stone" above), the properties of material at any given outcrop limit the possible choices to those where the rock can be successfully extracted, worked, and employed as dimension stone. The process of finding and then choosing a quarry depended first on discoverability, i.e., being able to see the material on the surface either through some type of exposure of bedrock or as loose boulders that allowed the Incas to gauge the suitability of the stone. The Incas then had to determine whether enough rock was accessible through their quarrying techniques to produce enough blocks to create one or more entire structures. In many cases, finding quarries probably entailed testing numerous outcrops or loose boulders through flaking off material to examine its properties. In essence, the Incas had to employ geological prospectors. But in some cases, the Incas may have simply utilized existing quarries. For example, it is thought that the Wari Empire may have been exploiting the Rumiqolqa andesite, as suggested by structures that appear to be of Wari style within the quarry area (McEwan 1984; Fig. 3.8). Taking over known resources was of course a typical imperial practice. But this was not necessarily a common option for the Incas when it came to building stone quarries because few groups in the Andes were actively using cut stone at the time of Inca expansion. Instead, it is possible that the Incas sometimes re-used quarries that had been long-abandoned, as there were other societies that had quarried cut stone in earlier eras. Nonetheless, that building technique was still rather limited geographically. Moreover, existing quarries for rough stone construction may have been tested, but many may not have been suitable for high quality dressed stone. If Rumiqolqa had been exploited by Wari, it was most likely for rough or partially worked stone, as they did not utilize fancy cut stone in their sites in the Cusco region.

Religion was another cultural element that could have influenced which places were chosen to be quarried as well as which stone to use in specific buildings. A basic religious connotation stems from the fact that the stone from these quarries was used to build the most sacred buildings in the empire, including the Qoricancha, as well as the palaces for the Incas themselves. Beyond that, some, if not all quarries were themselves considered sacred places, and by extension, the material from them would have been deemed sacred as well (this also applied to mining metals: see Chap. 12). For example, quarries featured as three of the shrines in Cobo's (1990 [1653]) catalog of the Cusco *ceque* system (a series of paths radiating from Cuzco that each contained a sequence of sacred places and objects). Guayrangallay, the seventh shrine along the second Chinchaysuyu *ceque*, was a quarry above Sacsayhuaman where the Incas made various sacrifices. Guayrangallay may have been located on a hill to the northeast of Sacsayhuaman (Bauer 1998: 54), and it could have been the primary source of the megalithic limestone blocks used at that site. Curovilca, the fourth shrine on the second Antisuyu *ceque,* was a quarry for which Cobo noted that "they sacrificed to it so that it might not give out, and so that the buildings built from it might not fall" (Cobo 1990 [1653]: 64). The fourth shrine on the fourth Antisuyu ceque was a quarry called Viracocha. It was said that a stone that looked like a person came out of the rock they were cutting out to build a house for the Inca, and that the Inca declared that the quarry should be worshipped as a *huaca* (an important place or thing held as sacred by the Incas).

Specific religious meanings could well have been attached to places after quarrying activity had commenced, as illustrated by the *ceque* shrine called Viracocha. But in other cases, places on the landscape where quarries were established may have been significant pre-existing *huacas* themselves. For example, the hill of Cojitambo was one of the major sacred places of the Cañaris. Also in Cañari territory, the Labrashcarrumi quarry was located next to the important sacred lake of Culebrillas (Fig. 3.4). The mountain of Acacana, quarried to produce the ceremonial site of Inkapirka of Saraguro, was a major *huaca*, noted by Albornoz to be the *pacarisca* (sacred place of origin) of the Paltas (Duviols 1967). The sacred nature of these places could have provided a compelling reason for the Incas to quarry stone from them, an act that could serve to appropriate native shrines in the provinces and put an indelible imperial stamp on them. Moreover, through the extraction of cut stone, the essence of those places was incorporated into major state temples and other buildings.

The religious dimensions of these sacred locations, the initiation of quarrying activity, and the routes of major imperial roads may have been closely intertwined.

In some cases, the Incas may have chosen to build a road next to major *huacas* as a strategy of appropriating their power (Ogburn 2010); quarry access may have been an added benefit to such a practice. But in other cases, did the Incas decide to build roads by places that could serve as quarries, or did they instead seek quarries that were close to the roads after the latter had been built? It is feasible they followed both strategies, but outside of the Cusco region it must have depended on the unfolding of imperial strategies as a province was conquered and integrated into the empire. Roads were essential to the movement of armies and the process of conquest as well as the primary needs of communication and transport of food and other resources. Thus they were likely to be constructed first, though desire for access to and appropriation of sacred places on the landscape could have influenced the planning of specific routes. The decisions to create imperial buildings with cut stone could well have come afterward as part of Inca strategies to further consolidate and maintain their control over the province.

Finally, there may have been cultural significance attached to the color of some of the stone extracted from these quarries. Given that the Incas often considered unusual things to be sacred or otherwise significant (Cobo 1990 [1653]), those stones that had remarkable coloration may have been considered especially sacred. Thus, the rose colored stone of Kachiqhata and the green stone of Cajón Tambo may have been quarried and utilized for the temples of Ollantaytambo and Ingapirca, respectively, because of their distinctive colors. Likewise, the rich blue gray color of Rumiqolqa stone may have been important, and its variability may have also been significant. Rumiqolqa stones that exhibit flow banding may have been particularly sacred, as many were selected to create the base of the Qoricancha (Protzen 1983). It is also possible that the Incas chose to exploit Rumiqolqa for the Qoricancha and other important buildings in Cusco because the stone approximated the color of the gray andesite from the Ccapia quarry that was used to create gateways and monolithic sculptures at Tiwanaku. Selecting similar material in terms of both color and type may well have been part of the emperor Pachacuti's program for re-building Cusco as supposedly inspired by his visit to Tiwanaku. While the Incas did not strive for uniform color of cut stone throughout the empire, it is interesting that the material from Cojitambo used in cut stone construction of Tomebamba was also a gray colored andesite. This suggests that perhaps the Incas were trying to approximate the look and feel of Cusco architecture in a site that they were building up to be their second capital. Unfortunately, it is likely impossible for us to know what significance particular stone colors had for the Incas, as I know of no specific reference to this aspect of cut stone in historical documents.

Summary/Conclusions

To quote Protzen (1983: 184) once again, it is clear "that quarrying was a very important operation to the Incas, and not a routine matter." There was a high degree of variability in the location of quarries, material extracted, distance from sites, extent of quarrying, and associated infrastructure. Yet in all cases, the quarrying, shaping, and transport of cut stone required a significant investment in labor. The choices the Incas made in terms of material and quarry location show that in some cases convenience and labor efficiency may have been a concern, whereas in other cases, the desirable qualities of the stone or its sacred associations may have been paramount, resulting in operations that required far more labor than if they had chosen closer sources of rock.

In all, it appears the varied cultural meanings associated with the stone from these quarries played a significant, if not primary, role in Inca decisions regarding where and what to quarry; these decisions were not entirely economic ones oriented toward minimizing labor expenditure. This is not surprising given that we are dealing with the cut stone that served as a symbol of Inca power through its use in imperial temples, shrines, and palaces—the most important buildings in the empire. Even at Machu Picchu, the use of on-site granite may have had a very significant meaning, and convenience of access may have been coincidental. A different set of criteria may have applied in the construction of lesser buildings out of field stone, where the Incas were communicating different meanings and messages. I have no doubt that there was even more variability between Inca cut stone quarries, and perhaps there were other dimensions to what the Incas were doing with those quarries. On the whole, understanding how the Incas quarried material for cut stone architecture is valuable not just as a case study, but also because it is necessary for fully understanding the nature of their imperial architecture and because it is integral to the broader questions of understanding how the Incas carried out their imperial enterprise.

Acknowledgments The fieldwork that served as a basis for this chapter benefited from collaboration with and participation of a number of people, and in particular I would like to thank José Luis Espinosa, Julio César Sierra, and Bill Sillar. Funding for my initial visits to Rumiqolqa and Cojitambo was provided by the Archaeological Research Facility of the University of California, Berkeley. I would like to thank Nico Tripcevich and Kevin Vaughn for their feedback on this chapter and efforts in creating this volume. I also appreciate the anonymous reviewer comments, which contributed to improving the chapter.

References

Bauer, B. S. (1998). *The Sacred Landscape of the Inca: The Cusco Ceque System*. Austin: University of Texas Press.

Betanzos, J. de (1996 [1557]). *Narrative of the Incas*. [R. Hamilton and D. Buchanan, Trans. and Eds.]. University of Texas Press, Austin.

Cieza de León, P. (1984 [1553]). *La crónica del Perú primera parte*. Pontificia Universidad Católica del Perú Fondo Editorial, Lima.

Cieza de León, P. (1985 [1553]). *La crónica del Perú segunda parte (El señorío de los Incas)*. Pontificia Universidad Católica del Perú Fondo Editorial, Lima.

Cobo, B. (1990 [1653]). *Inca Religion and Customs* [R. Hamilton, Trans.]. University of Texas Press, Austin.

D'Altroy, T. N. (1992). *Provincial Power in the Inka empire*. Washington/London: Smithsonian Institution Press.

Dean, C. (2010). *A Culture of Stone: Inka Perspectives on Rock*. Durham, NC: Duke University Press.

Duviols, P. (1967). Un inédit de Cristóbal de Albornoz: la instrucción para descubrir todas las guacas del Pirú y sus camayos y haziendas. *Journal de la Société des Americanistes, 55*(1), 7–39.

Garcilaso de la Vega, el Inca. (1966 [1609]). *Royal Commentaries of the Incas and General History of Peru* [H. V. Livermore, Trans.]. University of Texas Press, Austin.

Gasparini, G., & Margolies, L. (1980). *Inca Architecture* [P. J. Lyon, Trans.]. Indiana University Press, Bloomington.

Gomis, D. (2003). Prospección y excavaciones en la cantera de Cajón Tambo (provincia de Cañar). *Revista de Antropologia, 17*, 37–67.

Gregory, H. E. (1916). A Geologic Reconnaissance of the Cuzco Valley, Peru. *American Journal of Science, 41*, 1–100.

Harth-terré, E. (1965). Técnica y arte de la cantería incaica. *Revista Universitaria* (Cuzco) 51–52 (122–123 and 123–125):152–168.

Heizer, R., & Williams, H. (1968). Archaeological Research in Peru and Bolivia. In *National Geographic Society Research Reports for 1963* (pp.127–134). National Geographic Society, Washington, DC.

Heldal, T., & Bloxam, E. (2008). QuarryScapes Guide to Ancient Stone Quarry Landscapes: Documentation, Interpretation, and Statement of Significance. Electronic Document. http://www.quarryscapes.no/guide_content_text.php, accessed May 22, 2012.

Hunt, P. N. (1990). Inca Volcanic Stone Provenance in the Cuzco Province, Peru. *Papers from the Institute of Archaeology, 1*, 24–36.

Kalafatovich, C. (1970). Geología del grupo arqueológico de la Fortaleza de Saccsayhuaman y sus vecinidades. *Revista Saqsaywaman, 1*, 61–68.

McEwan, G. F. (1984). *The middle horizon in the valley of Cuzco, Peru: The Impact of the Wari occupation of Pikillacta in the Lucre Basin* (Unpublished Ph.D. dissertation). University of Texas, Austin.

Menzel, D. (1959). The Inca Occupation of the South Coast of Peru. *Southwest Journal of Anthropology, 15*(2), 125–142.

Ogburn, D. (2004a). Evidence for Long-distance Transport of Andesite Building Blocks in the Inca Empire. *Latin American Antiquity, 15*(4), 419–439.

Ogburn, D. (2004b). Dynamic Display, Propaganda, and the Reinforcement of Provincial Power in the Inca Empire. In K. Vaughn, D. Ogburn, & C. Conlee (Eds.), *Foundations of Power in the Prehispanic Andes. Archeological Papers of the American Anthropological Association*, Vol. 14, pp. 225–239.

Ogburn, D. (2010). Inca Manipulation of the Sacred Landscape of Saraguro, Ecuador. *Ñawpa Pacha, 30*(2), 165–186.

Protzen, J.-P. (1983). Inca Quarrying and Stonecutting. *Ñawpa Pacha, 21*, 183–214.

Protzen, J.-P. (1993). *Inca Architecture and Construction at Ollantaytambo*. New York: Oxford University Press.

Protzen, J.-P., & Nair, S. (1997). Who taught the Inca Stonemasons Their Skills? A Comparison of Tiahuanaco and Inca Cut-Stone Masonry. *Journal of the Society of Architectural Historians, 56*(2), 146–167.

Rapoport, A. (1993). On the Nature of Capitals and Their Physical Expression. In J. Taylor, J. G. Lengellé, & C. Andrew (Eds.), *Capital Cities: International Perspectives* (pp. 31–67). Ottawa: Carleton University Press.

Schreiber, K. J. (1992). *Wari Imperialism in Middle Horizon Peru*. Anthropological Papers, No.87. Museum of Anthropology, University of Michigan, Ann Arbor.

Squier, E. G. (1877). *Peru: Incidents of Travel and Exploration in the Land of the Incas*. London: Macmillan.

Trigger, B. G. (1990). Monumental Architecture: A Thermodynamic Explanation of Symbolic Behavior. *World Archaeology, 22*(2), 119–132.

Van Buren, M., & Presta, A. M. (2010). The Organization of Inka Silver Production in Porco, Bolivia. In M. Malpass & S. Alconini (Eds.), *Distant Provinces in the Inka Empire: Toward a Deeper Understanding of Inka Imperialism* (pp. 173–192). Iowa City: University of Iowa Press.

Wright, K. R., Zegarra, A. V., & Lorah, W. L. (1999). Ancient Machu Picchu Drainage Engineering. *Journal of Irrigation and Drainage Engineering, 125*(6), 360–369.

Chapter 4
Building Taypikala: Telluric Transformations in the Lithic Production of Tiwanaku

John Wayne Janusek, Patrick Ryan Williams, Mark Golitko, and Carlos Lémuz Aguirre

Stone was elemental to Tiwanaku identity. One name for Tiwanaku, and possibly a key epithet during its pre-Inca apogee, was *taypikala*, or "central stone" (Cobo 1990 [1653]: 100). The employment of impeccably carved stones and the elaboration of stonework in Tiwanaku are among the most notable physical aspects of monumental construction at the site. Tiwanaku monumentality emphasized the permanence, mass, color, and texture of stone. It showcased stone's materiality. Yet the stones used to build Tiwanaku, and thus the materiality and technology of Tiwanaku monumental construction, shifted dramatically between AD 500 and 700. Tiwanaku's employment of new stone sources and technologies distinguished it from other centers in the Lake Titicaca Basin. What was the character of this shift? What were its lithic sources, and why did they change?

Stone remains critical for shaping native identities in the south-central Andes. Today, a person raised in Tiwanaku is considered *kalawawa*, or born of stone. Although partly in jest, this identification has deeper resonance in the region. This was made clear at a celebration of Machac Mara (or "New Year") on the June solstice in 2002, on the ruins of the site of Khonkho Wankane in the upper Desaguadero basin. The ritual is a recently established tradition designed by native-identifying communities as they began to garner power in Bolivian national politics.

J.W. Janusek (✉)
Anthropology, Vanderbilt University, Box 6050 Station B, Nashville, TN 37235, USA
e-mail: john.w.janusek@vanderbilt.edu

P.R. Williams • M. Golitko
Field Museum of Natural History, 1400 S Lake Shore Dr, Chicago, IL 60605, USA
e-mail: rwilliams@fieldmuseum.org; mgolitko@fieldmuseum.org

C. Lémuz Aguirre
Sociedad de Arqueologia de La Paz, y Universidad Mayor de San Andrés,
Casilla postal, 5294 La Paz, Bolivia
e-mail: clemuzaguirre@gmail.com

N. Tripcevich and K.J. Vaughn (eds.), *Mining and Quarrying in the Ancient Andes*, Interdisciplinary Contributions to Archaeology, DOI 10.1007/978-1-4614-5200-3_4,

Initiated in the focal blood sacrifice of a carefully selected camelid as the sun ascends the horizon to the east, it continues as a lively event full of vendors selling wares, competitive local dances, and political discussion. At one point in 2002, a local political aspirant made a particularly dramatic gesture. In the midst of an animated speech on the importance of fostering a native Aymara identity, he bent over and patted his hand vigorously on one of Khonkho's carved stone monoliths, while proclaiming "we are this stone, this stone is us."

The relationship of humans and their landscapes in the highland Andes is deeply political. Hilltops and mountain peaks are particularly important places invoked in local prayers, libations, and sacrificial offerings in earnest bids to ensure health, community well-being, and the success of particular projects (archaeological and otherwise, see Abercrombie 1998). So are local springs and certain other prominent landscape features. Hilltops and mountain peaks are animated as powerful ancestral persons in native ritual practices such that they come to support, and indeed represent, the local communities who pay tribute to them. Hilltops and peaks also punctuate the landscapes of local communities and larger political organizations. Fundamental components of local ritual practices, they define sociopolitical territories while grounding communities to the altiplano and its agropastoral and mineral resources. The essential component of mountains, stone was a particularly important class of resources in the altiplano.

Given the profound long-term importance of stone in the Andean highlands, it is striking that so little systematic research has been dedicated to understanding quarrying and preparation practices for monumental construction. A notable example of recent research includes Dennis Ogburn's analysis of carved andesite ashlars in Saraguro, Ecuador, that date to the Inca period (Ogburn 2004a, b). Through XRF spectrometry, he determined that those volcanic blocks had been quarried in the Rumiqolqa quarry some 35 km from Cuzco, Peru; approximately 1,600 km from Saraguro. Ogburn concludes that it was important for the Inca to transfer volcanic material from a particular source and of a particular material type over an incredible distance. The Inca employed the same andesite quarry to construct many of the most important structures in Cuzco (Ogburn 2004a; Protzen 1983). Ogburn considers the movement of massive andesite stones a "transfer of sacredness," and suggests that the last legitimate Inca ruler, Wayna Capac, had the Saraguro andesite moved just as he was establishing a new center in the northern Andean highlands.

This chapter explores the shift from sandstone to volcanic stone at Tiwanaku. It explores (1) the shift from exclusive quarrying in the local Kimsachata mountain range to the inclusion of sources in more distant, ancient volcanoes and volcanic outcrops; and (2) the significance of this shift in relation to Tiwanaku's increasing regional prestige and sociopolitical centralization in the Tiwanaku Period. We present new data related to Tiwanaku's stone sources recovered through reconnaissance and analysis conducted in 2009–2011. Our conclusions center on these data and their significance in relation to previous research and thought on Tiwanaku stone sourcing. Secondarily, the paper proposes some preliminary hypotheses regarding the significance of the shift from sandstone to volcanic stone, emphasizing the ritual and political dimensions of this shift. We conclude that the full materiality of stone—its source, texture, color, relative hardness, mineral qualities, and so forth—were made to be *essential* in defining what Tiwanaku was, ritually and politically, and who "the Tiwanaku" were, personally and collectively.

Lithic Transformations in the Production of Tiwanaku Monumentality

Tiwanaku emerged as a major ritual–political center in the southeastern lake Titicaca Basin during the Late Formative Period (100 BC–AD 500) and expanded greatly in extent and monumentality during the latter generations of that period, or Late Formative 2 (~AD 250–500). Tiwanaku transformed into what most archaeologists agree was an expansive urban center after AD 500, during the Andean Middle Horizon, known locally as the Tiwanaku Period (which comprises early and late phases, known as Tiwanaku IV and V). Stone was a critical fundamental component of Tiwanaku monumentality from the early Late Formative (Fig. 4.1).

Sandstone and Tiwanaku's Late Formative Monumentality

During the Late Formative, Sandstone was the principal lithic component of Tiwanaku monumental architecture. Most of it varied from yellowish to reddish brown. Sandstone was carved to form "parallelepiped" ashlars for architectural revetments, paved floors, terrace foundations, subterranean canals, and a host of other structural features of Tiwanaku's monumental landscape. Most of Tiwanaku's sculpted monolithic stelae, which depict personified ancestral deities and animate natural forces (Janusek 2006, 2008) also consisted of sandstone (Ohnstad and Janusek n.d.).

Monumental Structures

Employing sandstone for monumental construction began early in the Late Formative Period. Many of Tiwanaku's ceremonial spaces and monolithic sculptures date to this period (Ohnstad and Janusek n.d.). The earliest extant monumental structure at Tiwanaku, the 'Semi-subterranean Temple,' most likely dates to Late Formative 1 (Fig. 4.2) (Bennett 1934; Ponce Sanginés 1981, 1990; Janusek 2004, 2008). We refer to this as Tiwanaku's Sunken Temple. Its walls consisted mostly of relatively small, roughly hewn red sandstone blocks supported by deeply set, vertical pilasters. Carved heads tenoned in the wall—which may postdate the temple's construction—consist of light sedimentary rock from chalky outcrops

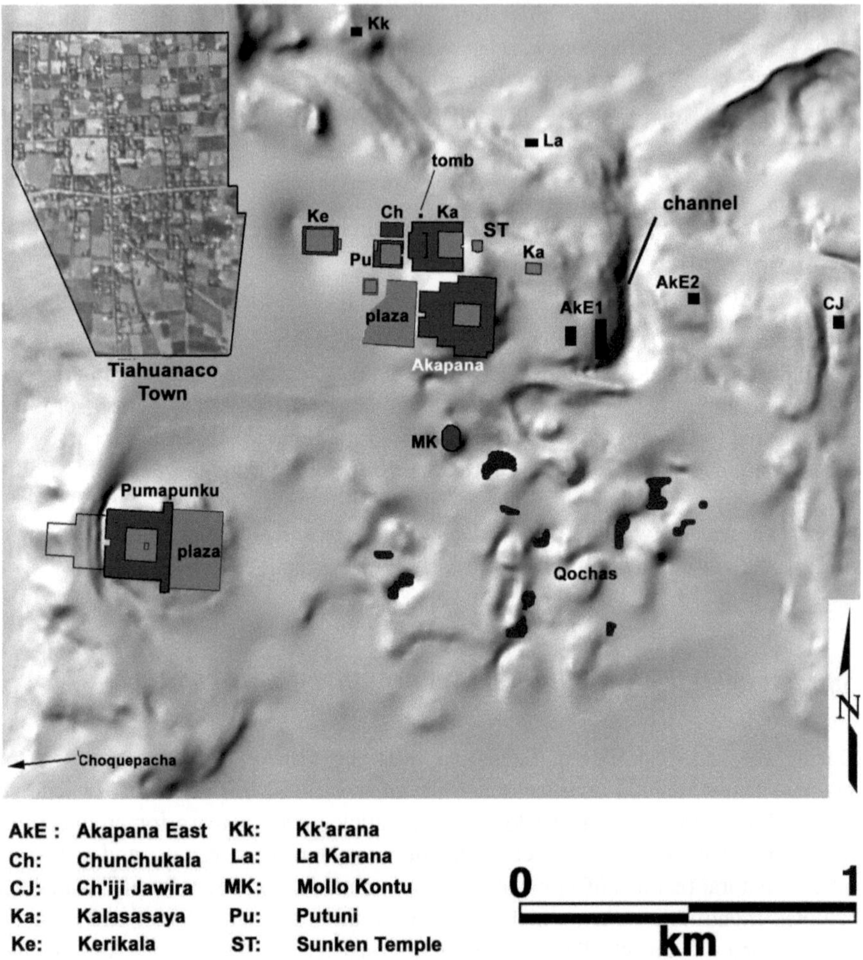

Fig. 4.1 Location of key structural features in Tiwanaku (from Janusek 2008: Fig. 4.1)

scattered across the region. A single south-facing stairway entrance consists of massive, rectangular sandstone blocks.

Construction on the Kalasasaya Platform Complex, which bounds the west side of the Semi-subterranean Temple, also began during the Late Formative (Ponce Sanginés 1981, 1990; also Janusek 2004, 2008). Kalasasaya dates later than the Semi-subterranean Temple and most likely to Late Formative 2 (Fig. 4.3). Its engineering drew on that of the Sunken Temple, yet manifested on a grander scale. The outer revetment mimicked the architectural construction of the Semi-subterranean temple's interior walls, but they were far more extensive. The deeply set vertical pilasters of Kalasasaya's platform revetment, which buttress its wall segments, are massive. The monumental east entrance to the Kalasasaya consisted purely of carved sandstone blocks.

Fig. 4.2 Tiwanaku's Late Formative Semi-Subterranean or Sunken temple (photo by J. Janusek)

Fig. 4.3 Tiwanaku's Late Formative Kalasasaya platform (photo by Wolfgang Schüler)

Monoliths

Sculpted sandstone monolithic stelae punctuated Tiwanaku's expanding monumental landscape. Late Formative monoliths consisted primarily of reddish-brown sandstone in Tiwanaku and at Late Formative centers across the southern Lake Titicaca Basin (Ohnstad and Janusek n.d.; Ohnstad n.d.). At Tiwanaku, as at Khonkho Wankane, each stela depicts a single anthropomorphic personage. The personage wears distinctive facial ornaments or painting, including suborbital lightning bolts

Fig. 4.4 Late Formative sandstone anthropomorphic stelae from Tiwanaku. Depicted are the "Barbado," "Dezcabezado," and the "Ídolo plano". Photos by Arthur Posnansky (1945:Figure 87), J. Janusek, and Carlos Ponce Sanginés.

or tears, and bears a distinctive gesture: arms folded over the chest, one hand above (and possibly over) the other. Most likely, this stylized gesture codified the pose of an interred, yet animate, mummified ancestor (Janusek 2006; Ohnstad n.d.; Ohnstad and Janusek n.d.).

Each Late Formative monolithic personage wore minimal clothing (e.g., sash, skirt, and headgear) and was decorated with sinuous living creatures that evoked earthy and watery domains (Fig. 4.4). Frequently, such images were rendered symmetrically, most commonly in pairs. Prominent among the latter are paired felines and neonate catfish, often presented as creatures slithering, swimming, or climbing up and down the sides of the stela. To the extent the stelae depicted animate ancestral persons, they gathered—in Heidegger's sense—multiple allusions to generative landscape features and resources. In critical sociospatial moments (e.g., communal ceremonies), stelae helped produce an understanding of the world that linked humanly produced stone monoliths to deceased relatives-cum-ancestors and to the natural landscapes that people inhabited. Each stela was a corporeal landscape in stone.

Volcanic Stone and Tiwanaku Monumental Construction

Tiwanaku monumentality was always being transformed. A particularly significant shift occurred at approximately AD 500–600. At this time, the massive Akapana and

Pumapunku complexes were initiated. Radiocarbon dating suggests that construction in the two structures began at roughly the same time (Janusek 2003a; Vranich 1999). After AD 600–700, during the so-called Tiwanaku period, the Kherikala and Kantatayita structures were built. Also during this dynamic Tiwanaku IV phase, the Kalasasaya was fitted with a west balcony wall (Ponce Sanginés 1990) and the Putuni complex was constructed on its west side (Couture 2002; Janusek 2004). All of these structures manifested an important material shift; unlike Late Formative construction, each structure now incorporated significant construction in volcanic stone.

Monumental Structures

Volcanic stone, principally bluish- to greenish-gray andesite, was selectively employed for Tiwanaku monumental construction. The superimposed terrace revetments of the Akapana and the structures they supported consisted primarily of sandstone. Akapana's basal terrace incorporates particularly massive, cyclopean sandstone ashlars. Andesite ashlars were incorporated strategically into its design, and possibly after the terraced structure had been originally completed in sandstone (Fig. 4.5). Voluminous, regularly spaced andesite blocks served as architectonic pilasters in the front (west) facade wall of the basal terrace of the Akapana. Andesite blocks formed the base of Akapana's central stairway entrance and a series of pedestals that stood in front of the terrace. At least some of these pedestals likely supported carved basalt *chachapumas*—or human-feline figures—associated with sacrificial imagery and practices. Crowning the top of the stairway was a (now-shattered) massive trilithic portal hewn of dark andesite.

Andesite was less regularly employed elsewhere in the Akapana. It was employed as pilasters in sections of the lower terrace revetments of Akapana's south side, in the wall facing the Kalasasaya; as pilasters in sections of the back (east) side of the Akapana, and as remnants in structures that once stood on top of the Akapana. In many of the latter cases, the blocks were clearly recycled from prior architectonic contexts in other parts of the Akapana or other structural complexes.

Employment of volcanic andesite in the Pumapunku complex was just as strategic. Most of Pumapunku's structure foundations themselves consist of carved sandstone ashlars. This includes the foundation of a platform on Pumapunku's east side, which incorporates some of the largest sandstone blocks employed in the center. The platform has been completely destroyed, and no convincing, systematically researched reconstructions have yet been published. This platform supported several—as many as eight—andesite portals (Ponce Sanginés 1971), at least the extant remains of five of which consisted of andesite (Fig. 4.6). Architectural blocks carved with decorative "portal windows" and with "blind portals," also found in association with the platform, all consisted of andesite (Ponce Sanginés 1971).

The Pumapunku structure itself was primarily a sandstone and earth construction (Escalante Moscoso 1997; Ponce Sanginés 1971; Vranich 1999). Like the Akapana, in Pumapunku andesite was built into particularly visible and evocative places. This included a central sunken patio, which was paved with both sandstone and andesite

Fig. 4.5 The west or "front face" of the Akapana, highlighting the strategic placement of andesite veneer pilasters in a sandstone terrace foundation. Andesite constitutes the large *rectangular* blocks in the *upper image*, which depicts the *inset* north "wing" of the second terrace west façade, and the intermittent pilasters in the *lower image*. Note: the sandstone facing of the west lower terrace facade in the lower image has been covered with mud facing for purposes of conservation) (photos by J. Janusek)

Fig. 4.6 Andesite monolithic portal located in Pumapunku's east portico. The iconography of its frieze replicates the "serpent band" on the better-known Sun Portal ("Gate of the Sun") (photo by J. Janusek)

Fig. 4.7 Totora-reed lintel from Pumapunku's primary west entrance (photo by Wolfgang Schüler)

ashlars. The fragment of an andesite portal that may have led into this court was found during the excavation of this central ritual space (Vranich 1999). As in the Akapana, volcanic stone figured prominently in the entrance to the Pumapunku. Here, relatively small andesite blocks were incorporated into the staircase leading up onto the platform. More dramatic were the lintels that covered the stairways leading up onto the Pumapunku platform (Fig. 4.7). They consisted in several fragments of andesite blocks with distinctively carved surfaces and sides (Nestler 1913; Ponce Sanginés 1971; Posnansky 1945). The volcanic stone lintel fragments appear to depict the reed thatch of roofs covering contemporaneous houses and other Tiwanaku buildings (Ponce Sanginés 1971: 56–57; also Janusek 2008; Kolata 1993), possibly including sections of the Pumapunku and other ceremonial structures.

Kalasasaya incorporated some of the largest carved andesite blocks known in Tiwanaku. The most prominent use of andesite was in its western extension, which was built or at least expanded during the Tiwanaku period (Ponce Sanginés 1990). It consisted of a raised earth platform bounded on its west side by a "balcony" supported by ten massive andesite pilasters (Fig. 4.8). A prominent gap indicates where a pilaster was later removed during one of many past events in which Tiwanaku monuments were "quarried" to facilitate post-Tiwanaku construction (Benitez 2009; Posnansky 1945). The pilasters supported a massive wall consisting of smaller andesite ashlars. Most of these had been quarried by the time archaeologists began working at Tiwanaku. The Bolivian archaeologist Carlos Ponce Sanginés had andesite ashlars from Viacha, to the southeast, quarried and carted to Tiwanaku by rail to complete the wall. Only small sections of the original wall remain.

A revetment on andesite foundations bounded the east side of Kalasasaya's west platform extension. A small central andesite staircase led select persons up and onto the platform from its east side, facilitating authoritative views over the Kalasasaya and across the east and west horizons. This prominent place facilitated

Fig. 4.8 Kalasasaya's West Platform and balcony wall, prior to their reconstruction (photo by Gregorio Cordero Miranda, courtesy of Alexei Vranich)

the observation of particularly important solar rise and set points at specific times, which may have correlated with the solar iconography carved onto the frieze of the nearby andesite Sun Portal, and in particular, the solar almanac depicted in its serpent band (Benitez 2009: 68). The Sun Portal, Tiwanaku's focal icon today, most likely originally stood on the Kalasasaya west platform extension.

Although Putuni postdates Kalasasaya's initial construction (Couture 2002; Couture and Sampeck 2003; Janusek 2003b, 2004), the two structures came to form an integrated monumental complex (Vranich 2009). Putuni consists of a platform-courtyard complex bounded to the northwest by elite residences that may have held purchase over it and the ceremonies held there. Built relatively late in Tiwanaku's urban history, the Putuni platform revetments consist entirely of andesite.

Nevertheless, a series of niches built into the east and west platform walls, seven on each side, incorporated internal sandstone foundations and walls. Building the Putuni consisted of digging wall trenches and then placing revetment stones in them. Recent excavations revealed sandstone flakes in the sides of many revetment trenches (Luis Calisaya, personal communication, July 2009), indicating that at least some of the foundations had originally consisted of sandstone blocks that were later switched out for the elegantly carved andesite blocks visible today. This conjunction of architectural data may indicate that Putuni was originally built of sandstone, and that sometime later, and most likely in Tiwanaku V (AD 800–1000), its more publically visible components, notably its revetments, were rebuilt of andesite ashlars.

The Putuni platform wall that faced the Kalasasaya was a crown jewel of Tiwanaku fitted andesite stonework (Fig. 4.9). Its construction mimicked archetypical earlier technologies; relatively tall and bulky pilasters supported wide areas of smaller

Fig. 4.9 The impeccable outer east platform wall of Putuni (photo by J. Janusek)

blocks. Yet in this wall, blocks were uniquely, sometimes trapezoidally carved and impeccably fitted. Carved andesite stones of 1 cm^2 filled some small gaps.

This wall was built to be admired. Perforating its center was Putuni's primary entrance, consisting of an elaborate multilithic portal and a polychrome central staircase (Créqui-Monfort 1904; Posnansky 1945). The corridor between the west side of Kalasasaya and east side of Putuni formed a key ritual passage (Vranich 2009). This passage led ritual participants from around the north side of the old Kalasasaya, now newly refitted with an elaborate west balcony, toward the south and into the central east entrance to Putuni. The materiality of the corridor was impeccable: to the left (east), one witnessed Kalasasaya's massive balcony wall supported by towering andesite pilasters; to the right (west), one witnessed, in anticipation of impending ceremony and feasting, the impeccably constructed andesite walls of Putuni's east platform wall. Beyond doubt, the artisanry of this stonework was touted by those who led groups of pilgrims, ritual participants, diplomats, and others through this potent space.

Sandstone, Volcanic Stone, and Tiwanaku Period Lithic Monumentality

Sculpted monolithic personages remained focal icons in Tiwanaku period ritual spaces. During Tiwanaku's generations of political hegemony in the southern Lake Titicaca Basin, they were restricted to Tiwanaku itself. The strategic use of sandstone and andesite in Tiwanaku monumental construction was mirrored in the use of sandstone and volcanic stone for the production of Tiwanaku monoliths, and in at least one place, the juxtaposition of Formative and Tiwanaku style stelae.

Fig. 4.10 Tiwanaku style monoliths as imposing invitational personages (photos by Clare Sammels and J. Janusek)

If Late Formative monoliths personified community ancestors, new monolithic sculptures presented community ancestors as elite persons (Fig. 4.10). Impassive, abstract faces still marked defied status, but corporeal decoration shifted from generative landscape beings to the elaborately woven tunics, sash, and headgear of an elite person (Janusek 2006). Notably, the focal gesture of each monolith changed markedly. If Late Formative personages depicted interred, animate mummies, Tiwanaku period personages were living ancestral "hosts" offering a complementary pair of mind-altering consumable substances to the person apprehending them; in their left hands, a *kero* full of fermented *chicha*, and in their right hands, a snuff tablet for hallucinogenic resins (Torres and Repke 2006: 40–42). This ubiquitous gesture was overtly invitational (Janusek 2006). Tiwanaku's dominant ethos emphasized consumption and mind alteration as means to becoming an ideal Tiwanaku ritual person. It was an open invitation to all.

Elements of dress and body decoration, such as face markings and hair tresses, varied markedly among monolithic personages. Iconographic differentiation may have been significant for multiple reasons; perhaps it indexed different ruling leaders or dynasties, each associated with a unique ancestral complex; perhaps it indexed

Fig. 4.11 Bennett's excavation of the Bennett and one of its flanking Late Formative-style monoliths during the early 1930s (Bennett 1934: Fig. 31)

different cults, each focused on a specific Tiwanaku temple and its ancestral complex; or, it may have indexed different landscape features outside of Tiwanaku—mountains, lakes, rivers, or other generative places—and their associated defied ancestors, communities, and productive activities. We opt for the latter hypothesis (Janusek 2006, 2008; Ohnstad n.d.), while noting that elements of the other hypotheses may help explain this variability.

Stelae of sandstone and andesite presented this charged imagery. They occupied key places in Tiwanaku's architectural complexes; notably sunken courtyards. The Kalasasaya, Akapana, and Pumapunku each incorporated a central sunken courtyard, each of which likely housed a central monolithic icon. Kalasasaya's sunken court housed the andesite Ponce Monolith, which was found in situ. Akapana's sunken court was dismantled during the Spanish Colonial Period. The andesite "Gigantic Monolith," known only from its decapitated head of hard greenish-gray andesite, may have once stood in Akapana's upper court, but other scenarios are possible. A sandstone monolithic stelae (Stela 4) found just west of Pumapunku was likely the focal monolithic icon that once stood in Pumapunku's central patio (Ponce Sanginés 1971).

Tiwanaku's Sunken Temple stood out. It incorporated a cluster of in situ sandstone stelae (Bennett 1934; Ponce Sanginés 1990). One of the four principal stelae lying supine in the court was carved in a style often termed "Yayamama" (Chávez and Chávez 1970; cf. Browman 1972, Ohnstad and Janusek n.d.) that was distinctive of the Late Formative period. This stela, the so-called Bearded Stela, may have been brought to Tiwanaku from the lake shore near Lukurmata. It was dwarfed by the massive sandstone Bennett Stela (Fig. 4.11). Standing over three meters high, the sandstone Bennett Monolith, like the andesite Ponce Monolith, presented important elements of Tiwanaku religious iconography. The Bennett embodied an imposing yet "invitational" anthropomorphic personage wearing a tunic decorated with dressed llamas and mind-altering plants.

Tiwanaku Stone Sourcing: Sandstone and Andesite

Previous Stone Sourcing at Tiwanaku

Early visitors noted that Tiwanaku monumental architecture comprised two very different types of stone, and often commented on their possible sources. As David Forbes put it in 1870 (p. 256), "the one is a light red sandstone…The other stone, however, is very different in nature, being a hard, tough, and compact volcanic rock, precisely the same as what was originally called Andesite." He continues (1870: 257), "although the sandstone has evidently been taken from the hills seen but a few miles distance from Tiahuanaco, the volcanic stone of which the two great monolithic portals, etc., have been constructed, has been conveyed a very great distance from the volcanic mountains on the other or western side of Lake Titicaca, where the quarries are still visible."

Travelers and researchers over the ensuing century brought more detailed analyses to bear on the origins of these two stones at Tiwanaku. While most agreed that the sandstone had derived largely from the Kimsachata mountain range south of Tiwanaku (Fig. 4.12), it remained unclear from which specific sectors, and how it was quarried and ported.

The first systemic studies were conducted and published in the early 1970s (Ponce and Mogrovejo Terrazas 1970; Ponce Sanginés 1971). In the late 1960s, a group of researchers headed by the fiery young Bolivian archaeologist Carlos Ponce Sanginés set out to determine the source of the sandstone employed to construct the Pumapunku complex. They conducted reconnaissance in several drainage valleys of the Kimsachata range south of Tiwanaku, isolating several possible sandstone quarries. A combination of comparative investigations that included X-ray diffraction (Avila Salinas 1971), geochemical analysis (Urquidi Barrau 1971), and most diagnostically, lithic petrography (Castaños Echazú 1971), indicated that Pumapunku sandstone came from the Kimsachata range, and most likely from the Kausani *quebrada* southeast of Tiwanaku (Mogrovejo Terrazas 1970).

The same authors present the photo of a possible sandstone source "near Cerro Muiundani" (Ponce Sanginés 1971: Fig. 128) in the Kausani *quebrada* (Fig. 4.13). This is a photo of the quarry that Janusek, Lemuz, and Julio Condori identified as a primary sandstone quarry in 2010 and 2011. Ponce and colleagues present photos showing the base of Kausani from the valley bottom, or pampa, to the north (Ponce Sanginés et al. 1971: Figs. 114 and 129). Yet they say little else about this possible

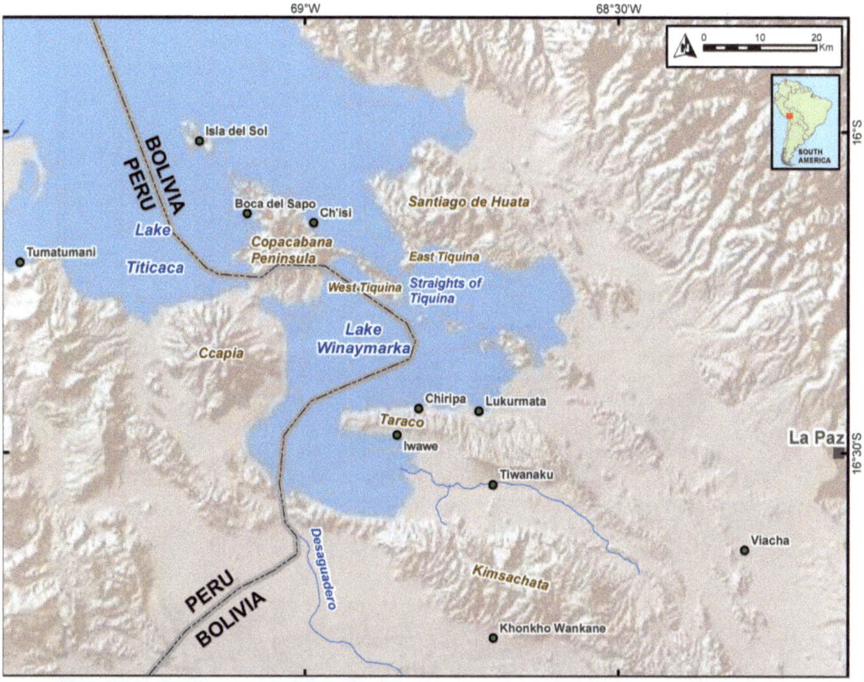

Fig. 4.12 Location of Tiwanaku in relation to the Kimsachata range and other geographical features, including Lake Titicaca, Cerro Ccapia, and Copacabana (Image courtesy of Nicholas Tripcevich)

source. It appears that they did not recognize the source as a "quarry" per se. The main photo depicts a field of jumbled, angled stones, and the caption reads:

> Typical process of disintegration by way of geological weathering on a mountain near Miundani, in the meridional [Kimsachata] range of Tiwanaku. Presumably, no sandstone quarries per se were exploited by the Tiwanaku, whether exposed cuts or galleries. Rather [the Tiwanaku] took advantage of blocks already separated by [geological processes] of fracturing[1] (Ponce Sanginés 1971: Fig. 128; translation by Janusek).

The authors did not consider this field of sandstone blocks the result of human activity. Our research determined that it was an anthropogenic production.

Locating Tiwanaku's volcanic stone quarries has been far more problematic.

Volcanic rock formations shape much of the southern Lake Titicaca basin, including the low mountainous chains bounding Lake Titicaca and much of the irregular chain that separates most of it from its smaller southern portion, or Lake

[1] "Tipico proceso de desintegracion por intemperismo mecanico en un cerro proximo a Muiundani en la sierra meridional [Kimsachata] de Tiwanaku. Se supone que no existio canteras propiamente dichas de areniscas aprovechadas pos las tiwanacotas, trabajo a tajo abierto o galleria, sino se acudio a bloques separados por disclasa."

Fig. 4.13 Ponce and colleagues' photographs of a "natural" stone source near Cerro Miundani, in the Kausani Valley of the Kimsachata range [(**b**) Ponce Sanginés et al. 1971: Fig. 128; (**a**) Ponce and Mogrovejo Terrazas 1970: Fig. 23]

Wiñaymarka. Even portions of the Kimsachata range, and particularly Mounts Chilla and Gloriakollu, respectively, southwest and southeast of Tiwanaku, have volcanic components.

Archaeologists and others honed in on a few likely sources for Tiwanaku's volcanic stones. As early as 1894, the German archaeologist Max Uhle took comparative samples for petrographic analysis, some of which indicated a strong correspondence between andesite ashlars at Tiwanaku and andesite outcrops at the foot of Mount Ccapia, an extinct volcano southwest of Yunguyu, Peru, and north west from Tiwanaku across Lake Wiñaymarka (Fig. 4.12) (Bergt 1894). Macroscopic analysis indicated to the French geologist George Courty (1907) that most of

Tiwanaku's volcanic stone derived from outcrops on the Huatta Peninsula east of Lake Titicaca, northeast of Tiwanaku. A few years later, Arthur Posnansky, an eccentric Austrian who spent half a century studying Tiwanaku, suggested that its volcanic stone derived primarily from Ccapia (1904, cited in Ponce and Mogrovejo Terrazas 1970: 35). Ironically (it turns out), he later changed his mind based on a small-scale petrographic analysis, opting for outcrops near Comanche, far southeast of Tiwanaku, as a more likely source.[2]

Ponce Sanginés and other young geologists set out to determine the source of Tiwanaku's andesite stonework. While the results were far less definitive than those for sandstone, the group made progress in eradicating several igneous outcrops as possible primary sources for Tiwanaku's volcanic stone constructions (Mille and Carlos Ponce 1968; Ponce and Mogrovejo Terrazas 1970). Mineralogical and petrographic analysis marked the volcanic outcrops of Chilla and Gloriakollu in the Kimsachata range south of Tiwanaku as clear nonprimary sources (Mogrovejo Terrazas 1970: 247–248). A key conclusion of the collaboration of Ponce and Mogrovejo, published in 1970, was that Tiwanaku's primary volcanic sources were nonlocal (Ponce and Mogrovejo Terrazas 1970). Somewhat earlier, Carlos Ponce and Max Mille had determined, based on reconnaissance and petrographic analysis, that Tiwanaku volcanic stone had not originated from sources southeast of Tiwanaku. This effectively discounted Viscachani, Comanche (Posnansky's favored source), and Viacha (Ponce Sanginés 1968: 37), from which Ponce had hundreds of andesite ashlars carved to reconstruct the west balcony wall of Tiwanaku's Kalasasaya in the 1970s (Marcelino Lopez, personal communication, July 2010).

For Ponce and colleagues, more likely sources for Tiwanaku's andesitic stone included: Mount Ccapia, across Lake Wiñaymarka to the west; the Copacabana Peninsula, to the northwest; and other volcanic outcrops along the shores of Lake Titicaca further north. Considering Tiwanaku's distance from the shore, they hypothesized that Tiwanaku leaders, artisans, and stoneworkers took advantage of lake shore sources so that stone traveled relatively minimal overland distances. Impeccably engineered rafts crafted of locally harvested *totora* reeds, they suggested, served to cart heavy volcanic stones across long stretches. Recent projects in experimental archaeology have developed multiple proposals for how such heavy-duty *totora* reed rafts—essentially barges—might have been constructed (Vranich, personal communication, 2006).

---

[2]Posnanksy commissioned petrographic analyses that were conducted by a certain Dr. Schneiderhohn at the Petrographic Institute of the University of Berlin (Ponce and Mogrovejo Terrazas 1970: 36). He published the results of 23 petrographic analyses, two of them conducted on volcanic stone samples from Ccapia (samples Q and R). At least ten samples derived from Tiwanaku monumental constructions, and two others from Tiwanaku basalt monuments. (Posnansky 1945, II; Figs. 162–180; Ponce and Mogrovejo Terrazas 1970: 36–37). The two samples from Ccapia (Q and R) did not provide a good match for the Tiwanaku samples. Of course, this is not surprising. Mineralogical and chemical signatures vary significantly in volcanic sources due to variable local formation conditions. Posnansky made the critical mistake of second guessing his initial instinct. As it turns out, most Tiwanaku andesitic stone derived from Mount Ccapia.

Fig. 4.14 Abandoned, roughly carved andesite block near the lake shore at Iwawe (photo by J. Janusek)

Ponce and colleagues' interpretation of stone quarrying and transport for Tiwanaku hinged on the discovery of andesite "tired stones" at Iwawe (Fig. 4.14) (Ponce Sanginés 1968: 38; Ponce Sanginés 1970: 60, 145–146), located on the shores of Lake Wiñaymarka and at the south edge of the Taraco Peninsula (Fig. 4.12). There are no nearby volcanic quarries. The abandoned stones, Ponce and colleagues surmised, had been ported from elsewhere. Their archaeological reconnaissance located both (1) finely and (2) roughly carved stones (Ponce Sanginés 1970: 146), indicating a range of quarrying techniques. Further, excavations supervised by Gergorio Cordero Miranda in 1968 (Ponce Sanginés 1970: 146) and, much later, by William Isbell and Juan Albarracin-Jordan (Isbell and Burkholder 2002) revealed Tiwanaku occupations at Iwawe. Ponce thus interpreted Iwawe as a port that received monolithic volcanic stones destined for Tiwanaku.

Building on prior analysis, and taking into consideration (1) the quality of andesitic stone, (2) overall distance to Tiwanaku, and (3) proximity to the lake shore, Ponce and colleagues considered two places as particularly likely sources for Tiwanaku andesitic stone. One was Mount Ccapia. Forbes (1870) and others had noted the presence of "tired stones" (*piedras cansadas*) at the foothills of Ccapia, southwest of Yunguyu, Peru. These were quarried stones that had been carved into ashlars and other rectilinear forms and then left en route to the lake shore in destination for lacustrine travel (Forbes described one such stone as a "sofa"). A second was the Calvario of Copacabana. Ponce noted (1970: 60), "in Mount Calvario we have recovered samples very similar to those of the archaeological monuments of Tiwanaku."

Ultimately, Ponce and colleagues favored Copacabana as the likely source for Tiwanaku's monolithic andesite. Ponce surmised in 1970 (p. 146), "Iwawe…was

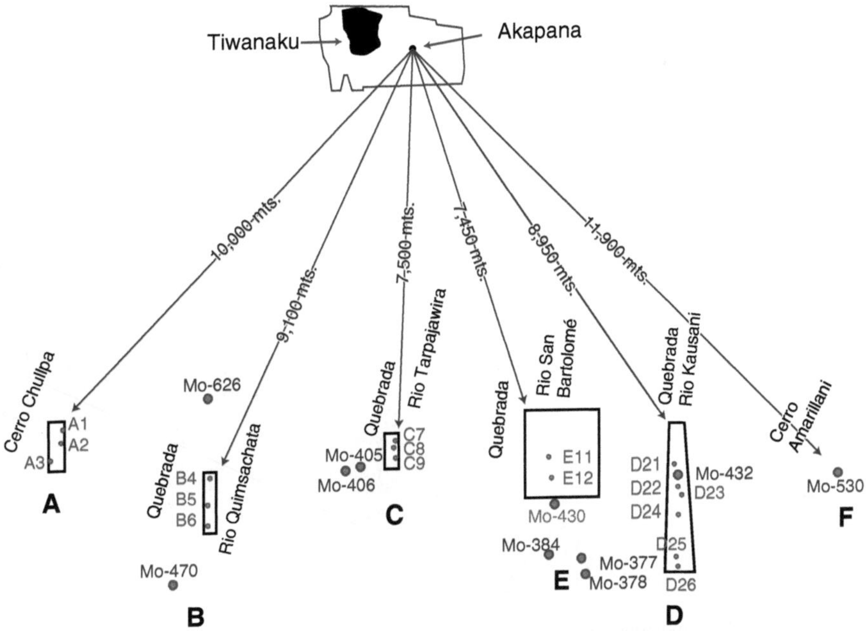

Fig. 4.15 The six possible sandstone sources that Ponce and colleagues identified in the Kimsachata range south of Tiwanaku (redrawn by Nicholas Tripcevich from Ponce Sanginés et al. 1971: Fig. 115). Kausani is D

the port where rafts unloaded their material carried from the Copacabana Peninsula." He settled on this source, in part, for reasons of political expediency. Ponce explains (1968: 37): "It appears currently indispensible to concentrate on the possibility of the quarry of Mount Capira [Ccapia]…[yet], For us, it was not possible to conduct even a simple reconnaissance in the region, because it is located on the border of Peru and Bolivia, in the territory of our brother country, where for obvious reasons difficulties arise for even the most summary explorations." Serious consideration of Ccapia as a source for Tiwanaku's volcanic stone apparently was out of the question.

Results of Stone Sourcing, 2010–2011

Sandstone

Our research confirms some of Ponce and colleagues' conclusions and challenges others. We confirmed that Kimsachata was the principal source of sandstone for Tiwanaku architectonic construction. Ponce and colleagues had reconnoitered the range and identified six possible valleys with sandstone outcrops (Ponce Sanginés 1971): Cerro Chullpa (A), Quebrada Kimsachata (B), Quebrada Tarpajawira (C), Quebrada San Bartolome (E), Quebrada Kausani (E), and Cerro Amarillani (F) (Fig. 4.15).

Fig. 4.16 Kaliri sandstone quarry. Location on Google Map (*inset*) and view from the south (photo by J. Janusek)

Petrographic (Castaños Echazú 1971), X-ray diffraction (Avila Salinas 1971), and geochemical (Urquidi Barrau 1971) analysis converged on the conclusion that Kausani Valley was the principal sandstone source for Pumapunku (Ponce Sanginés 1971). The co-editors present a photo of the base of the valley, just as the stream enters the altiplano pampa (Ponce Sanginés et al. 1971: 316), and a photo of the upper valley that shows a dense field of large, rectangular sandstone blocks (Ponce Sanginés et al. 1971: 324). The caption concludes that Tiwanaku took advantage of geologically fractured sandstone outcrops to mine stones for Pumapunku (Ponce Sanginés et al. 1971: Fig. 128). Knowing Ponce's acuity in other aspects of Tiwanaku archaeology, we suspect that he never visited—or at least never more than cursorily inspected—this quarry. The quarry is known locally as Kaliri.

Janusek initiated reconnaissance in Kimsachata with Julio Condori and Manuel Choque in 2010, and continued in 2011 with Carlos Lémuz and Andy Roddick (McMaster University). We located only one positive source of sandstone for Tiwanaku in the Kimsachata range; that is, only one source where partially carved stones had been left behind. Kaliri was likely the major source of Tiwanaku's sandstone (Fig. 4.16). Reconnaissance located several potential minor sources in the range, including distant outcrops in the San Bartolome Valley southeast of Tiwanaku, nearer outcrops in the Tarpa Jawira Valley, and outcrops in the large Kimsachata Valley southwest of Tiwanaku. Each valley presented relatively small outcrops that may have been quarried in the past. Few of them revealed extant quarried stones. None of them revealed extant quarried stones to the extent located in Kaliri.

Fig. 4.17 Fissures in the Kaliri sandstone outcrop that Tiwanaku quarry masons exploited to pry open naturally formed, yet roughly quadrangular sandstone blocks (photo by J. Janusek)

The Kaliri quarry is located in the northern foothills of the Kimsachata mountain range and at an altitude of 4,200–4,300 m above sea level. It comprises one of several geologically uplifted and vertically tilted sedimentary exposures on the northern flanks of this range. Traveling up into the range southward from Tiwanaku, it is the first sedimentary exposure that naturally produces relatively large and nonfoliated, contiguous blocks of finely laminated reddish-brown sandstone. Tiwanaku stonemasons opportunistically employed this sandstone outcrop. They did not simply "grab" geologically fractured blocks, as Ponce seems to suggest; and the Kaliri quarry was not simply a natural phenomenon. It was a humanly created quarry, and the "jumble" apparent in Ponce's photo resulted from Tiwanaku-related quarrying activities. Humans took advantage of natural fractures in embedded sandstone outcrops and produced the disarrayed stones visible in Ponce's and our more recent photos.

As Ponce (1971: 176) notes, Kaliri was not a quarry to the extent that stonemasons carved stones "whole cloth" from natural bedrock. They took advantage of natural, geological joints to disembed large natural blocks from a naturally fragmented outcrop (Fig. 4.17). Natural blocks deemed inadequate to a particular task were cast aside; those deemed useful were moved to other areas for fine carving (Fig. 4.18). The quarry revealed several hundred sandstone blocks that had been moved by human force from their geological locations.

Means of transporting massive sandstone blocks from Kaliri to Tiwanaku has been one of the fraught questions of stone production in the region. It may be one of the simplest to answer. Many of the largest blocks left behind at Kaliri and along the road leading down alongside the Kausani River present carved grooves on opposing corners that Janusek refer to as "rope holds" (Fig. 4.19). Stone blocks with carved grooves comprise approximately seven percent of the carved blocks left behind at the Kaliri quarry and 90% of the sandstone blocks left behind on the path that leads down the Kausani Valley toward Tiwanaku. Creating a rope hold was apparently

Fig. 4.18 Relatively flat platform in the Kaliri quarry that was possibly used for fine carving. Photo shows its surface littered with sandstone flakes (photo by J. Janusek)

Fig. 4.19 Quarried sandstone block on the path between Kaliri and Tiwanaku, demonstrating rope holds (photo by J. Janusek)

key to preparing a stone for transport to Tiwanaku and demonstrates that rope—woven of whatever durable material—was critical to monumental construction. Several sandstone blocks incorporated into Akapana's foundation terrace maintain remnant rope holds (Julio Condori, personal communication, 2010).

Specific technical sequences Tiwanaku stonemasons and laborers employed to port sandstone to Tiwanaku from Kimsachata are not entirely clear. Available evidence and felicitous analogy offer probable scenarios. First, it is likely that blocks were removed from joints by rope, roughly worked, and then carted down to Tiwanaku on consistently recycled beds of wooden logs, employing mud, logs, and might (Ponce Sanginés 1970). The location of a mountain spring just below the quarry may have aided in lubricating the road down the valley. In this scenario, two or more people were in charge of consistently taking logs from behind the train and placing them in front of it as they carted a block downward along the Kausani River. Although wood is a premium resource in the altiplano, the tough and century-long *kiswara* does well, and other wood may have been imported from eastern valleys. Still, substantial scraping on some *tired stones* indicates that porters dragged some blocks directly over the soil, perhaps primarily where mud-and-log porting technology was not topologically feasible.

Volcanic Stone: Andesite and Basalt

We reconnoitered the places Ponce and colleagues considered likely sources of volcanic stone for Tiwanaku. Using geological maps, Janusek, Williams, Lemuz, and Roddick collected samples from various places in the southern Lake Titicaca Basin. We collected source materials from numerous localities and analyzed them using an Innov-X Alpha Series portable X-ray fluorescence spectrometer.[3] In 2009, Williams and Janusek used the same spectrometer to measure elemental concentrations in 122 architectural stones and carved monoliths located at the sites of Tiwanaku, Lukurmata, and Iwawe (Fig. 4.20). At these "target sites," we identified five different chemical source groups—Group 1, 2, 3, 7, and 8—that we then sought to link to their specific volcanic sources.

We confirmed that Ccapia was a principal source of andesite for Tiwanaku monumental construction, but that other sources, including Tiquina, and as yet undetermined sources, served as secondary and tertiary quarries. The presence of "tired stones" near the shores of Lake Titicaca at the foothills of Ccapia had suggested this conclusion (Stanish et al. 1997). Nevertheless, it had not been clear to what extent Ccapia stones were employed to build Tiwanaku monuments and monolithic stelae in relation to volcanic stones quarried from other likely locales, including Viacha, Copacabana, Tiquina, Santiago de Huata, and other volcanic outcrops located on the eastern and western sides of the lake. Determining where Tiwanaku's volcanic stone was quarried, and the relative importance of Tiwanaku's different volcanic sources, were key goals.

Our research confirms that volcanic outcrops on the slopes of Ccapia, in contemporary Peru, were the primary sources of volcanic stone at Tiwanaku (Fig. 4.21). We term this core source Group 1. Group 1 accounts for all volcanic stone tested at Iwawe and all volcanic stone tested at Lukurmata. The core group is

[3] Quantitative results were calculated, using fundamental parameters, by the software provided with the instrument. Previous comparison of quantitative data derived from this instrument with data from obsidian standards and with samples measured by LA-ICP-MS is reported in Williams et al. (2012). See Grave et al. (2012) for a comparable use of pXRF quantification to characterize volcanic stone in another world region.

Fig. 4.20 P. Ryan Williams trains his portable X-ray fluorescence spectrometer on Tiwanaku's iconic Sun Portal (photo by J. Janusek)

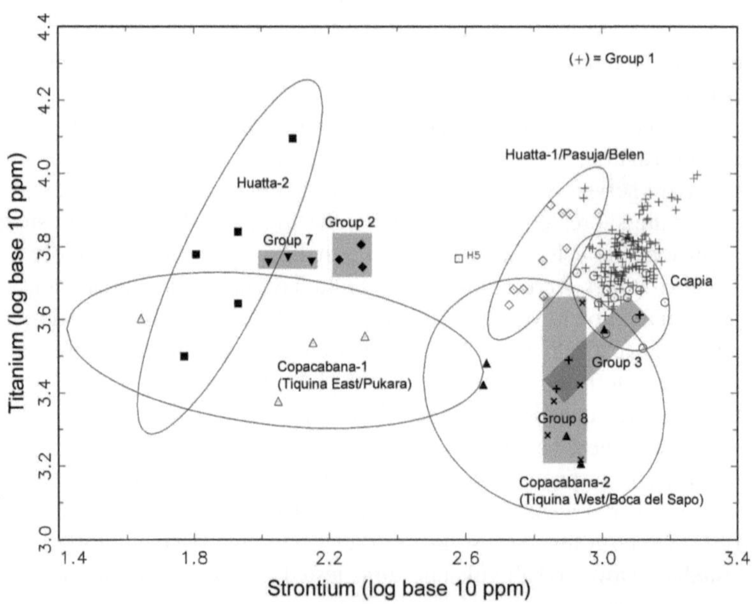

Fig. 4.21 Bivariate plot of logged (base 10 ppm) concentrations in geological samples and archaeological samples, comparing geological chemical profiles to identified archaeological chemical profiles. 90% confidence ellipses are drawn around raw material chemical profiles. Geological samples: Ccapia (*open circle*), Huatta-1/Pasuja Belen (*open diamond*), Hutta-2 (*filled square*), Copacabana-1 (*open triangle*), Copacabana-2 (*filled triangle*). Geological sample H5 is a chemical outlier sample collected near Huatta

characterized by moderate levels of iron (2–4%) and relatively enriched levels of strontium (1,000–1,600 parts per million). It is also characterized by moderate levels of titanium (4,000–8,500 ppm) and low concentrations of zircon (200–250 ppm).

Four minor groups from Tiwanaku and related sites best match source outcrops located near the Straits of Tiquina on the Copacabana peninsula. The source data collected from these locales are limited, so we reserve final judgment pending further sampling.[4] However, target groups 2 and 7 appear to chemically match source samples from the east side of the Straits of Tiquina and Tiquina Pukara (collectively, Copacabana-1 in Fig. 21). These groups are characterized by low concentrations of zircon and strontium (between 100 and 200 ppm each). Group 2 is manifested only in volcanic blocks incorporated into the Pumapunku temple and in two ex situ pieces in the Tiwanaku Regional Museum: a *chachapuma* (were-feline) sculpture and window-lintel architectural element. Group 7 is manifested in the one in situ basalt stone incorporated into the Sunken Temple,[5] and in two ex situ basalt *chachapuma* sculptures located at the Tiwanaku Regional Museum.

Target Groups 3 and 8 differ chemically from all other stones. Both groups have very low concentrations of titanium (2,000–3,000 ppm) and iron (1–2%). They match source materials recovered from the west side of Tiquina and from Boca del Sapo on the Copacabana peninsula (collectively, Copacabana-2). Two distinctive blocks of Tiwanaku's Kantatayita structure represent Group 3. One was the massive *maquette* or temple effigy that occupies the center of Kantatayita's east structure.

Group 8 included the highly eroded, white diorite pilasters that bolster Kalasasaya's east façade. Current accounts of Tiwanaku construction suggest that Kalasasaya's east wall was built early in Tiwanaku's urban history, specifically in early Late Formative 2 (Janusek 2004: 108–111). It is perhaps significant, then, that the volcanic stone comprising its pilasters was so different in appearance—and so much coarser in texture and friable in endurance—than all later volcanic stone employed for Tiwanaku monumental construction. Kalasasaya's east wall may incorporate the first igneous stone on a major scale at the center. Quarrying volcanic pilasters from the Copacabana peninsula may have been an early, experimental stage in the monumental production of Tiwanaku.

By the time architects were redesigning structural details on Akapana and Pumapunku, and sculptors were crafting monoliths and lithic portals in "Classic" Tiwanaku style, most volcanic stone was being quarried on the north edge of Ccapia (Group 1). Most of the stone consisted of bluish-gray andesite (though color and texture varied dramatically). Most Akapana volcanic blocks, Kalasasaya's west balcony wall, the Sun Portal, and the Ponce and Suñawi Monoliths all consisted of particularly fine-grained, bluish-gray andesite from Ccapia quarries.

---

[4]We recovered several more samples from Tiquina in summer, 2011, but they have not yet been analyzed.

[5]This is a vertically placed flat stone located at nearly the exact center of the north wall of the Sunken Temple. Benitez (2009) argues that the oddly placed stone was critical in orienting the structure to celestial observations and calendrical calculations.

Other sources provided stones for specific places and purposes. Kantayita's miniature effigy temple, surrounded by stones quarried from Ccapia, was quarried somewhere on the Copacabana peninsula. This monolithic sculpture was a key focus of ritual activity in the temple, raising the question of whether its source, or other aspects of its materiality, was important to its being chosen as Kantayita's focal carved stone.

The mountain range east of Tiquina, including our East Tiquina and Tiquina Pukara (Copacabana 1) sources (Groups 2 and 7), provided some of the heavier, denser basalt for Tiwanaku's fine architectural elements and sculptures. Basalt was quarried to form very particular architectural elements; notably architectural stones in Pumapunku's east portico and the narrow, vertically oriented stone located in the center of the Sunken Temple's north wall (Benitez 2009). Basalt was also quarried to form several of the *chachapuma*, or were-feline, personages that punctuated Tiwanaku's monumental core.

Discussion: Sandstone, Andesite, and the Shifting Production of Taypikala

Stone was essential to Tiwanaku. It was essential to the ballooning fame of the center and to coordinating the actions and identities of the many communities that identified with it. Stone for monumental construction and monolithic erection had a long history in the southern Lake Titicaca Basin. It began during the Early or Middle Formative (800–100 BC; Bennett 1936; Cohen and Roddick 2007; Hastorf 2003; Lemuz 2001; Chávez 1988). It became a primary component of monumental construction during the Late Formative (100 BC–AD 500), at sites such as Khonkho Wankane (Ohnstad n.d.) and Tiwanaku (Janusek 2008). Local stone fueled an increasing appetite for monumental permanence and mass. It served to embody key ancestral personages, crafted as monolithic sculptures to which local populations made offerings and attributed causality to ongoing events in the world (Janusek 2006, 2008, 2010). Stone continued to be quarried, carved, and incorporated into monumental structures and monolithic sculptures throughout the Tiwanaku Period.

Tiwanaku's incorporation of stone changed dramatically at the beginning of the Tiwanaku Period (AD 500–1100), just as its regional prestige began to explode. Sandstone quarrying and sculpting practices, focused on the Kimsachata range south of Tiwanaku, shifted as stoneworkers and masons began to incorporate heavier volcanic stone from more distant ancient volcanoes and igneous outcrops in Ccapia, Copacabana, and beyond. This was a momentous transformation with technical, esthetic, and ultimately, spiritual dimensions (Janusek 2006).

Sandstone was quarried in the Kimsachata range south of Tiwanaku. Our reconnaissance located one clear sandstone quarry in the upper Kausani Valley of that range and a number of other possible quarry sites. Our research confirms some of the results of analyses conducted under the direction of Carlos Ponce Sanginés in the 1960s and 1970s (Ponce Sanginés and Mogrovejo Terrazas 1970; Ponce Sanginés et al. 1971). Ponce and colleagues determined, first, that Tiwanaku's sandstone derived from sandstone quarries located in narrow valleys of the Kimsachata range south of the site. Based on a conjunction of analyses, they suggested that sandstone for the Pumapunku derived from the Kausani Valley southeast of Tiwanaku (Ponce Sanginés et al. 1971).

Our reconnaissance confirmed that the Kausani Valley quarry was Tiwanaku's primary sandstone quarry. We located several other possible secondary quarries in nearby narrow valleys in the Kimsachata range. We can also update Ponce's thoughts about the process of sandstone quarrying (Ponce Sanginés et al. 1971: 324). Ponce concluded that Tiwanaku builders took advantage of naturally fissured sandstone outcrops to mine blocks for Tiwanaku constructions. He suggests that Tiwanaku builders did not quarry in the strict sense. We find that this is only partly true. Tiwanaku quarrysmiths took advantage of natural outcrop fissures to mine sandstone blocks, as Ponce argues, but the disarrayed stone blocks currently found in the Kaliri quarry of Kausani are the product of long-term, intensive stone quarrying. Tiwanaku quarrysmiths mined blocks from natural fissures, leaving behind a disarrayed field of blocks of varying size. The lithic landscape Ponce depicts was not naturally produced; it was the product of human stone quarrying and carving.

We also recovered evidence pertinent to the technical practice of moving stone blocks from Kaliri down the narrow Kausani Valley, across a wide stretch of altiplano pampa, and ultimately into Tiwanaku. Stonemasons carved sandstone blocks to roughly desired rectangular shapes on flat areas near the fissured sandstone outcrops. They then carved 7–10 cm wide divets into the edges of the blocks, usually located on opposing edges just below (~10–15 cm) the corners of the block. Janusek terms these "rope holds." Some blocks had two opposing divets, others had four. Less than ten percent of the carved sandstone blocks found in preparation areas of Kaliri yielded rope holds. Approximately ninety percent of the blocks found along the road leading down from Kaliri to the altiplano pampa yielded rope holds. We believe this is compelling evidence that multiple persons used strong hemp to transport each block from Kaliri to Tiwanaku.

Volcanic andesite began to be employed in Tiwanaku construction early in the Tiwanaku period. Our analysis determined that, by far, most of Tiwanaku's andesite derived from Mount Ccapia, across Lake Wiñaymarka and northwest of the site. Smaller proportions of andesite derived from other sources, including volcanic outcrops near the straits of Tiquina and on the Copacabana peninsula. Ponce considered the latter Tiwanaku's primary sources. We believe he did this, in part, because as he notes, he was unable to conduct research in the Ccapia region just across the border in Peru. But we believe he did this also *because* Mount Ccapia is located in Peru. Ponce's Bolivian archaeology was vehemently nationalistic (1978a, b). It would have been distasteful for him to acknowledge that stone for Tiwanaku's monumental constructions, which he was simultaneously reconstructing and recreating as Bolivian national patrimony (Ponce Sanginés 1961, 1990), derived from sources currently located (however circumstantially) in Peru.

Volcanic andesite was employed in Kalasasaya's West Balcony, Akapana's front west face, and in Pumapunku's east portico. These structures all date to the early generations of the Tiwanaku period, sometime between AD 500 and 700. Later structures such as the Putuni and Kerikala continued to employ volcanic stone.

Stonemasons employed volcanic stone, and principally andesite, strategically in Tiwanaku's early monumental structures. Andesite pilasters punctuated Akapana's façade, andesite comprised mono- and multilithic portals in Pumapunku's portico, and massive andesite blocks bolstered Kalasasaya's impressive west balcony wall.

This balcony wall, a structural addition to a Late Formative structure (Ponce Sanginés 1981), may have been a critical turning point in the use of quarried andesite (Janusek 2010). Its 11 andesite pilasters apparently marked solar setting points, solstice to solstice, on the distant west horizon as viewed from a monolithic andesite staircase on the east edge of the west platform extension (Benitez 2009). Viewed from the staircase, the sun on June solstice set over the northernmost andesite pilaster, and simultaneously over the peak of Mount Ccapia, its source. The very materiality of andesite tied Tiwanaku monumental construction to key terrestrial features—Ccapia—and celestial movements. It linked Tiwanaku to major cosmic forces and rhythms.

Monolithic sculptures also manifested a dramatic shift in iconographic composition. Most Tiwanaku sandstone monoliths depict ancestral personages rendered in Late Formative style (Browman 1972; Janusek and Ohnstad n.d.). These can be easily cross-dated to monoliths at other Late Formative sites, in particular Khonkho Wankane. Monoliths feature impassive personages with arms crossed over chest, high-relief carving, and serpentine zoomorphs. All Tiwanaku andesite monoliths depict personages rendered in Classical Tiwanaku style. Many depict impassive personages with arms presenting two objects—a *kero* and a probable snuff tablet—fine low-relief carving, and profile "attendants." Sandstone monoliths embody personages with minimal clothing (emphasizing headdress and sash), while andesite monoliths embody personages wearing elaborate garments. Most sandstone monoliths are quadrangular in form, while andesite monoliths emphasize three-dimensional shoulders, head, arms, legs, and feet (Janusek and Ohnstad n.d.).

There are, of course, many exceptions to this chronological/material distinction. Standing on the Kalasasaya platform, the Fraile monolith, carved of sandstone, depicts a three-dimensional personage in high relief who grasps a *kero* and snuff tablet and wears minimal clothing including a sash decorated with crustaceans. Pumapunku's Stela Four, also carved of sandstone, depicts a more rigidly carved personage in high relief who grasps a *kero* and snuff tablet in his hands (Ponce Sanginés et al. 1971: Figs. 84–86). This personage wears far more elaborate clothing, as depicted in its three-dimensional, high-relief sculpture; a squared headdress topped with feathers, ear ornaments, and almost fully clothed torso (now highly eroded). Just as andesite came to accompany sandstone construction in Tiwanaku's later monumental buildings, sandstone monoliths continued to be crafted and stood alongside andesite monolithic personages in these buildings during the Tiwanaku Period.

Yet relative significance in the differentiated materiality of sandstone and andesite may be more complex. Most impressive among monoliths is the well-known Bennett Monolith, which until 2002 stood in an effigy sunken temple designed by Arthur Posnansky in the city of La Paz. Bennett, crafted of sandstone, is the largest extant monolith known from Tiwanaku. Though crafted of sandstone, the personage presents a *kero* and snuff tablet, wears elaborate clothing, much of its body is three-dimensionally rendered, and is decorated with profile attendants and other Classical Tiwanaku lithic imagery. Like Kalassaya's Fraile and Pumapunku's Stela Four, the Bennett Monolith is a Tiwanaku Period sculpture. Yet its context may help explain its sandstone materiality.

The Bennett Monolith stood inside the Sunken Temple (Bennett 1934; Ponce Sanginés 1981, 1990). This structure dated to the Late Formative and is Tiwanaku's earliest documented monumental construction. Unlike all of Tiwanaku's monumental structures, the Late Formative Sunken Temple was aligned north–south, with a north-facing staircase entrance that framed Mount Kimsachata to the south (Benitez 2009). All later monumental buildings, including Kalasasaya, Akapana, and Pumapunku, were aligned east–west. Thus, the Sunken Temple was "retro-fitted" with a classical Tiwanaku period sculpture; the most immense sculpture in the southern Lake Titicaca Basin. It stood beside, and essentially dwarfed, two smaller sandstone sculptures depicting Late Formative personages. Why was the Bennett placed in the Sunken Temple? Perhaps, sandstone attracted sandstone in the early Tiwanaku period. That is, sandstone monoliths were carved in an emergent Tiwanaku style early in the Tiwanaku period, and they were placed in key ritual locales between AD 500 and 700. Most of these locales still maintained substantial sandstone foundations, and some Late Formative sandstone monoliths (such as the Sunken Temple). The Sunken Temple, in particular, appears to have been converted into a place for recreating Tiwanaku "historical patrimony" in the Tiwanaku Period (Janusek 2006). Emplacing the massive Bennett in the Temple, thereby dwarfing its predecessor monoliths "in the eyes" of surrounding tenoned head sculptures, may have been a critical architectonic movement in this direction.

Archaeologists cite the Bennett Monolith to demonstrate that the sandstone/volcanic stone distinction has no chronological significance. The caution is well taken. Sandstone and andesite are not absolute determinants of Late Formative or Tiwanaku Period lithic production, as witnessed in Tiwanaku architectural construction over the long term. During the Tiwanaku period, sandstone and andesite came to serve complementary, highly strategic purposes in monumental construction and monolithic production. Stonemasons integrated andesite blocks into particularly visible facades or public spaces such as Akapana's front and Pumapunku's portico. Excavations in Putuni indicate that early sandstone blocks forming the inner courtyard façade were later swapped out for andesite blocks. Volcanic andesite came to complement sandstone as Tiwanaku's novel public esthetic and spiritual technology.

In other venues, Janusek (2006, 2008: 134–135) suggests that during the Tiwanaku period, sandstone and andesite accrued complementary material semiotic qualities. Relatively friable, sandstone from local Kimsachata outcrops was red in color. More durable, andesite from more distant volcanic outcrops was bluish-gray in color. Andesite production demanded an entirely new technological sequence (or *chaîne opératoire*) of production: from quarrying, through movement across the lake, to final fitting in structures such as Akapana. Examining this sequence is a primary objective of our ongoing research. Andesite came from more distant volcanic sources that had been more recently incorporated into Tiwanaku's ritual–political sphere. Red, friable sandstone came to index Tiwanaku's telluric ancestors and Late Formative heritage, as perfectly manifested in the Sunken Temple with its imposing sandstone Bennett Monolith. Bluish-gray, durable andesite came to index Tiwanaku's newly incorporated landscapes, communities, and sacred places, as strikingly manifested in the Kalasasaya west balcony, which created visual relations to Mount Ccapia, the origin of most of Tiwanaku's andesite, and the andesite Ponce Monolith newly emplaced in Kalasasaya's sunken courtyard. The juxtaposition of the Sunken Temple and Kalasasaya, and the mutual visual relation of their central monoliths - Bennett and Ponce - spatially configured Tiwanaku's emergent historical narrative, rendering it concise and eminently legible. Part of the message must have been: Tiwanaku has arrived, and we have a mission.

Conclusions

Whether or not *taypikala* was an original name for Tiwanaku, the term captures its lithic essence. Indeed, those who identify with Tiwanaku—those who were born there, and even politicians who (despite well-reasoned commentary to the contrary) legitimize their power there—continue to affirm the power of the site and its stone-founded constructions. People continue to identify as *kalawawa*, authentic persons "born of stone." Visits to Tiwanaku have become analogous to Catholic or Muslim pilgrimage but grounded in contemporary interpretations of native Andean faith (now usually meshed with fantastic Atlantean and other New Age interpretations of the past).

We identified Tiwanaku's primary sandstone and andesite sources. Tiwanaku's primary sandstone source was the Kaliri quarry located in the upper Kausani Valley, some 15 km southeast of Tiwanaku, and its primary andesite source consisted of (as yet undocumented) quarries located on Mount Ccapia, across the small lake to the west. The shift from the Late Formative to the Tiwanaku Period was, in part, a shift from primarily sandstone to the strategic addition of volcanic stone in Tiwanaku stone construction and monolithic production. In the Tiwanaku Period, stonemasons employed andesite for relatively public facades and communal spaces in otherwise largely soil and sandstone buildings. Use of andesite and other volcanic stone, notably basalt for small stone sculptures, corresponded with tectonic sociopolitical shifts. Volcanic stone quarrying and construction corresponded with the incorporation of more distant landscapes on Lake Titicaca, including those that produced volcanic stone. Volcanic stone production enacted and propelled Tiwanaku's emergent imperialism and its ascent to power.

Acknowledgments We thank Nico and Kevin for inviting us to present a paper in the SAA panel that generated this volume. The fieldwork on which this paper is based is indebted to the help of many people. Andy Roddick (the archaeologist, not the tennis player) helped during most phases of the project. Others include Manuel Choque, our guide in the Kausani Valley, Julio Condori, Charee Peters, and Michelle Young. Project funding derived in part from a Vanderbilt University

Discovery grant and a grant from the Curtiss T. and Mary G. Brennan Foundation. The Innov-X PXRF device used in this study (project EAF076) was purchased with a grant from the Field Museum's Grainger Fund for Scientific Research.

References

Abercrombie, T. A. (1998). *Pathways of memory and power: Ethnography and history among an Andean people*. Madison: University of Wisconsin Press.

Avila Salinas, W. (1971). Estudio comparativo por diffraccíon de rayos X de las areniscas de Pumapunku. In *Procedencia de las areniscas utilizadas en el temple Precolombino de Pumapunku* (Tiwanaku), No. 22 (pp. 221–230). La Paz: Academia nacional de Ciencias de Bolivia.

Benitez, L. (2009). Descendants of the sun: Calendars, myth, and the Tiwanaku State. In S. J. Chávez & K. L. Mohr Chávez (Eds.), *Tiwanaku: Papers from the 2005 Mayer center symposium at the Denver Art Museum*, (pp. 49–81), Edited by Margaret Young-Sanchez. Denver, CO: Denver Art Museum.

Bennett, W. C. (1934). Excavations at Tiahuanaco. *Anthropological Papers of the American Museum of Natural History, XXXIV*(Part III), 359–494.

Bennett, W. C. (1936). Excavations in Bolivia. *Anthropological Papers of the American Museum of Natural History, 35*, 329–507.

Bergt, W. (1894). Die Gesteine der Ruinenstatte von Tiahuanaco im Alten Peru (Bolivia). *Abhandlungen der Naturwissenschaftlechen Gesselschaft ISIS, 5*, 35–52.

Browman, D.L. (1972). Asiruni, Pucara-Pokotia and Pajano: Pre-Tiahuanaco South Andean monolithic stone styles. *Paper presented at the 34th Annual Conference of the Society for American Archaeology,* St. Louis.

Castaños Echazú, A. (1971). Estudio petrográfico comparativo de las areniscas de Pumapunku. In *Procedencia de las areniscas utilizadas en el temple Precolombino de Pumapunku* (Tiwanaku), No. 22 (pp. 209–220). La Paz: Academia nacional de Ciencias de Bolivia.

Chávez, K. L. M. (1988). The significance of Chiripa in Lake Titicaca basin developments. *Expedition, 30*, 17–26.

Chávez, S. J., & Chávez, K. L. M. (1970). Newly discovered monoliths from the highlands of Puno, Peru. *Expedition, 12*(4), 25–39.

Cobo, F. B. (1990 [1653]). *Inca religion and customs (Selections from Historia del Nuevo Mundo) (R. Hamilton, Trans. and ed.)*. Austin: University of Texas Press.

Cohen, A., & Roddick, A. (2007). Excavations in the AC (Achachi Coa Kkollu) sector. In M. S. Bandy & C. A. Hastorf (Eds.), *Kala Uyuni: An early political center in the Southern Lake Titicaca Basin* (pp. 25–34). Berkeley: University of California Archaeological Research Facility.

Courty, G. (1907). *Explorations géologiques dans l'Amérique du Sud. Mission Scientifique G. der Créqui-Monfort et E. Sénéchal de la Grange*. Paris: Impremerie Nationale.

Couture, N. C. (2002). *The construction of power: monumental space and an elite residence at Tiwanaku, Bolivia* (Unpublished Ph.D. dissertation). University of Chicago.

Couture, N. C., & Sampeck, K. (2003). Putuni: A history of palace archietcture in Tiwanaku. In A. L. Kolata (Ed.), *Tiwanaku and its hinterland: Archaeology and paleecology of an Andean civilization* (Vol. 2, pp. 226–263). Washington, DC: Smithsonian Institution Press.

Créqui-Montfort, Count G. de (1904). Fouilles de la Mission Scientifique française à Tiahuanaco. *Internationaler Amerikanisten-Kongerss. Verzehnte Tagung Stuttgart* 1904, 531–550. Stuttgart: Druck und verlag von W. Kohlhammer.

Escalante Moscoso, J. F. (1997). *Arquitectura Prehispanica en los Andes Bolivianos*. La Paz: CIMA.

Forbes, D. (1870). On the Aymara Indians of Bolivia and Peru. *Journal of the Ethnological Society of London, 2*(3), 193–305.
Grave, P., Attenbrow, V., Sutherland, L., Pogson, R., & Forster, N. (2012). Non-destructive pXRF of mafic stone tools. *Journal of Archaeological Science, 39*, 1674–1686.
Hastorf, C. A. (2003). Community with the ancestors: ceremonies and social memory in the middle formative at Chiripa, Bolivia. *Journal of Anthropological Archaeology, 22*, 305–332.
Isbell, W. H., & Burkholder, J. (2002). Iwawi and Tiwanaku. In H. Silverman & W. H. Isbell (Eds.), *Andean archaeology I: Variations in sociopolitical organization* (pp. 199–242). New York: Kluwer Academic / Plenum Publishers.
Janusek, J. W. (2003a). Vessels, time, and society: Toward a chronology of ceramic style in the Tiwanaku heartland. In A. L. Kolata (Ed.), *Tiwanaku and its hinterland: Archaeology and paleoecology of an Andean civilization* (Vol. 2, pp. 30–92). Washington, DC: Smithsonian Institution Press.
Janusek, J. W. (2003b). The changing face of Tiwanaku residential life: State and social identity in an Andean city. In A. L. Kolata (Ed.), *Tiwanaku and its hinterland: Archaeology and paleoecology of an Andean civilization* (Vol. 2, pp. 264–295). Washington, DC: Smithsonian Institution Press.
Janusek, J. W. (2004). *Identity and power in the ancient Andes: Tiwanaku cities through time*. London: Routledge.
Janusek, J. W. (2006). The changing 'nature' of Andean religion and the rise of an Andean state. *World Archaeology, 38*(3), 469–492.
Janusek, J. W. (2008). *Ancient Tiwanaku*. Cambridge: Cambridge University Press.
Janusek, J. W. (2010). El surgimiento del urbanismo en Tiwanaku y del poder político en el altiplano andino. In K. Makowski (Ed.), *Señores de los Imperios del Sol* (pp. 39–56). Lima: Banco de Crédito de Peru.
Janusek, J. W., & Ohnstad A. T. (n.d.). Stone Stelae of the Southern Lake Titicaca Basin: A stylistic chronology of ancestral personages. In W. A. Isbell (Ed.), *The Southern Andean iconographic series*. Washington, D.C.: Dumbarton Oaks (in press).
Kolata, A. L. (1993). *Tiwanaku: Portrait of an Andean civilization*. Cambridge: Blackwell.
Lemuz, A. C. (2001). Patrones de Asentamiento Arqueologico en la Peninsula de Santiago de Huata, Boliva (Unpublished Licenciatura thesis). Universidad Mayor de San Andres.
Mille, M., & Carlos Ponce, S. (1968). *Las andesitas de Tiwanaku*, No. 18. La Paz: Academia Nacional de Ciencias.
Mogrovejo Terrazas, G. (1970). Estudio Geológico petrográfico. In *Acerca de la Procedencia del material lítico de los monumentos de Tiwanaku*, No. 21 (pp. 189–258). La Paz: Academica Nacional de Ciencias de Bolivia.
Nestler, J. (1913). Beitrage zur Kenntnis der Ruinenstatte von Tiahuanaco. *Mitteilungen der Kais. Konigl. Geographischen Gesselschaft in Wein, 56*(4), 267–291.
Ogburn, D. (2004a). Power in stone: the long-distance of building blocks in the Inca empire. *Ethnoshistory, 51*, 101–135.
Ogburn, D. (2004b). Evidence for long-distance transportation of building stones in the Inka Empire, from Cuzco, Peru, to Saraguro, Ecuador. *Latin American Antiquity, 15*(4), 419–439.
Ohnstad, A. T. (n.d.). Monoliths and Monolithic Iconography at Khonkho Wankane. In J. W. Janusek (Ed.), *Khonkho Wankane and its Hinterland: Early Complexity in the South-Central Andes*. Los Angeles: Cotsen Archaeological Institute, University of California at Los Angeles.
Ohnstad, A. T., & Janusek, J. W. (n.d.). The development of 'Tiwanaku Style' out of the ideological and political landscapes of the formative lake Titicaca basin. Paper presented at the conference, *The Southern Andean Iconographic Series: A Colloquium in Pre-Columbian Art and Archaeology*, Santiago, Chile.
Ponce Sanginés, C. (1961). *Informe de Labores*. La Paz: Centro de Investigaciones Arqueológicas en Tiwanaku.
Ponce Sanginés, C. (1968). Perspectiva arqueológica. In *Las andesitas de Tiwanaku*, No. 18 (pp. 25–43). La Paz: Academia Nacional de Ciencias.

Ponce Sanginés, C. (1970). Examen Arqueológico. In *Acerca de la Procedencia del material lítico de los monumentos de Tiwanaku*, No. 21 (pp. 11–188). La Paz: Academica Nacional de Ciencias de Bolivia.

Ponce Sanginés, C. (1971). Exam arqueológico de las ruinas Precolombinas de Tiwanaku. In *Procedencia de las areniscas utilizadas en el temple Precolombino de Pumapunku* (Tiwanaku), No. 22 (pp. 13–206). La Paz: Academia nacional de Ciencias de Bolivia.

Ponce Sanginés, C. (1978a). *El Instituto Nacional de Arqueologia de Bolivia: su organizacion y proyeccione*. La Paz: Instituto Nacional de Arqueología.

Ponce Sanginés, C. (1978b). *Apuntes Sobre Dearrollo Nacional y Arqueologia*. La Paz: Instituto Nacional de Arqueología.

Ponce Sanginés, C. (1981). *Tiwanaku: Espacio, tiempo, y cultura*. La Paz: Los Amigos del Libro.

Ponce Sanginés, C. (1990). *Descripción sumaria del templete Semisubterraneo de Tiwanaku* (6th ed.). La Paz: Juventud.

Ponce Sanginés, C., Castaños Echazu, A., Avila Salinas, W., & Urquidi Barrau, F. (1971). *Procedencia de las areniscas utilizadas en el temple Precolombino de Pumapunku (Tiwanaku)*, No. 22 . La Paz: Academia Nacional de Ciencias de Bolivia

Ponce Sanginés, C., & Mogrovejo Terrazas, G. (1970). *Acerca de la Procedencia del material lítico de los monumentos de Tiwanaku*. La Paz: Academia nacional de Ciencias de Bolivia.

Posnansky, A. (1904). Petrografía de Tiahuanacu. *Boletín de la Sociedad Geográfica de La Paz, 18–20*, 207–211.

Posnansky, A. (1945). *Tihuanacu: The cradle of American man* (Vol. I and II). New York: J. J. Augustin.

Protzen, J.-P. (1983). Inca quarrying and stonecutting. *Ñawpa Pacha, 21*, 183–214.

Stanish, C., Edmundo de la Vega, M., Steadman, L., Chávez Justo, C., Frye, K. L., Onofre Mamani, L., Seddon, M. T., & Calisaya Chuquimia, P. (1997). *Archaeological survey in the Juli-Desaguadero region of the Lake Titicaca Basin, Southern Peru*. Chicago: Field Museum of Natural History.

Torres, C., M., & Repke, D. B. (2006). *Anadenanthera: Visionary Plant of Ancient South America*. New York: Haworth Herbal Press.

Urquidi Barrau, F. (1971). Geoquímica de las areniscas de Pumapunku. In *Procedencia de las areniscas utilizadas en el temple Precolombino de Pumapunku* (Tiwanaku), No. 22 (pp. 231–240). La Paz: Academia nacional de Ciencias de Bolivia.

Vranich, A. (1999). *Interpreting the meaning of ritual spaces: the temple complex of Pumapunku, Tiwanaku, Bolivia* (Unpublished Ph.D. dissertation). University of Pennsylvania.

Vranich, A. (2009). The development of the ritual core of Tiwanaku. In M. Young-Sanchez (Ed.), *Tiwanaku: Papers from the 2005 Mayer Center Symposium at the Denver Art Museum* (pp. 11–34). Denver, CO: Denver Art Museum.

Williams, P. R., Dussubieux, L., & Nash, D. J. (2012). Provenance of Peruvian Wari Obsidian: Comparing INAA, LA-ICP-MS, and portable XRF. In I. Liritzis & C. Stevenson (Eds.), *Obsidian and ancient manufactured glasses* (pp. 75–96). Albuquerque, NM: University of New Mexico Press.

Chapter 5
Arcillas and *Alfareros*: Clay and Temper Mining Practices in the Lake Titicaca Basin

Andrew Roddick and Elizabeth Klarich

Introduction

This chapter aims to integrate archaeological and ethnoarchaeological research on ceramic production in the Lake Titicaca Basin. Drawing on over 60 years of scholarship exploring the early stages of ceramic manufacture, we examine the acquisition of clays at quarries and the subsequent processing of these raw materials. Investigations into clay quarries have often focused on the availability of raw materials appropriate for pottery production. This research has included pedestrian survey for clays and sediments, and geochemical and mineralogical work on the quality of clays (Bishop et al. 1982; Neff et al. 1992). While such work is unquestionably useful (and unfortunately still rare in some regions), the dynamic nature of clays makes defining historic and prehistoric sources difficult. As a result, many archaeologists have considered these early technical stages through other means. For instance, research on prehistoric ceramics has long included careful analysis of ceramic pastes—the mixture of the aplastic inclusions and the plastic clay components of ceramics (for a good summary, see Arnold 2000). These findings have permitted for variability in local recipes to be correlated with regional and sometimes local deposits. In this work some have deployed sophisticated analytical tools in the laboratory to examine the techno-functional aspects of particular technological choices at quarry sites. This research has tended to focus on the relative performance of particular materials under a range of conditions (Bronitsky and Hamer 1986; Skibo et al. 1989; Summerhayes 1997).

Arcillas and *Alfareros*: Clays and potters

A. Roddick (✉)
Department of Anthropology, McMaster University, 1280 Main Street West, Hamilton, ON L8S4L8, Canada
e-mail: roddick@mcmaster.ca

E. Klarich
Department of Anthropology, Smith College, Green St. Classroom Annex 104c, Northampton, MA 01063, USA

N. Tripcevich and K.J. Vaughn (eds.), *Mining and Quarrying in the Ancient Andes*, Interdisciplinary Contributions to Archaeology, DOI 10.1007/978-1-4614-5200-3_5,

Other researchers have conducted ethnoarchaeological studies of clay quarrying, tempering choices, and paste preparation within contemporary societies. Here scholars working in a range of contemporary cultural contexts are able to investigate the role of particular choices across wider cultural fields, including the relationship of clay and temper choices to linguistic, ethnic, or political affiliation, and to other types of quotidian practice (Gosselain 1992; Herbich 1987; Neupert 2000; Sillar 2000). A recent overview by Olivier Gosselain and Alexandre Livingstone Smith summarizes some of the social mechanisms that underlie the spatial and temporal variations in clay selection and processing. They note that patterns may be the result of "[a]ccidental discovery; competition over land use; competition between artisans; individual or collective conceptions regarding the quality of raw materials; habits and traditions in technical behaviour; and social interactions at a local or regional level" (Gosselain and Livingstone Smith 2005: 34). The authors stress that while environmental and technical *constraints* of raw materials are important, the vast *possibilities* in social practice are equally important.

This chapter focuses on some of these social mechanisms, drawing on insights from active ethnoarchaeological research to guide the interpretation of archaeological data. Careful ethnoarchaeology, such as that presented by Gosselain and Livingstone Smith and their colleagues of the "ceramics and society" project in sub-Saharan Africa (Corniquet 2011; Gosselain 1992, 1999, 2008, 2011; Livingstone Smith 2000; Wallaert-Pêtre 2001, 2008), challenges archaeologists to consider such social mechanisms involved in the initial steps of pottery production, including quarrying and paste preparation. They encourage a consideration of the cultural logics behind technical practice in their wider project of "decomposing traditions and their dynamics" (Gosselain and Livingstone Smith 2005: 45). Ongoing research in the South-Central Andes by the authors is guided by similar goals.

Investigations of cultural logics behind ceramic production are certainly not new to the Andes, and a number of prominent Andean ethnoarchaeologists and archaeologists have focused on such technical grammars (what some call "emic" perspectives) within wider social contexts (Hosler 1996; Lechtman 1977; K. Chávez 1992; Sillar 2000). However, the implications of both Andean and global ethnoarchaeology are not always considered in wider archaeological research. For instance, Gosselain and Livingstone Smith express frustration that while a wide range of patterns of quarrying and processing practices are repeatedly observed, archaeologists seldom consider them in interpreting variations in paste recipes. In fact, "variations in paste composition continue to be used mostly as chronotypological markers or interpreted in techno-functional terms" (Gosselain and Livingstone Smith 2000: 34).

Pastes are increasingly playing a key role in building regional chronologies in the Lake Titicaca Basin of the *altiplano* (high plains) of Bolivia and Peru (Fig. 5.1), yet few scholars have considered the processes that produce such variability. There are two main reasons for this particular, and somewhat limited study of ceramic pastes and raw materials. First, the assemblages that are characteristic of Titicaca Basin sites are highly fragmented. In the absence of clearly identifiable iconography or vessel forms in many collections, paste recipes have been key in the defining of periods and horizons, occasionally used for techno-functional interpretations, and

Fig. 5.1 Map of the Lake Titicaca Basin with relevant archaeological sites and modern communities

in some cases serve as the uncritical basis for delimiting ethnic or political identities and boundaries. The second reason may be methodological; there are few ceramicists in the region, and surveys for raw materials, with a few exceptions, have rarely been conducted. In other words, this lack of engagement with larger trends within the ethnoarchaeological literature is due to research orientation. So while Titicaca Basin archaeologists use varying paste recipes for chronological purposes, they rarely ask "why" questions concerning this variability.

This chapter represents an early effort by the authors to ask such questions of ceramics of the Late Formative Period and to consider the social dynamics of prehistoric quarrying, raw material choice, and paste preparation through ethnoarchaeological means. First, details of the paste types within the regional chronology are provided, which serve to frame the first author's work on Middle and Late Formative pottery production in the southern Titicaca Basin. Second, a brief summary of existing ethnoarchaeological research from the region is presented, which provides a context for the second author's preliminary research on the mining of raw materials in the traditional pottery producing community of Pucará in the northern Basin. In both the archaeological and ethnographic cases, specific "quarry quandaries" are examined, focusing in particular on the characteristic primary materials in these areas and the possibilities for distinguishing between "types" or boundaries of these sources. The relationship of particular quarries to technological choices is also briefly explored. In closing, the authors revisit the challenge proposed by Gosselain and Livingstone Smith in their work in sub-Saharan Africa and highlight the difficulties in studying source landscapes in both the present and the past. To this end, the authors attempt to bring together the results of recent ethnoarchaeological and archaeological research, drawing on wider Andean scholarship to consider the significance of shifting raw materials and sources, concepts of ownership, and technological choice.

Prehistoric Quarry Quandaries: The Late Formative Taraco Peninsula

The Lake Titicaca Basin is a high plateau approximately 80,000 km^2 in area and 4,000 m above sea level in the Central Andes of South America. Flanked by several mountain chains, it is an internal drainage basin centered on Lake Titicaca, but continuing south on the Desaguadero River. Twenty kilometers from the edge of Lake Titicaca is the Middle Horizon (AD 400–950) site of Tiwanaku. This large urban center located at high altitude has intrigued tourists and scholars alike with its remains of large cut stones, carved monoliths, beautiful pottery, and a diversity of long-distance trade items. Up until fairly recently, the understanding of the Formative Period was limited to the Middle Formative occupation of Chiripa located on the shores of the smaller, southern Lake Wiñamarka (Bennett 1948; Browman 1978; K. Chávez 1988; Hastorf 1999; Ponce Sanginés 1970), and the Late Formative site of Pukara in the northern Titicaca Basin (S. Chávez 1992; Kidder 1948; Klarich 2005; Stanish 2003; Tantaleán 2010; Wheeler and Mujica 1981).

The last 20 years, however, have seen an explosion of Formative Period research on both the Peruvian and Bolivian sides of the lake. Much of this research has been focusing on the Late Formative Period (200 BC–400 AD), a 600-year period before the emergence of the city and polity of Tiwanaku (see Hastorf 2005; Janusek 2004; Stanish 2003 for overviews). Unfortunately, chronological hiccups and a lack of excavation have forestalled much of this research. In some parts of the region, it has been particularly difficult to distinguish this long period from the earlier Middle

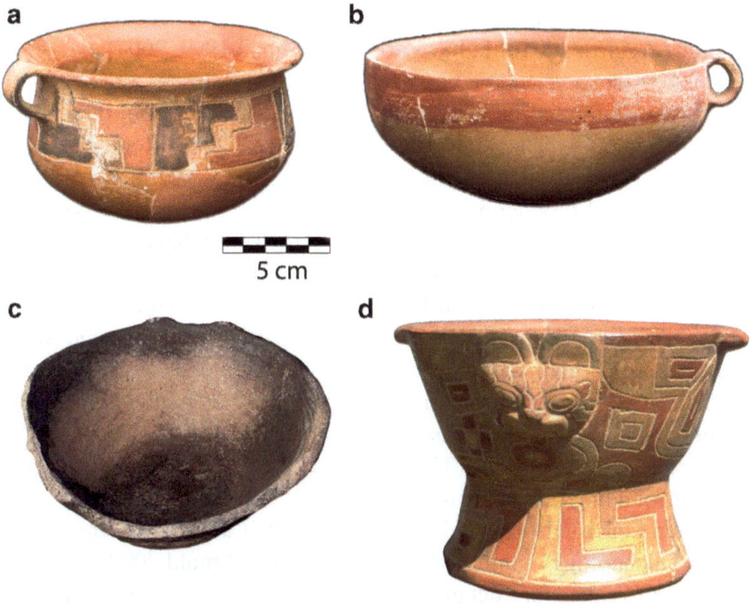

Fig. 5.2 Late Formative ceramics. Vessels made of fine-compact pastes (often with red hematite inclusions) from the southern Titicaca Basin: (**a**) Kalasasaya "zonally-incised" bowl from Tiwanaku; (**b**) Kalasasaya red rimmed bowl from Tiwanaku; and (**c**) a more typical coarse and micaceous tempered vessel from the site of Kala Uyuni, on the Taraco Peninsula. (**d**) Polychrome Pukara *incensario* from the northern Titicaca Basin, not to same scale as (**a–c**). Images compiled by Andrew Roddick

Formative (800 BC–200 AD) and later Tiwanaku (400 AD–950 AD) phases. There is some continuity in technology and production sequences resulting in subtle chronological boundaries. Defining these chronological boundaries has also been impacted by the predominance of survey projects and lack of systematic deep excavation in the region. Initially, Titicaca Basin cultural chronologies were constructed based on changes in surface decoration. For example, the Late Formative in the northern Basin was defined by the development of a new style of polychrome pottery with standardized iconography (S. Chávez 1992). Southern Basin decorated pottery includes red-banded and incised "Kalasasaya" vessels (some of which draw on styles from Pukara; Fig. 5.2) and a later Late Formative phase includes a rare polychrome style commonly known as Qeya (Janusek 2003; Ponce Sanginés 1971, 1993; Wallace 1957).

More recent research has focused on other attributes in Formative Period ceramic manufacturing sequences. In these highly fragmented assemblages, the paste recipes of nondecorated vessels have been particularly useful in creating chronological divisions in the Formative Period. In the northern Basin, Late Formative nondecorated pottery is distinguished from that of previous and subsequent periods by the presence of relatively large flakes of mica used as tempering material (Carlevato 1988; S. Chávez 1992; Klarich 2005; Oshige Adams 2010). In the southern Basin tempering is also key to chronology building, where similar shifts have been observed (Janusek 2003; Lemuz Aguirre 2001; Mathews 1992; Steadman 1999, 2007).

Since 2005, the first author has been investigating issues of Late Formative ceramic production based on a large assemblage of sherds from three sites excavated by the Taraco Archaeological Project (Roddick 2009; Roddick and Hastorf 2010). While it appears that production was occurring on the Taraco Peninsula (a variety of tools of production have been recovered) few primary typical indicators of ceramic production (i.e., wasters, ash, burned clay) have been identified. As such, attention has been on technological choices in ceramic production (as seen in manufacturing attributes) across the three settlements through the Formative Period. The archaeological assemblage is defined by particular diachronic shifts. For instance, the Late Formative is defined by the appearance of a new decorated form—the red-banded Kalasasaya bowl discussed above—that is ubiquitous on Taraco sites. Other changes concern the use of raw materials, in particular related to clay and temper mixtures, or pastes. The interest in quarries and sources stems from questions surrounding these clay recipes.

A first step consisted of examining the "potential and limitations of available material and technology" (Shimada 1998a: 7, see also Arnold 2000). Over 40 years ago, Mohr (1966) called for a detailed survey of available resources on the Peninsula, and both she and Bandy (2001) have suggested that raw materials were quarried from the Taraco Hills. In 2007 a survey was conducted to investigate the possibility of ancient quarrying activities by testing the nature of locally available raw materials and their relationship to Late Formative ceramics. Unlike many other areas in the circum-Titicaca region, there are no active potters on the Peninsula today. Nevertheless, the ubiquitous clay deposits are still being used as a building and flooring material in many households (Goodman-Elgar 2008) and geophagy is a common practice.[1] The first author worked with a Tiwanaku potter in collecting the raw materials of the region, which were sampled from the surface, from areas dug by property owners, deep gallery-like deposits, and (most commonly) deep cuts through the *quebradas* of the Taraco hills (Fig. 5.3). A total of 112 clays, 14 possible temper sediments, and 5 colorful mineral pigments were collected.

A series of subjective tests similar to those used by a variety of contemporary potters were then deployed to gauge the plasticity and water retention of these local materials. These tests included a "coil test," a "loop test," and a "ball test," which allowed for plasticity, stiffness, and strength to be examined (McReynolds and Herbert 2004; Roddick 2009: 275–277). It was found that they are all high-quality potting clays that would have been well suited for pottery manufacture. X-ray diffraction supported these more subjective claims, as the clays were rich in smectites, illites, and kaolonite, making a particularly good potting mineralogical mix. The clays collected from the Taraco Formation included a range of colorful iron oxides, including white, light brown, red brown, orange, and yellow clays. These clays

[1] Like elsewhere in the Andes (including Pucará), Taraco villagers consume clay as a condiment during communal potato roasts called watiyas.

Fig. 5.3 Clay collection on the Taraco Peninsula, southern Titicaca Basin. Photo by Andrew Roddick

refired to colors in the same range as Late Formative ceramic samples. While detailed analysis is ongoing, petrographic and elemental analysis of Late Formative ceramics and the refired briquettes suggests Late Formative inhabitants on the Taraco Peninsula used local materials (Roddick 2009: 263–299).

Unsurprisingly, there is little direct evidence for clay quarrying 2,000 years ago on the Taraco Peninsula. However, there is plenty of evidence of sources that would have been ideal for mining clay and analysis has identified very little diversity among these sources. Clays from sources all along the Peninsula up to the site of Tiwanaku are quite similar, and true diversity in quality and geochemistry are not apparent until some distance away. There is significant color variability across otherwise similar (mineralogically, chemically, and subjectively) sources (Roddick 2009: 263–299). The ubiquity, but also homogeneity of possible sources in such a dynamic environment suggests that the entire peninsula could be defined as a potential prehistoric quarry.[2]

The recipes, or pastes, utilized during the Formative Period appear to have always included a temper mixing stage. If local Taraco materials were indeed being used in the Late Formative, tempering would have been essential for these fine clays to withstand the firing process. The paste mixing stage has proven to be vital for defining diachronic shifts. As noted briefly above, late Middle Formative pottery is defined by the presence of grass temper and often included large opaque quartz fragments. In contrast, Late Formative ceramics lack organic inclusions and are

[2] Dean Arnold noted deposits of clay and tempers that varied both vertically and over great horizontal distances in Quinua (Peru). His work in Ticul (Mexico) found raw materials that were much more homogenous, while not as widely distributed (Arnold 2000: 340). The Taraco case here appears to offer both homogeneity and fairly wide horizontal distribution.

defined by a high density of micaceous sands. Petrographically, grains of quartz, biotite, muscovite, plagioclase and sanidine feldspars, hornblende and both sedimentary and igneous rock fragments have been identified. The larger fragments are angular, yet unaltered, and clean looking compared to the smaller, nonclay minerals, suggesting the larger were temper elements (Roddick 2009: 310–314). There is also the appearance of several other paste recipes that were not present in earlier phases, including compact pastes. At magnification one group has characteristic hematite aggregates and another has high densities of volcanic ash (Roddick 2009: 314–316). The survey for tempering material did not find many sources of micaceous or ashy materials. So while ideal potting clay is ubiquitous in the deep rivers cuts on the current day Taraco Peninsula, the particular tempering material, used by generations of potters, is not as commonly available.

Nevertheless, researchers working in other regions of the southern Basin have noted similar shifts from fiber to micaceous sands in a number of pottery producing regions (e.g., Janusek 2003; Lemuz Aguirre 2001; Steadman 1995). Is this shift based on the constraints of raw materials and the intended purpose of the vessels? Do recipes change based on the regional variability of raw materials over time? Or are social decisions being made at quarry sites a key element? The rich tradition of ethnoarchaeological studies from the Andean region (e.g., Arnold 1984, 1993; Bowser 2000; Cleland and Shimada 1998; Donnan 1971; Druc 2001; Hagstrum 1988, 1989; Hosler 1996; O'Neale 1976; Ravines 1978; Ravines and Viller 1989; Shimada 1998a, b; Shimada and Wagner 2007) may offer some perspective on such prehistoric quarry quandaries. After a brief overview of studies focused in the circum-Titicaca Basin, contemporary clay quarrying and preparation practices in the community of Pucará, Peru are explored.

Modern Quarry Quandaries: Ethnography in Pucará, Peru

In order to frame the methods, data sets, and goals of recent ethnographic work in Pucará, it is important to first review the extant ethnoarchaeological literature from the Lake Titicaca Basin. Formal ethnoarchaeological research on ceramic production in the region began with Tschopik's (1950) "An Andean Ceramic Tradition in Historical Perspective." The article traces "the Aymara ceramic tradition" (Tschopik 1950: 196), which he argues remained relatively stable from the Inca period, through the colonial period, and into contemporary times (from 1940 to 1942) in three communities near the large town of Chucuito in the western Basin. He provides a detailed inventory of locally produced vessels, discusses regional specialization in various crafts, and includes a number of details on ceramic technology pertinent to the present discussion.

In terms of procuring raw materials, Tschopik describes one potter's process of creating *ñeq'e,* or paste, after moving from the countryside to Chucuito (ibid: 209). Because of a lack of suitable local materials in town, this potter returns to his home village and collects two types of clays (translated as red earth and purple earth) plus tempering materials (white sand and "gold sand" with mica) from a dry riverbed. This potter's particular recipe involved combining an *olla* sized amount of purple earth, an *olla* of sand, and a bowl of red earth to create the paste used in the production of a wide variety of vessel forms. While Tschopik briefly mentions the well-known pottery centers of Pucará and neighboring Santiago de Pupuja (1950: 201) in the northern Basin, he focuses on workshop production contexts from this area, not the early stages of material procurement or processing as discussed for the Chucuito case study.

More detailed information about contemporary ceramic production in Pucará and neighboring potting communities appears several decades later in a series of publications written not by archaeologists or ethnographers, but by several scholars interested in documenting and promoting Andean folk or popular art traditions. Studies by Spahni (1966) and Litto (1976) provide valuable information on local clay sources and tempering materials, forming techniques, firing technology, and often exchange, which serve as a foundation for all subsequent ethnoarchaeological studies in the region (e.g., Chávez 1987 and Sillar 2000 citing Litto 1976).

The ethnoarchaeological studies most influential in framing the Pucará project are K. Chávez's (1987, 1992) work at Raqchi (Department of Cuzco, 150 km northwest of Pucará) and Sillar's (e.g., 1997, 2000) research in a number of Peruvian communities surrounding Cuzco (including Raqchi) and potting communities in Bolivia (in the southern Lake Titicaca Basin, Northern Potosi, and Cochabamba). At Raqchi, Chávez provides detailed descriptions of the stages of pottery production, consumption, and distribution, including valuable information on clay and temper procurement practices in the region. Her primary goal was to determine if the household- and community-level relationships concerning the organization of pottery production had "natural or artificial causes" (1987: 186). For example, did communities specialize in certain types of vessels because of limited access to appropriate materials (a technological constraint) or due to a shared interest in maintaining interdependency between those communities (a social constraint)?

This study is of particular interest because Pucará is noted as one of Raqchi's exchange partners (Chávez 1987: 186); Pucará offered the more reliable shale or slate-tempered *ollas* in contrast to vessels manufactured with volcanic temper in Raqchi. Raqchi potters provided Pucará potters with large *rakis*, a vessel type that would not be durable if produced in Pucará, according to local sources. While noting that her research was preliminary and limited to a single community, Chávez proposed that various patterns of complementarity—seen at the household level in the sexual division of labor and at the local level through community specialization—were maintained artificially for a variety of social and economic reasons. Clearly, this is a subject that merits further exploration through expanded interviews and analysis of raw materials used by potters producing for intercommunity exchange.

Lastly, Sillar's (2000) study of ceramic production, distribution, and consumption provides a wide range of information about the diversity of vessel types utilized in the processes of cooking, storing, serving, and celebrating in the region today. In all the potting communities Sillar visited, he observed production at the household level, making "the economics of pottery manufacture…inseparable from the domestic economy" (Sillar 2000: 125). Unlike many ethnoarchaeologists, Sillar is not as concerned with using contemporary practices to model material signatures of ancient craft production, but instead focuses on "how the production, exchange and use of pottery is embedded in the wider activities and cultural values of Andean people" (ibid: 125). The work of Sillar not only provides excellent comparative data on the technical aspects of pottery production in the circum-Titicaca region (raw material procurement and processing for a variety of vessel types), but like Chávez offers insights into a range of social factors influencing decision making by contemporary potters.

Inspired by this rich tradition of ethnoarchaeological research in the region, the second author began to explore clay quarries in and around the town of Pucará in 2006. This work was associated with a full-coverage mobile-GIS survey of the archaeological site of Pukara (Klarich and Román 2012), where Klarich has been working since 2000. Casual surveys during multiple field seasons at Pukara and discussions with local potters suggested that there were numerous local and nonlocal clay sources that continued to be exploited for a variety of purposes. During the 2006 survey, which was conducted with a lifelong potter on the crew, clay samples were collected from four "defined" modern quarries and will be integrated into future studies of Late Formative pottery production.

The ethnographic data presented in this chapter were collected during the summer of 2010 as part of a multiyear project to develop a new *sala* (exhibition space) about modern craft production at the Museo Lítico Pukara, which opened in early 2011. As 80–90% of the few thousand residents of Pucará are part or full-time potters, there was quite a bit of interest in participating in the project. The co-directors of the Pukara Archaeological Project were able to conduct interviews in over 20 household-level ceramic workshops. Guided by both content needed for the new *sala* and previous ethnoarchaeological research in the region, the interviews primarily explored the organization of craft production, including the technological, economic, and social aspects of pottery production, consumption, and distribution.

Based on the interview responses, Pucará potters exploit a wide range of sedimentary and residual clays from meandering riverbank sources, deep galleries, and extensive surface deposits across a variety of settings (Fig. 5.4). When asked which clay sources are used to make their pots, potters responded using categories of color primarily, and then referred to location, or, less frequently, differentiated sources based on physical properties. In terms of color categories, there are two main local sources in Pucará: the *chocolate* sources about 2 km east of the town center near the river and the red source located on the northern edge of town. The other major clays used by Pucará potters (brown and red) come from sources approximately 7 km to the east in the district of Santiago de Pupuja. Minor sources include a yellow clay found along the margins of the river owned by the neighboring town of Choquehuanca; a blueish source that co-occurs with the *chocolate* source; a gray source; and an opaque white source located "in its place of origin" in the hills about 4 km south of Pucará.

Within color categories there are distinctions based on source location. For example, when discussing *chocolate* sources potters always distinguished between *relave*, which is mined along the river cut on the east edge of town or *hatun pampa*, mined from galleries several meters below the modern surface of the oxbow

Fig. 5.4 *Chocolate* clay sources on the edge of Pucará, northern Titicaca Basin: (**a**) *relave* and (**b**) *hatun pampa* (with David Oshige). Photos by Matthew Wilhelm

(see Fig. 5.4). As illustrated in Fig. 5.4, the *relave* sources are small, informal, and easily accessible pits along the river that fluctuate annually with shifts in water levels. In contrast, obtaining *chocolate* clays from galleries excavated into the *hatun pampa* requires entering deep, unstable pits that expand with use; the clays are typically removed by experienced men and often transported by the truck or tricycle load. The surface remains of previously used galleries are apparent across the pampa, which often fill with trash after the source has been exploited. These two sources of *chocolate* clays are only separated by a 5-minute walk, but were noted as distinct sources by potters interviewed.

Red is also a tricky category, as it includes both local and nonlocal clay sources used for distinctive productive activities. At the local level, Pucará red is a transported clay deposit that is mined from shallow, informal pits on the *pampa* (open plain) just north of town. The nonlocal red source is mined from transported surface deposits at Santiago de Pupuja, which is discussed below. The other named sources appear to be from either smaller deposits or ones not exploited with frequency. The consistent use of clay color categories across household workshops to define sources (potters always corrected interviewers who misspoke when referring to different clays) most likely reflects direct knowledge about the quarries from which potters select and/or buy their clays. Potters clearly know their landscape well; for example, they have no trouble differentiating between numerous clay sources located in the small, neighboring communities within the district of Santiago de Pupuja, which they would not actually visit with any regularity.

How do color categories correlate to the physical properties of the clay? Unlike the clays collected from the Taraco peninsula discussed above, the second author has yet to conduct formal analyses of the clays collected in and around Pucará. However, local potters were very clear and consistent about the properties (or "personalities") of these different sources, which are combined in a range of paste recipes for different vessel types. For example, *chocolate* is characterized as being very plastic, but with little "resistance to firing" (Interview #2). It is typically mixed with the red clay from Santiago de Pupuja, which is characterized as being resistant to higher temperatures, but with little plasticity and a tendency to dry too fast if used

Fig. 5.5 Honorato Ttacca sampling clays at Santiago de Pupuja, a nonlocal clay source used regularly by Pucará potters. Clay to left of the white dashed line is bright red and to the right is a deep grey. Photo by Matthew Wilhelm

alone (#2). When discussing the main source in Santiago de Pupuja, local potters would refer interchangeably to "red" and "brown" or "light chocolate" clays, which initially confused the interviewers as to whether the interviewees were referring to one or multiple sources. However, after reviewing photos from the 2006 survey, it was clear that this particular clay deposit was quite variable in color (Fig. 5.5) and that potters were selecting from different areas within the same source. In terms of less common colors, yellow is sometimes added to make finished objects "more resistant to firing" and is described as being "like simple dirt…it doesn't have much plasticity" (#2). Gray, an "antiplastic," is described as "sandy" and is mixed with plastic clays to make them stronger and increase the range of possible firing temperatures (#14). The white clay is also mixed with other clays to "raise their firing temperatures" (#14). Blueish clay is never used for pottery as it sticks to molds and is hard to fire; however, it is used to paint houses.

Interesting to note is that the red clay from Pucará, which is mined from shallow pits on the *pampa* just north of town, was only mentioned in detail by a potter who continues to hand build low-fired, large utilitarian vessels like *ollas* (cook pots) and *phuos* for making *chicha* corn beer. He mixes this red clay with nonlocal clay from Azángaro (up to 30 km away) and the local yellow clay for these coil-built pots, which he constructs on a circular, ceramic or stone *molde* (disk), which rests on the floor (Fig. 5.6).[3] It is interesting to note that during the 2006 survey there were stacks of drying adobe bricks near the shallow, informal pit quarries for this red clay

[3] Similar stone or ceramic disks have not been recovered from prehistoric ceramic production areas at Pukara. However, large stone slabs may have been used as bases or platforms for producing coil-built pottery (e.g. Klarich 2005, Block 3 excavations).

Fig. 5.6 Photograph of semi-finished vessel on a *molde* (disk) used by Pucará potter. Photo by Elizabeth Klarich

on the *campo ferial* (market grounds) just north of town. During the interviews in 2010, some potters also referred to this clay as *sañu*, which means both "earth" and "roof tile" in Quechua, indicating a combination of color, location, texture, and even use in the varieties of names for this source.

Clearly, there remains much to do to tease apart the myriad categories of local and nonlocal clay sources based on the initial interviews of Pucará potters. While there are some trends across all workshops (e.g., every family uses some type of local clay), it is very difficult to generalize about clay recipes for a particular type of vessel (cooking pot or *olla* vs. a bowl) or for a certain production technique (mold vs. wheel vs. hand built). Most households use three to four different types of clays and, as summarized by one of the potters, for each type of pot "we have to prepare different materials—*ollas* are different, grotesque art is different and then there is the liquid form of clay used for mold-made" (#12) objects. Most potters did not offer specific information about their clay recipes; however, one potter (#2) explained that she made grotesque figurines by mixing 60% *chocolate* from Pucará, 30% red/brown from Santiago, and 10% yellow from Choquehuanca plus small amounts of "fine construction sand."

In general, few potters spoke of adding temper to their clays, even when asked directly if their ceramic recipes included anything besides "just clay." This was also noted in the work of Spahni (1966: 64), where he documented that Pucará potters used 75% "plastic clay" and 25% "antiplastic clay" in their paste recipes, but there was no mention of adding temper.[4] However, one potter (#14), a teacher at the local

[4] Although archaeologists continue to use standardized categories when examining raw materials, paste preparation often "does not conform nicely to immutable definitions of "clay" and "temper" as plastic and added non-plastic respectively. Rather potters are interested in modifying the paste so that they can successfully make pots with it" (Arnold 1998: 355).

school, had participated in a number of workshops sponsored by NGOs and was very familiar with the technical terminology surrounding both modern and ancient pottery production. For example, he explained that during ancient times the Pukara, Inca, and other cultures had used mica as an additive within their clay recipes, but that was no longer the practice. Today in Pucará, potters who make *ollas* use slate (*piedra pizarra*) temper, which comes from nonlocal, flat, soft stones of red, silver, and black, and is ground and added to clays to improve consistency and increase resistance to firing temperatures. The use of slate is a long-standing tradition in some parts of the central Andes (Druc 2001). Silica is also used as temper for strength and added to glazes to keep them from cracking. This is an interesting, if unfortunate, case of a source becoming "local"; it is commonly known that potters today acquire this material by knocking chunks off the vertically placed, worked stone slabs that line the Classic Pukara sunken courts on the Qalasaya complex. The original source has yet to be determined, but several people have mentioned the possibility that the slabs came from Asillo in the Province of Azángaro. It may be that in addition to the technological properties of these stones their position within the sunken court holds some significance for modern potters, which is a topic that merits further exploration.

A Social Orientation to Ceramic Raw Materials in the Lake Titicaca Basin?

We began this chapter noting that archaeological studies in the Lake Titicaca Basin are increasingly constructing cultural chronologies from the basis of variability of paste recipes. Contemporary studies of material culture in the Titicaca Basin (and perhaps the broader South-Central Andes) have been attempting to move beyond cultural historical models, prioritizing explanatory models to diachronic cultural patterns. Andeanists are enthusiastic about discussing the broad social, political, and economic processes that may correlate with shifts in productive practice. Yet there has been little engagement with the social aspects that lay behind minor ceramic variations, relations that may be key in explaining the subtle elements that are increasingly essential in fine-grained chronological frameworks. At this early stage the ethnographic data discussed above cannot speak to long-term trends directly (in the manner of Arnold's (2000) inspiring work), yet the ethnoarchaeology discussed above can reveal potential social practices behind paste recipes in the Lake Titicaca Basin. In this brief discussion, we focus on three aspects—material constraints and technological choice, social and technical boundaries, and raw materials on a dynamic landscape—all of which reveal future research questions and the need for engagement with larger issues of the social aspects of clay quarrying and paste preparation.

Material Constraints and Technological Choice

As stated in the introduction to this chapter, functional aspects are a key interpretive aspect to exploring manufacturing choice. On the Taraco Peninsula choices of raw materials do appear to correlate somewhat with the actual use of finished vessels. While compact pastes were used for decorated vessels, the more friable micaceous pastes have been identified for a variety of forms, including charred vessels used directly over the fire. While it certainly is possible that micaceous materials were employed to manufacture a better cooking pot (Roddick 2009; Steadman 1995), it is not clear how this explains the process of adoption. As Sillar (1996: 260) argues, functional interpretations should be made very cautiously, and efficiency models should be constantly evaluated within the framework of cultural choices and local representations. So while mica may have been deployed for functional use initially, it does not preclude a range of future uses, meanings, and landscape associations. In the case of Pucará, it is worth investigating whether the wide range of paste recipes used for different production techniques (wheel, mold made, or hand-built) are mutually exclusive as most potters note, or if they vary based on family traditions, access to nonlocal clay sources, or other factors.

Gosselain and Livingstone Smith (2005) observe that potters in sub-Saharan Africa are willing to use a wide range of clays, including low percentage clay mixtures, to produce their pots and that particular mixtures infrequently correlate with the intended functions of the vessels. They have found that the physical attributes of raw materials are rarely the key elements in deciding which clay deposits will be quarried (Gosselain and Livingstone Smith 2005: 39). Rather the *potters' perception* of what is considered proper materials are more important: "When discussing the matter with them, one gets the feeling that each clay used locally is thought to be so "appropriate" as to be non-substitutable" (ibid). This observation has also been made for the Andes, where potting clay is often widely available (Litto 1976). It is reflected in Chávez's (frustrated) observation about the "artificial" nature of potters' explanations of their inabilities to make particular forms (K. Chávez 1992: 82, discussed in Sillar 1997).

As the second author's research shows, the wide availability of high-quality raw materials does not preclude culturally meaningful categories for clay sources, which can in turn have real social and material consequences. For instance, at Pucará color, as a referent to particular quarries, is the main way of judging or naming clay qualities and clay ownership. These clay qualities may in fact communicate a wide range of contextual information about both the raw materials and the finished pots. Sillar (1997: 12, drawing substantially on K. Chávez 1992) has discussed how a good, or "reputable pot," may incorporate several interrelated contextual factors, including "serviceability, aesthetics and exchange value." Like "reputable pots," "reputable quarries," such as those found around modern day Pucará might carry significant weight, serving as a trademark, and creating a type of bond between producers and consumers.

A consideration of such reputable quarry sites could be revealing in both ethnographic and archaeological contexts. For example, one may ask how having a vessel made from the famous "Pucará clay" mined from the various quarries around the town might enhance the prestige of the item; it has been documented that thieves from Puno, Cuzco, and Arequipa illegally mine community-owned clays from Pucará because of the fame of the source across the region. Such a perspective also introduces questions concerning similarities in manufacturing material in the Late Formative Period that go beyond performance requirements. Regardless of potential cooking advantages of micaceous pastes there may be broader social significance to particular technological choices in micaceous pastes, including the cultural value placed on the shiny particles by both producers and consumers (Arnold 1993; Lunt 1988; as discussed by Sillar 1996: 273). Wider regional survey for both clay and temper quarries and further detailed compositional research is necessary before such questions may be explored.

Social and Technical Boundaries

This relationship between "reputable quarries," producers and consumers also involves social boundaries, senses of place that can also create conflict and battles over property rights and use of clay (Arnold 2000: 357–359). Legal battles over ownership and access have been documented as early as the 1680s for nearby sources in the northern Lake Titicaca Basin (Spurling 1992: 244). At Pucará, different types of access impact the nature of boundaries, with some resources controlled by community members, and others purchased from private landowners in neighboring communities, while before land reform the local clay sources were owned by the *haciendas*. At this point it is unclear as to whether Pucará potters always pay for access to nonlocal clays (most responded that they did purchase nonlocal clays plus pay labor and transportation costs). Potters mentioned half a dozen sources they use from distinct communities within the nearby district of Santiago de Pupuja (e.g., Llallahua, Coqra, Checca) and it is unclear with the present data how the value of these sources vary both from a social and technological perspective.

This investigation of social boundaries around these quarries has introduced a number of questions to be investigated in the future: When potters exploit nonlocal sources owned by relatives or by communities in which they have close kin do they incur the same costs as other Pucará potters (clay, transportation, and labor)? Do the modern boundaries used today by potters (e.g., yellow clay is from Choquehuanca, the neighboring community, even though it is just on the other side of the riverbank) apply the same way across all families? How do intercommunity relationships and patterns of land ownership, for example, affect perceived physical and social boundaries of clay sources? Defining the range of boundaries for local and nonlocal clay sources clearly merits further attention in future interviews.

This work, along with other Andean ethnoarchaeology, suggests that raw material choice "may be heavily influenced by the particulars of local land ownership and/or political control" (Sillar 2000: 69). This relationship between technological choice and social boundaries would manifest itself in some subtle variability across archaeological landscapes, which would create the patterns that archaeologists recognize as regional traditions (Livingstone Smith 2000: 27). The nuanced social boundaries that are being revealed through ethnoarchaeological research are evocative, and certainly require finesse to develop insight into archaeological research. An initial stage is to examine the technological boundaries that result from these social boundaries in both the ethnoarchaeological and archaeological contexts.

Roddick's (2009) dissertation work has clearly defined common technical practice on the Taraco Peninsula, with identical diachronic shifts occurring across the three sites. Although future publications will present the geochemical and mineralogical details of these technological styles, there are some brief trends to be noted. There was little statistically significant difference in the categories of paste recipes being deployed across these sites. Qualitative petrography on these pastes suggests that not only are similar quarries being used (likely located in the fairly homogeneous geological Taraco hills) but also that similar learned practices of processing are being deployed. All three sites had a consistent presence of the paste groups, including the micaceous sand subgroups, and the two compact paste groups characterized by the presence of either ash or red hematite inclusions. The micaceous pastes appear to be similarly ground up and prepared, while the fine compact pastes all appear to have either been accessed as fine, well-sorted material, or carefully prepared. In sum, there is no indication of a technological boundary associated with the quarrying and preparing of pastes. We still do not have a sense for Late Formative social boundaries in ceramic production on the regional scale. On the one hand, this is due to little baseline research on the geological diversity in the broader region. On the other, since pastes have predominantly been used to construct chronologies, the disjunctures in recipes are often explained away, rather than engaged with in greater detail. Recent findings suggest such questions surrounding technological choice and social boundaries merit further investigation. For instance, John Janusek (personal communication, 2009) has observed ceramic assemblages in the Tiwanaku Valley with red inclusions, descriptively comparable to the hematite aggregates found in both clay deposits and fired Late Formative ceramics of the Taraco Peninsula. Janusek notes that these ceramics are not evenly distributed across the landscape, and apparently some communities had differential access to either the finished products or the raw materials (Janusek 2003). One working hypothesis, based on the finding of clays with such hematite inclusions locally (discussed above), is that these materials were associated with a "reputable quarry" on the Taraco Peninsula. Ongoing work seeks to identify possible disjunctures in ceramic assemblages in the region, paying particular attention to the quarrying and processing stages of particular production sequences.

Raw Materials on a Dynamic Landscape

The final element to be considered in this discussion is the nature of clay quarries themselves. As briefly mentioned in the beginning of this chapter, the clay–water system is particularly dynamic, both in a seasonal sense but also over the long term. In fact, "source locations of ceramic raw materials are not immutable and may change over time" (Arnold 2000: 346). As Arnold (2000: 348–351) points out, clay sources can come and go for a variety of reasons, including environmental processes, land ownership, and the exhaustion of sources. This contributes to the difficulty in clearly defining quarry sites; unlike many of the chapters in this volume, it is quite rare to find primary archaeological evidence for clay mining practices.[5] This is the elephant in the room, a troublesome issue within both the ethnographic and archaeological research projects outlined in this chapter that has concerned both of the authors.

While certainly a methodological concern, this dynamism is also a key element of the social practices surrounding raw materials. Like the Bariba potters that Gosselain and Livingstone Smith (2005) interviewed, the northern Andean potters of Las Animas (a pseudonym) stress that clay is alive, "it is sensitive, delicate and gets upset easily; it responds to the mood of the potter. Clay knows when the potter is ill or inexperienced, and becomes impossible to work. Clay changes state, formed and shaped by the potter's idea, but sometimes, they told me, clay has its own idea" (Hosler 1996: 83). While speaking directly to the materiality of clay, Hosler's observations also speak to the seemingly living nature of clay quarries themselves.[6] These sources are places with dynamic histories, which may include the mundane disappearance of a source through generations of use, to more dramatic and dangerous cave-ins (Arnold 2000: 348).

The shifting nature of such sites would have had a dramatic impact on the synchronic and historical processes seen at both the ethnographically visible time scales in contemporary potting communities and at archaeologically visible time scales in ancient potting assemblages of the Titicaca Basin. Such dynamism produces important questions for both research agendas. For instance, how might this impact naming conventions and debates over community ownership and access? One must imagine that named mines exploited over centuries (such as that of Santiago de Pupuja) must fluctuate across the landscape as they are exploited and erode.[7]

[5] Some historical archaeology projects have had some success. For instance, Stahl et al.'s (2008) compositional work in Ghana included raw materials from abandoned clay pits and galleries. These were likely accessed in the last 100 years and could not be accurately dated (Stahl, personal communication 2011). See also Arnold and Bohor (1977).

[6] This may draw comparison to the ethnographic work of Cruikshank (2005), who has written of the dynamic geological places associated with Northern Canadian communities. Cruikshank writes of First Nations' oral histories of "surging glaciers," shifting ice flows that are named animistic entities with cultural histories.

[7] It could also be expected that changes in larger scale sociopolitical organization (e.g. from hacienda-controlled to community-owned sources) would impact how sources are named, who accesses them, and their relationship to community (vs. individual) identity (also discussed in Sillar 1997). Clearly a much more dynamic understanding of larger political and environmental processes is required both in long-term archaeological and short-term historical processes.

From an archaeological perspective, how is it that particular quarrying practices could be maintained for many generations? In the Lake Titicaca Basin there are pastes that do not appear to change for over 200 years in a dynamic environment, which suggests potters are making an effort to maintain such stability. The transition from tempering with ubiquitous *ichu* grass during the Middle Formative to the Late Formative practice of tempering pots with more limited resources (volcanic ash and micaceous materials) suggests not only particular mining practices and potential trading relationships, but also a specific relationship to the landscape. While the lake levels of Lake Titicaca rose and fell, and while the Pucara River changed course, potters for several generations sought out (and likely named) particular sources. In the relatively geologically homogeneous Taraco Peninsula, this waxing and waning of sources may not impact the clay matrix of the pottery. More intriguing, perhaps, are the temper types that last across generations of practice. While not dynamic in the same sense as clays, particular temper deposits (such as the volcanic ash and mica discussed here) were important points on the broader Titicaca landscape, and likely were enmeshed in complex social histories of exploitation, much like other mineral extraction practices discussed in this volume.

Conclusions

This chapter represents an ongoing effort of two projects in the Lake Titicaca Basin to confront temporally disparate, but interconnected issues related to prehistoric and contemporary clay selection on the landscape. While these projects are different in scale and methodology, they find common ground in regards to perceptions of boundaries (both social and physical) and raw material qualities (or "appropriate" materials). There are two distinctive research trajectories as the authors continue to engage with "quarry quandaries." First, work on the Taraco Peninsula will continue to focus on the elemental variability of tempering materials, moving outward to a more regional approach to source materials. Second, continuing research in the northern Basin will examine the time depth of associations between Pucará and quality clays, taking a diachronic approach not typically employed in the ethnoarchaeology of pottery production (see Sillar 1997 and Arnold 2000 for a similar call for such archival research).

This chapter also speaks to the current state of ethnoarchaeology in the Andes, particularly the recent shifts to focus on consumptive practices, such as *chicha* (corn beer) production and consumption or feasting practices more generally.[8] These are welcome and much needed areas of study, and may impact our understanding of a range of technical sequences (e.g., Sillar's (2006: 274) discussion of large jars for

[8] E.g. Bray 2003 (editor); Hayashida 2008; Jennings and Bowser 2009 (editors); and Klarich 2010 (editor).

beer preparation being reused for clay processing). However, as this chapter demonstrates, a consideration of the processes behind patterns in raw materials reveals a rather astonishing number of potential research questions.

This leads to a final conclusion concerning the rich tradition of Andean ceramic ethnoarchaeology, which is fairly specialized scholarship rarely considered in wider archaeological practice in the region. As Andeanists continue to build local and regional cultural chronologies, a perspective on the social practices behind such elements would appear essential. Ethnoarchaeological perspectives offer important reminders of the complexities behind patterns in ceramic attributes, practices that occur at dynamic quarry sites as well as in potters' backyards.

Acknowledgments Thank you to the editors for their invitation to participate in the SAA symposium and this publication. As the only contributors dealing with clay and temper sources we often felt a little out of our element, but their encouragement motivated us to think more deeply about shared concerns relating to ancient mining practices across material types. Roddick would like to thank the communities of the Taraco Peninsula (particularly San Jose and Coacollu) and all members of the Taraco Archaeological Project (directors Dr. Christine Hastorf and Dr. Matthew Bandy) within which this ongoing research is based. Research in raw materials on the Taraco Peninsula could not have proceeded without support from the National Science Foundation, the Canadian Social Sciences and Humanities Research Council, and the Dirección Nacional de Arqueología y Antropología de Bolivia (DINAAR). Klarich would like to thank the following: the many generous potters of Pucará (particularly the Ttacca family); the Instituto Nacional de Cultura office in Puno; David Oshige, Barbara Carbajal, Matthew Wilhelm, Luis Flores, and Nancy Román; and Roberto Ramos and his students at the UNA-Puno. Funding was provided by the Heinz Foundation (2006), Wenner Gren (2008–2010), the Cotsen Institute of Archaeology at UCLA (2009–2010), and Smith College (2010).

References

Arnold, D. E. (1984). Social interaction and ceramic design: Community-wide correlates in Quinua, Peru. In P. Rice (Ed.), *Pots and potters: Current approaches in ceramic archaeology* (pp. 133–161). Los Angeles: UCLA Institute of Archaeology.

Arnold, D. E. (1993). *Ecology and ceramic production in an Andean community*. Cambridge: Cambridge University Press.

Arnold, D. E. (1998). Ancient Andean ceramic technology: An ethnoarchaeological perspective. In I. Shimada (Ed.), *Andean ceramics: Technology, organization, and approaches* (pp. 353–367). Philadelphia: University of Pennsylvania Museum of Archaeology and Anthropology.

Arnold, D. E. (2000). Does the standardization of ceramic pastes really mean specialization? *Journal of Archaeological Method and Theory, 7*(4), 333–375.

Arnold, D. E., & Bohor, B. F. (1977). An Ancient Clay Mine at Yo'k'at, Yucatan. *American Antiquity, 42*(4), 575–582.

Bandy, M. S. (2001). *Population and history in the ancient Titicaca Basin* (Unpublished Ph.D. dissertation). University of California – Berkeley.

Bennett, W. (1948). A revised sequence for the south Titicaca Basin. *Memoirs of the Society for American Archaeology, 4*, 90–92.

Bishop, R. L., Rands, R. L., & Holley, G. R. (1982). Ceramic compositional analysis in archaeological perspective. *Advances in Archaeological Method and Theory, 5*, 275–330.

Bowser, B. (2000). From pottery to politics: An ethnoarchaeological study of political factionalism, ethnicity, and domestic pottery style in the Ecuadorian Amazon. *Journal of Archaeological Method and Theory, 7*(3), 219–248.

Bray, T. (2003). *The archaeology and politics of food and feasting in early states and empires.* New York: Kluwer/Plenum Press.

Bronitsky, G., & Hamer, R. (1986). Experiments in ceramic technology: The effects of various tempering materials on impact and thermal-shock resistance. *American Antiquity, 51*(1), 89–101.

Browman, D. L. (1978). The temple of Chiripa (Lake Titicaca, Bolivia). In M. M. Ramiro (Ed.), *III Congreso peruano "El hombre y la cultura andina" actas y trabajos* (pp. 807–813). Lima: Universidad Nacional Mayor de San Marcos.

Carlevato, D. (1988). Late ceramics from Pucara, Peru: An indicator of changing site function. *Expeditions, 30*(3), 39–45.

Chávez, K. M. (1987). Traditional pottery of Raqch'i, Cuzco, Peru: A preliminary study of its production, consumption and distribution. *Ñawpa Pacha, 22–23*(1984–1985), 161–210.

Chávez, K. (1988). The significance of Chiripa in Lake Titicaca Basin developments. *Expeditions, 30*(3), 17–26.

Chávez, K. M. (1992). The organization of production and distribution of traditional pottery in south highland Peru. In G. J. B. Bey III & C. A. Pool (Eds.), *Ceramic production and distribution: An integrated approach* (pp. 49–92). Boulder: Westview Press.

Chávez, S. J. (1992). *The conventionalized rules in Pucara pottery technology and iconography: Implications for socio-political development in the northern Lake Titicaca basin* (Unpublished Ph.D. dissertation). Michigan State University.

Cleland, K. M., & Shimada, I. (1998). Paleteada potters: Technology, production sphere, and subculture in ancient Peru. In I. Shimada (Ed.), *Andean ceramics: Technology, organization and approaches* (pp. 111–152). Philadelphia: MASCA, University of Pennsylvania Museum of Archaeology and Anthropology.

Corniquet, C. (2011). Cadres de pratiques et circulation des connaissances chez les potières de l'arewa (Niger). *Cahiers D'études Africaines, 1*, 87–114.

Cruikshank, J. (2005). *Do glaciers listen?* Vancouver, BC: University of British Columbia Press.

Donnan, C. B. (1971). Ancient Peruvian potters' marks and their interpretation through ethnographic analogy. *American Antiquity, 36*(4), 460–466.

Druc, I. C. (2001). Soil sources for ceramic production in the Andes. In I. C. Druc (Ed.), *Archaeology and clays* (pp. 95–105). Oxford: British Archaeological Reports International Series 942.

Goodman-Elgar, M. (2008). The devolution of mudbrick: Ethnoarchaeology of abandoned earthen dwelling in the Bolivian Andes. *Journal of Archaeological Science, 35*(12), 3057–3071.

Gosselain, O. P. (1992). Technology and style: Potters and pottery among Bafia of Cameroon. *Man, 27*(3), 559–586.

Gosselain, O. P. (1999). Not so much a pot, more a way of life: Current approaches to artefact analysis in archaeology. *American Journal of Archaeology, 103*(2), 382–384.

Gosselain, O. P. (2008). Mother bella was not a bella. Inherited and transformed traditions in southern Niger. In M. T. Stark, B. J. Bowser, & L. Horne (Eds.), *Cultural transmission and material culture: Breaking down boundaries* (pp. 150–177). Tuscson: University of Arizona Press.

Gosselain, O. P. (2011). Fine if I do, fine if I don't. Dynamics of technical knowledge in sub-Saharan Africa. In B. W. Roberts & M. V. Linden (Eds.), *Investigating archaeological cultures: Material culture, variability, and transmission* (pp. 211–227). New York: Springer.

Gosselain, O. P., & Livingstone Smith, A. (2005). The source: Clay selection and processing practices in sub-Saharan Africa. In A. Livingstone Smith, D. Bosquet, & R. Martineau (Eds.), *Pottery manufacturing processes: Reconstruction and interpretation* (pp. 33–48). Oxford: British Archaeological Reports International Series 1349.

Hagstrum, M. B. (1988). Ceramic production in the central Andes, Peru: An archaeological and ethnographic comparison. In C. C. Kolb & L. M. Lackey (Eds.), *A pot for all reasons: Ceramic ecology revisited* (pp. 127–143). Philadelphia: Temple University Press.

Hagstrum, M. B. (1989). *Technological continuity and change: ceramic ethnoarchaeology in the Peruvian Andes* (Unpublished Ph.D. dissertation), University of California – Los Angeles.

Hastorf, C. A. (1999). *Early settlement at Chiripa, Bolivia—Research of the Taraco archaeological project*. Berkeley: University of California Archaeological Research Facility.
Hastorf, C. A. (2005). The upper (middle and late) formative in the Titicaca region. In C. Stanish, A. B. Cohen, & M. S. Aldenderfer (Eds.), *Advances in Titicaca Basin archaeology,* (Vol. 1, pp. 65–94). Los Angeles: Cotsen Institute of Archaeology, University of California.
Hayashida, F. (2008). Ancient beer and modern brewers: Ethnoarchaeological observations of chicha production in two regions of the North Coast of Peru. *Journal of Anthropological Archaeology, 27*, 161–174.
Herbich, I. (1987). Learning patterns, potter interaction and ceramic style among the Luo of Kenya. *African Archaeological Review, 5*, 193–204.
Hosler, D. (1996). Technical choices, social categories and meaning among the Andean potters of Las Animas. *Journal of Material Culture, 1*(1), 63–92.
Janusek, J. W. (2003). Vessels, time, and society: Towards a chronology of ceramic style in the Tiwanaku heartland. In A. L. Kolata (Ed.), *Tiwanaku and its hinterland: Archaeology and paleoecology of an Andean civilization,* (Vol. 2, pp. 30–92). Washington, DC: Smithsonian Institution Press.
Janusek, J. W. (2004). Tiwanaku and its precursors: Recent research and emerging perspectives. *Journal of Archaeological Research, 12*(2), 121–183.
Jennings, J., & Bowser, B. J. (Eds.). (2009). *Drink, power and society in the Andes*. Gainesville: University Press of Florida.
Kidder, A. L., II. (1948). The position of Pucara in Titicaca basin archaeology. *American Antiquity, 13*(4), 87–89.
Klarich, E. A. (2005). *From the monumental to the mundane: Defining early leadership strategies at late formative Pukara, Peru* (Unpublished Ph.D. dissertation). University of California, Santa Barbara.
Klarich, E. A. (2010). *Inside ancient kitchens new directions in the study of daily meals and feasts*. Boulder: University Press of Colorado.
Klarich, E. A., & Román, N. (2012). Scale and diversity at Late Formative period Pukara. In A. Vranich, E. A. Klarich, & C. S. Stanish (Eds.), *Advances in Titicaca Basin archaeology, Vol. III*. Ann Arbor, MI: University of Michigan Press.
Lechtman, H. (1977). Style in technology: Some early thoughts. In M. Merril (Ed.), *Material culture: Styles, organization, and dynamics of technology* (pp. 3–20). St. Paul: American Ethnological Society.
Lemuz Aguirre, C. (2001). *Patrones de asentamiento arqueologico en la Peninsula de Santiago de Huata, Bolivia* (Unpublished Licenciatura thesis). Universidad de San Andres, La Paz, Bolivia.
Litto, G. (1976). *South American folk pottery*. New York: Watson-Guptill Publications.
Livingstone Smith, A. (2000). Processing clay for pottery in northern Cameroon: Social and technical requirements. *Archaeometry, 42*(1), 21–42.
Lunt, S. W. (1988). The manufacture of the Inca aryballus. In N. J. Saunders & O. de Montmollin (Eds.), *Recent studies in pre-Columbian archaeology* (pp. 489–511). Oxford: British Archaeological Reports 421(ii).
Mathews, J. (1992). *Prehispanic settlement and agriculture in the Middle Tiwanaku Valley, Bolivia* (Doctoral dissertation, University of Chicago). Available from ProQuest Dissertations and Thesis database (UMI No. 304041397).
McReynolds, T. E., & Herbert, J. M. (2004). *An Evaluation of the utility of Carolina clays for Woodland potters*. Paper presented at the Southeastern Archaeological Conference and the Midwestern Archaeological Conference, St. Louis, MO.
Mohr, K. L. (1966). *An analysis of the pottery of Chiripa, Bolivia: A problem in archaeological classification and inference* (Unpublished master's thesis). University of Pennsylvania, Philadelphia, PA.
Neff, H., Bove, F. J., Brenda, L. L., & Piechowski, M. F. (1992). Ceramic raw materials survey in Pacific coastal Guatemala. In H. Neff (Ed.), *Chemical characterization of ceramic pastes in archaeology* (pp. 59–84). Madison, WI: Prehistory Press.

Neupert, M. A. (2000). Clays of contention: An ethnoarchaeological study of factionalism and clay composition. *Journal of Archaeological Method and Theory, 7*(3), 249–272.

O'Neale, L. M. (1976). Notes on pottery making in highland Peru. *Ñawpa Pacha, 14*, 41–60.

Oshige Adams, D. (2010). *La secuencia más temprana en el sitio de Pukara, cuenca norte del Lago Titicaca* (Unpublished Licenciatura thesis in Archaeology). Pontificia Universidad Católica del Perú, Lima, Perú.

Ponce Sanginés, C. (1970). *Las culturas Wankarani y Chiripa y su relación con Tiwanaku.* La Paz: Academia Nacional de Ciencias de Bolivia.

Ponce Sanginés, C. (1971). *La cerámica de la época I de Tiwanaku.* La Paz: Academia Nacional de Ciencias de Boliva.

Ponce Sanginés, C. (1993). La cerámica de la época I (aldeana) de Tiwanaku. *Pumapunku: Nueva Época, 4*, 48–89.

Ravines, R. R. (1978). Cerámica actual de Ccaccasiri, Huancavelica. In R. R. Ravines (Ed.), *Tecnología Andina* (pp. 447–466). Lima: Instituto de Estudios Peruanos.

Ravines, R. R., & Viller, F. (Eds.). (1989). *La cerámica tradicional del Peru.* Lima: Editorial Los Pinos.

Roddick, A. P. (2009). *Communities of pottery production and consumption on the Taraco Peninsula, Bolivia, 200 BC-300 AD* (Unpublished Ph.D. dissertation). University of California – Berkeley.

Roddick, A. P., & Hastorf, C. A. (2010). Tradition brought to the surface: Continuity, innovation and change in the Late Formative period, Taraco Peninsula, Bolivia. *Cambridge Archaeological Journal, 20*(2), 157–178.

Shimada, I. (1998a). Andean ceramics: An introduction. In I. Shimada (Ed.), *Andean ceramics: Technology, organization and Approaches* (pp. 1–22). Philadelphia: MASCA, University of Pennsylvania Museum of Archaeology and Anthropology.

Shimada, I. (Ed.). (1998b). *Andean ceramics: Technology, organization and approaches.* Philadelphia: MASCA, University of Pennsylvania Museum of Archaeology and Anthropology.

Shimada, I., & Wagner, U. (2007). A holistic approach to pre-Hispanic craft production. In J. M. Skibo, M. W. Graves, & M. T. Stark (Eds.), *Archaeological anthropology: Perspectives on method and theory* (pp. 163–197). Tucson: University of Arizona Press.

Sillar, B. (1996). The dead and the drying: Techniques for transforming people and things in the Andes. *Journal of Material Culture, 1*(3), 259.

Sillar, B. (1997). Reputable pots and disreputable potters: Individual and community choices in present-day pottery production and exchanges in the Andes. In C. Cumberpatch & P. Blinkhorn (Eds.), *Not so much a pot, more a way of life* (pp. 1–20). Oxford: Oxbow Books.

Sillar, B. (2000). *Shaping culture: Making pots and constructing households: An archaeological study of pottery production, trade and use in the Andes.* Oxford: British Archaeological Reports International Series 883.

Skibo, J. M., Schiffer, M., & Reid, K. C. (1989). Organic-tempered pottery: An experimental study. *American Antiquity, 54*(1), 122–146.

Spahni, J. C. (1966). *La cerámica popular del Peru.* Lima: Peruano Suiza.

Spurling, G. E. (1992). *The organization of craft production in the Inka state: The potters and weavers of Milliraya* (Unpublished Ph.D. dissertation). Department of Anthropology, Cornell University.

Stahl, A. B., Dores Cruz, M., Neff, H., Glascock, M. D., Speakman, R. J., Giles, B., & Smith, L. (2008). Ceramic production, consumption and exchange in the Banda area, Ghana: Insights from compositional analyses. *Journal of Anthropological Archaeology, 27*(3), 363–381.

Stanish, C. (2003). *Ancient Titicaca: The evolution of complex society in southern Peru and northern Bolivia.* Berkeley, CA: University of California Press.

Steadman, L. H. (1995). *Excavations at Camata: An early ceramic chronology for the western Titicaca Basin, Peru* (Doctoral dissertation, University of California-Berkeley). Available from ProQuest Dissertation and Thesis database. (UMI No. 304153957).

Steadman, L. H. (1999). The ceramics. In C. A. Hastorf (Ed.), *Early settlement at Chiripa, Bolivia: Research of the Taraco Archaeological Project* (pp. 61–72). Berkeley, CA: Archaeological Research Facility.

Steadman, L. H. (2007). Ceramic analysis. In C. A. Hastorf & M. S. Bandy (Eds.), *The Taraco Archaeological Project's 2003 excavations at Kala Uyuni*. Berkeley, CA: Archaeological Research Facility.

Summerhayes, G. R. (1997). Losing your temper: The effect of mineral inclusions on pottery analyses. *Archaeology in Oceania, 32*(1), 108–117.

Tantaleán, H. (2010). *Ideología y realidad en al primeras sociedades sedentarias (1400 ANE- 350 DNE) de la cuenca norte del Titicaca, Peru.* Oxford: British Archaeological Reports International Series 2150.

Tschopik, H. (1950). An Andean ceramic tradition in historical perspective. *American Antiquity, 15*(3), 196–218.

Wallace, D. T. (1957). *The Tiahuanaco horizon styles in the Peruvian and Bolivian highlands* (Doctoral dissertation, University of California- Berkeley). Available from ProQuest Dissertations and Thesis database. (UMI No. 301929908).

Wallaert-Pêtre, H. (2001). Learning how to make the right pots: Apprenticeship strategies and material culture, a case study in handmade pottery from Cameroon. *Journal of Anthropological Research, 57*(4), 471.

Wallaert-Pêtre, H. (2008). The way of the potter's mother: Apprenticeship strategies among the potters from Cameroon, West Africa. In M. T. Stark, B. J. Bowser, & L. Horne (Eds.), *Cultural transmission and material culture: Breaking down boundaries* (pp. 178–198). Tuscson: University of Arizona Press.

Wheeler, J., & Mujica, E. (1981). *Prehistoric pastoralism in the Lake Titicaca basin, Peru, 1979–1980 field season*. Washington, DC: National Science Foundation.

Chapter 6
The Huarhua Rock Salt Mine: Archaeological Implications of Modern Extraction Practices

Justin Jennings, Félix Palacios, Nicholas Tripcevich, and Willy Yépez Álvarez

As many of the chapters in this volume demonstrate, there is much still to learn about mining and quarrying in the ancient Andes. This is particularly true in regards to the mining of halite or rock salt. Though one of the most widely traded and highly valued exchange goods among indigenous communities, rock salt was of little consequence to the metal hungry Conquistadores and has thus far managed to escape the interests of archaeologists. A tentative sketch can be made of early silver, gold, copper, obsidian, mercury, and hematite mining from historical, ethnographic, and archaeological resources (e.g., Berthelot 1986; Eerkens et al. 2009; Petersen 1994, 2010; Shimada 1994). Yet outside of a handful of references in early Spanish documents, we know almost nothing about the ancient exploitation and exchange of rock salt.

In this chapter, we hope to explore some aspects of ancient rock salt mining through the observations that each of us made during independent visits from 1999 to 2007 to a currently operating rock salt mine in southern Peru. We stress that our time in and around the mine was quite limited—our combined stay at the site was less than a month, but we had the opportunity to talk at different parts of the year

Mining and Quarrying in the Ancient Andes, edited by Nicholas Tripcevich and Kevin Vaughn.

J. Jennings (✉)
Royal Ontario Museum, Toronto, ON, Canada
e-mail: collota@gmail.com

F. Palacios
Universidad Nacional de San Agustín, Arequipa, Peru
e-mail: felpalacios@yahoo.com

N. Tripcevich
University of California-Berkeley, 2251 College Bldg, UC Berkeley, Berkeley, CA 94720-1076, USA
e-mail: nico@mapaspects.org

W.Y. Álvarez
Proyecto Arqueológico La Real, Av. Los Proceres N° 198, Arequipa, Perú
e-mail: tenakaka2007@gmail.com

N. Tripcevich and K.J. Vaughn (eds.), *Mining and Quarrying in the Ancient Andes*, Interdisciplinary Contributions to Archaeology, DOI 10.1007/978-1-4614-5200-3_6,

with the villagers, miners, traders, and customers that were affected by the rock salt trade. Although mining practices have undoubtedly changed through the centuries, we believe that what we learned about the methods and organization of modern halite extraction at this mine may provide some preliminary insights into how rock salt mines could be managed and controlled in the past.

The Huarhua Mine Today

The Mina de San Francisco de Asisi is located a few minutes walk from the village of Huarhua in the Cotahuasi Valley (Fig. 6.1). The mine is located within a salt dome found amid a series of sedimentary rocks that include limestone, a deep-water ocean sediment. The rock salt comes in a variety of colors and textures. Different inclusions, impurities, and structural defects in the salt's crystal lattice cause the various shades of black, red, pink, and white (e.g., Sonnenfeld 1995). Ground water seeping through the mine alters the salt's texture and the salt is often layered in alternating bands of color ranging from 0.5 cm to 0.75 m in thickness. The Huarhua salt is a highly valued commodity in the region's economy and has a wide variety of uses—from seasoning to preservative, saltlick, and medicine—that varies with the color and texture of the salt. Transported primarily through llama and burro

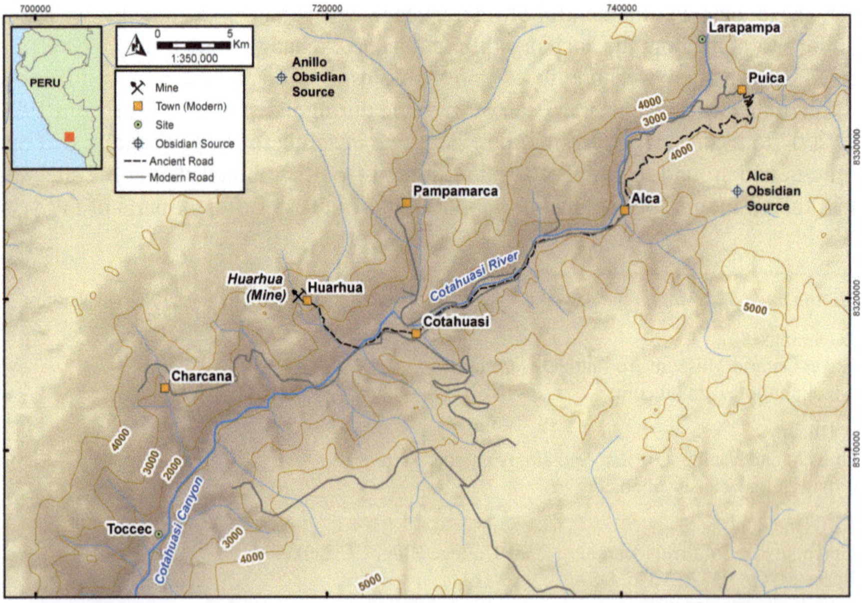

Fig. 6.1 Location of the Huarhua salt mine in the Cotahuasi Valley. Salt similar to that mined in Huarhua was found in tombs at Toccec and Larapampa. The town of Huarhua is adjacent to the precipitous Quebrada Espanja. From the community a trail traverses and then descends to the present entrance of the mine. Image courtesy of Bing Maps/Digital Globe, 16 April 2010

Fig. 6.2 Children in front of the gate of the Huarhua mine

caravans, the salt is exchanged in the southern Peruvian Departments of Arequipa, Cuzco, Apurimac, and Ayacucho (Flores Ochoa and MacQuarrie 1994: 125–127).

The Huarhua mine today is reached by a narrow trail that runs along a steep slope that separates the village from the mine entrance. A padlocked gate restricts access into the mine (Fig. 6.2). To reach the rock salt deposits, a shaft cuts approximately 300 m straight into the heart of the mountain. The mine then breaks into six tunnels, named Torro Moso, Choccon, Atutikana, Haiutaccanai, Pique, and Socoban, that meander through the deposit. All of the shafts come together at different levels into one central chamber that measures almost 60 m in height. The chamber, as well as some of the wider sections of tunnels, is supported by leaving in place pillars of unmined salt. The amount of salt extracted from the source fluctuates depending on the number of active miners. In 1999, there were 23–39 miners who worked the source actively during at least part of the year (our informants differed on the exact number of people who worked the mine). These miners and their families extracted approximately 100,000-kg of salt during that year.

Like other salt sources in Peru, the mine became the property of the state in 1969 with the formation of the Empresa Pública de la Sal (EMSAL) (Palomino Meneses 1985: 165). State control over the mine, however, did not seem to have markedly changed how salt was exploited according to villagers. The mine continued to be run by a cooperative made up largely, but importantly not exclusively, of people living in the nearby village of Huarhua for the next 3 decades. All associates of the mine paid an "exceso" of one arroba of salt (approximately 11.4 kg) to the cooperative for every quintal (around 45.5 kg) extracted. The arroba of salt was then sold to third parties to pay both the state taxes and fund the general maintenance of the mine and the cooperative. Associates were also granted a small amount of salt for personal consumption that was not taxed by the cooperative.

Fig. 6.3 Extracting rock salt in the mine

The cooperative would choose a head of operations for the mine and this individual held the key to the mine's entrance. Other fixed term positions were determined by the cooperative with expectations that each member would fulfill a variety of roles during their lifetimes. Cooperative members and their family members had rights to a particular part of the mine and each group worked the mine with very little hierarchical control. One of the few concessions made to the state was the hiring of an engineer to annually inspect the mine and mark the pillars of rock salt used to maintain the mine's integrity with large red "X"s and skulls and cross bones.

Our description of the collective is written in the past tense because in 2007 one of the more politically connected members of the collective staked a claim to the entire mine and changed the lock on the gate. Though many were afraid to speak out, some members of the collective have put forward a formal legal challenge asserting the long-standing communal ownership of the mine. Extraction has steeply declined in recent years as the case makes its way through the Peruvian judiciary system. The mine at least temporarily remains under private ownership and the exceso to work the mine has doubled to two arrobas. From a group of 23–31 associates just before the dispute, only 15 continue to remove salt from the Huarhua source. Mining log books that record activities at the mine over the last few decades are of critical importance in this dispute. Unfortunately, we have not yet been granted access to this invaluable source of data.

The change in ownership has not altered salt extraction techniques. The salt is still quarried with only the use of a hammer and chisel (often just a rock and a piece of rebar) (Fig. 6.3). A hole is burrowed deeper and deeper into the salt by banging the chisel into the rock. Chunks of rock salt are then obtained by pushing the chisel back and forth until the salt breaks off and fall to the ground. Ladders are used to reach higher areas (Fig. 6.4), while candles, and much more rarely flashlights, are used to

Fig. 6.4 Ladders used to reach higher areas in the mine

Fig. 6.5 Locals organizing bags of salt by candlelight

provide illumination for workers (Fig. 6.5). Most of the quarried salt is shoveled into bags that can hold about 11.5 kg of salt, and then loaded into wheelbarrows or on to one's shoulder. Particularly large chunks of salt are sometimes left intact to sell as saltlicks for animals.

Until the last few years, the salt was provisionally weighed at the mine's entrance, loaded on to burros or llamas for the short trip to Huarhua, and then brought to the cooperative office in the village to be re-weighed, officially recorded, and for the

Fig. 6.6 Llama caravan bearing rock salt near the mine's entrance

excise to be taken. Since 2007, all parts of this transaction occur at the mine's entrance. Miners, both under the cooperative and now private ownership, have been occasionally known to transport salt as well. In most cases, however, the mined salt is passed on to traders who bring their animal trains to the mine (Fig. 6.6). Though these traders are regular visitors to the mine, and sometimes have a relative among the members of the cooperative, they are not members themselves. Loaded on to burros and llamas, the salt begins its circulation through the barter networks that stretch far across the southern Peruvian highlands (Concha Contreras 1975: 71; Inamura 1981: 70–73; Trawick 2003: 48–69).

Ancient Use of the Huarhua Mine?

Although the Huarhua mine has likely long been an important source of salt for the people of the south-central Andes, there is no direct evidence for the early use of the mine. Mining is notorious for obliterating evidence of earlier exploitation (Stöllner 2008a: 73), and this is practically the case in rock salt mining since the best opportunity to date a mine often comes from examining the talus outside of its entrance. Rock salt mining usually leaves little talus (you can eat pretty much everything that you take out of the mine), and there is generally less of a mess within which artifacts can be more easily lost or buried. In the case of the Huarhua mine, there is also little chance of finding Pre-Columbian material on the steep, eroded slopes around the mine. A landslide buried one of the entrances in 1997, and we expect that similar geologic episodes in the past have destroyed any early evidence of the mine's use.

Nonetheless, villagers suggest that the mine dates to before the Spanish Conquest and their assertions are backed by at least three lines of indirect archaeological evidence. First, the modern village of Huarhua was likely founded in the Pre-Columbian period. The village sits on a small plateau just a few minutes walk from the mine, and ceramic sherds collected in the village date from the Middle Horizon through the Colonial Periods. Ten tombs dating from the Middle Horizon through Late Intermediate Periods were also found in the agricultural terracing above the site (Jennings 2002: 566).

The second line of indirect archaeological evidence for the salt mine's earlier use is the passage of a major pre-Columbian trail through the site (Fig. 6.1). Climbing up from the Cotahuasi River, the wide, well-maintained trail begins at a suspension bridge and then runs through terracing that abuts the trail on both sides. This trail was likely a major Inca road that connected Cuzco to the sea (Jennings and Yépez Álvarez 2008: 143–144). The empire likely built the road on top of a preexisting trail that led out of the valley—terracing, at least elsewhere in the valley, was likely constructed in the Middle Horizon (Jennings 2006: 360), but the placement of the road through the village of Huarhua might also reflect Inca interest in facilitating access to the rock salt mine.

Perhaps the best evidence for earlier use of the Huarhua mine is the documentation in looted tombs at two sites located elsewhere in the Cotahuasi Valley of two large chunks of rock salt that are similar in appearance to those mined today at Huarhua. A chunk of salt was found just inside a looted tomb at Larapampa, a funerary site dating from the Middle through Late Horizon that is almost 30 kg upriver from the mine. The second piece of salt was documented in the rubble of a destroyed tomb more than 20 kg down river at the Middle through Late Horizon site of Toccec (Fig. 6.7). Since the Huarhua mine is the only known rock salt source in this part of Peru, the salt found at both sites was most likely extracted from this mine. The location of the finds not only provides a minimum distribution radius, but also raises the possibility that salt was sometimes a ritually significant object for the pre-Columbian people of the Cotahuasi Valley.

Contextualizing Ancient Salt Production at Huarhua

Even if we can build a strong case for the long-standing exploitation of Huarhua's salt, we still unfortunately know nothing about the organization of production in the mine beyond people's recollections back to the mid-twentieth century. Yet, the few references to earlier salt production elsewhere in Peru leads us to believe that the cooperative practices followed at the Huarhua mine until the last few years may have been longstanding.

Fig. 6.7 The woman in the foreground holds a piece of rock salt found outside of a looted tomb at Toccec

The only rock salt source for which extensive early Spanish documentation exists is the Cerro de Sal in the Tarma area of southern Peru (Tibesar 1950; Verese 2002, 2006). In the seventeenth and eighteenth centuries, several groups from the Amazon and eastern Andes visited the site on annual trips that occurred usually in the months of July, August, and September. The salt was quarried from an exposed 18-m wide vein near Cerro de Sal's summit that ran for 16.5 km (Tibesar 1950: 103–104). As many as 500 people would work the vein using iron axes and river cobbles to carve out blocks of salt that weighed between 15 and 22 kg (Tibesar 1950: 103–106). While some of the salt was transported away in baskets, most of the salt was carried down to the Paucartambo River where it was loaded on to balsa rafts for passage into the jungle. Most of the rafts were made on the spot, packed from bow to stern with salt, and then loosely tied together into flotillas of 10–20 vessels (Tibesar 1950: 106).

The accounts are unclear regarding control of the salt quarry. Although people visiting the site paid the local cacique in clothing, feathers, pottery, and other goods, this seems to have been more for the privilege of moving through his lands rather than for his permission to access the source (Tibesar 1950: 107). The argument for a long-held tradition of open access to the vein is further strengthened by the 300 years of vigorous protests by indigenous groups in regards to the taxation and control of the salt source by the Spanish Crown and private companies (Brown and Fernández 1991: Verese 2002: 83, 132–133). In the winter of 1897, these clashes turned violent when groups of Campa from the eastern Andean foothills attacked and burned the farms of English settlers who were working for a Peruvian company that was preventing them from accessing the salt (Verese 2002: 132).

The exploitation of Cerro de Sal exemplifies rock salt's great appeal. The groups that visited said that they traveled for days to the mountain because the rock salt both tasted better than locally available brine salt and was easier to obtain in large quantities. After a few moments of work, a person could break off the salt required for a family's annual sustenance (Tibesar 1950: 104). As Michael Brown and Eduardo Fernández noted for Andean rock salt in general (1991: 18):

> …if veins of mineral salts are known, Indians are willing to travel hundreds of miles, risking privation and enemy attack, to unearth the precious substance in large quantities.

These colonial accounts suggest that the Cerro de Sal vein was considered an open access resource. This is not unusual (Weller 2002). Considered among the most basic necessities of life, access to salt has often been seen as an inalienable right—recall Gandhi's march to the ocean in protest of the British salt monopoly. This open access is in practice usually limited to those salt sources that require little effort to exploit. These sources, such as a deposit on the edge of the sea or an exposed vein, are perhaps seen as available to all because no person or group has invested heavily to make the salt accessible.

Huarhua is *not* one of these easily accessible sources. Though the basic tools of extraction used in the mine would have been similar to those used to pry chunks of salt off of the Cerro de Sal vein, considerable investment and coordination is required to hollow out a mountain using the techniques of chamber-and-pillar mining (Stöllner 2008b: 6). Cross-culturally, property rights are strongly linked with efforts to create, improve, or access a resource (Earle 2000). The Huarhua salt may have been considered an open access resource in theory, but the tunnels and other investments that were necessary to acquire this salt were likely privately held (just as the iron axes used to work the open access Cerro de Sal vein were privately owned).

Historical accounts of the brine salt makers of Maras and San Blas give a sense of how private ownership rights could occur within an open access salt source. Since at least the seventeenth century, the salt makers have successfully asserted ownership over the production infrastructure of saltpans, ovens, etc. that they have created over the centuries. Importantly, they do not make a claim on the salt source itself, nor do they claim that the source was the property of the local community (Espinoza Soriano 1984; Kumaki 2011; Palomino Meneses 1985). Though theoretically open to all, the salt makers effectively control the source since they own the means of production required to make salt out of the brine water.

Who then were the pre-Columbian salt makers at these sites and what was their relationship to local communities? Espinoza Soriano's work on the San Blas salt makers is particularly illuminating. He suggests that the saltpans were "occupied and controlled by multi-ethnic micro-colonies" that came from many regions in antiquity (Espinoza Soriano 1984: 192, first author's translation)—a claim supported by Daniel Morales' archaeological work at San Blas (1998) that showed salt production associated with a diversity of ceramic styles dating back to the second millennium BC. These micro-colonies of salt makers were part of the resilient vertical archipelagos that brought a wide variety of resources back to distant communities (Murra 1972). The colonists who lived in these archipelago communities had few local ties, and instead were provisioned through exchanges with their home regions. The San Blas salt makers were therefore a critical part of regional economies, and even the Incas chose not to interfere with traditional production and exchange at the site (Espinoza Soriano 1984: 199).

After Spanish Conquest, the San Blas salt makers were adamant that they should not be counted as locals. They possessed no fields or pasture land in the region, and argued that they were not obligated to pay the tribute or provide the labor service that was assigned to local groups. Instead, they claimed that their lone responsibility was to provide salt to the distant communities to which they belonged (Espinoza Soriano 1984: 201). As these connections withered during the colonial era, the salt makers found it more and more difficult to survive without reliable access to local food sources. Yet, the groups steadfastly refused efforts to integrate them into the surrounding communities of San Blas and continued to make their claims of independence until much of the source fell into private hands in the mid-nineteenth century (Espinoza Soriano 1984: 221).

The pre-2007 ownership nuances of the Huarhua rock salt mine are in line with what we know from these other salt sources. The salt in the Huarhua mine was seen in theory as an open access resource. Neither the cooperative nor the adjacent village owned the salt itself, and the state's universal claim to all salt sources was tolerated as long as their interference at the mine was minimal. In practice, however, the cooperative's claim to the salt was widely acknowledged because of its continual investment in the mine.

Though the Huarhua source was not owned by the cooperative, it was controlled by it. Mines have limited number of access points—a characteristic driven home by the miner who effectively blocked entry into the Huarhua mine by padlocking its only entrance—and so it was easy to restrict access to the salt. The nearby Alca obsidian source in contrast was nearly impossible to control since it could be easily quarried and was spread out across almost 50 km^2 (Jennings and Glascock 2002). Anyone could hypothetically come to Huarhua and dig into the mountainside for salt. Starting a new tunnel to reach the deeply buried deposits, however, would have been prohibitively costly for an individual or small group that could far more easily access the mine's salt through barter, purchase, or by joining the cooperative.

We can glean nothing from our meager sources about how a cooperative may have organized production during earlier periods. Each member of the cooperative took turns in positions of authority until quite recently at the Huarhua mine, and members were granted a great deal of autonomy in how and where they worked in the mine. The means of production (a hammer stone, rebar, and a wooden ladder) were not a barrier to participation in the collective, and traders from different groups were welcome to come to the mine to barter for salt. One explanation for the limited organizational hierarchies in the mine is that it was part of the wider decline is hereditary political power that occurred following the Spanish Conquest (e.g., Gose 2008). If Huarhua's early miners came from a variety of different communities like those documented at San Blas (Espinoza Soriano 1984), however, then it is more difficult to imagine standing centralized control over the source by a single stakeholder. Instead, earlier mining may have been organized in a manner similar to that used today.

Salt Mining, Common Resources, and State Control in the Ancient Andes

If we use the Huarhua salt mine as a tentative window into possible past practices, then what are some of the implications for understanding rock salt mining in pre-Columbian Peru? One implication of our work is to raise the possibility that rock salt mines, and perhaps other kinds of mines, could be worked communally. Mining demands, of course, depend upon the desired minerals and the matrix within which they are found (Stöllner 2003b). Rock salt is soft and easy to mine. It is largely impermeable, but also plastic enough so that shifting salt seals fractures created through tectonic shifts and mining activities. The stability and softness of rock salt would have been attractive to earlier miners, and this matrix would have been particularly conducive for people working within largely self-supervised groups. Miner could work a particular section independently, with little worry that work elsewhere in the mine would lead to cave-ins, gas blasts, or other hazards that are commonly associated with mining.

Though the massive rock salt mines of continental Europe were often run by the state during later periods, the earliest shafts were often created in periods that preceded the development of the state in a region (Alexianu et al. 2011; Barth 1982; Boenke 2005; Megaw et al. 2000; Stöllner 2003a). Centralized administration by the state or other authority is therefore not a necessary prerequisite to large-scale mining—Andean rock salt miners could have created an extensive, complex tunnel system without being organized by Inca, Wari, or Moche officials. Andean states almost certainly controlled some mining activities (e.g., Berthelot 1986), but the Huarhua data suggests that state control needs to be demonstrated rather than assumed.

A second implication of our work is to raise the possibility that rock salt, like other salt sources, was considered an open access resource that did not belong to the surrounding community. In the cases of a source where the salt was easily obtainable, anyone could come to the source to remove the resource. In the cases where significant investments to extract salt were necessary, then anyone that made these efforts was given exclusive rights to those areas where that person's investments occurred. Local groups would have had the same right to invest in extracting from the source, and so it makes sense that a resource was often worked in part by those that lived nearby. Whoever worked the source likely had a responsibility to share salt widely.

If a rock salt mine could have been worked by outside groups, then a foreign enclave of salt makers in a valley would not necessarily suggest foreign control of that valley—even indirect or hegemonic. A group of Wari colonists living near a mine, for example, did not necessarily signal control over the mine or surrounding groups. If a salt source was not inherently the property of local communities, then salt workers might actually have had few ties to nearby villages and instead been part of vertical archipelagos stretching out from distant communities like those documented by Espinoza Soriano at San Blas (1984).

We were only able to obtain a tentative glimpse into the operation of the Huarhua salt mine during our brief stays there. Nonetheless, we suggest that more intensive studies of cooperative mines like Huarhua can potentially provide important insights into the ways in which mining and quarrying were organized in the pre-Columbian Andes. Each year, however, more and more of these mines are transformed by the economic forces of globalization, the mandatory addition of iodine, and the penetration of the modern nation state. Our conversations at the Huarhua mine were often tinged with nostalgia as people lamented the mine's state, and now, private ownership, as well as the decline of the salt trade. They talked with a mix of optimism and fear about the construction of a road to Huarhua that is slowly making its way to the village, and they fantasized about iodine-injecting machines that could make their salt more marketable in Lima and abroad. In short, they recognized that the old ways of mining are swiftly crumbling—archaeologists should take notice of this vanishing source of information.

Acknowledgments The authors would like to thank Fidel Cruz Anco, Virginia Gutiérrez, Tadeo Ancco, Catalina Borda, Edison Mendoza Martínez, Cheyla Samuelson, and Amelia Argüelles Talavera. Support was provided by the Heinz Grant for Latin American Archaeology and the National Science Foundation (Award # 9903508 and # 0630081). Permission to work in the valley was granted by the Peruvian National Institute of Culture (Permits 977/INC and 828/INC).

References

Alexianu, M., Weller, O., & Curca, R.-G. (Eds.). (2011). *Archaeology and anthropology of salt: A diachronic approach*. Oxford: Archaeopress.

Barth, F. E. (1982). Prehistoric saltmining at Hallstatt. *Bulletin of the Institute of Archaeology, 19*, 31–43.

Berthelot, J. (1986). The extraction of precious metals at the time of the Inka. In J. Murra, N. Wachtel, & J. Revel (Eds.), *Anthropological history of Andean polities* (pp. 69–88). New York: Cambridge University Press.

Boenke, N. (2005). Organic resouces at the Iron Age Dürnberg salt-mine (Hallein, Austria)—long-distance trade or local sources. *Archaeometry, 47*(2), 471–483.

Brown, M. F., & Fernández, E. (1991). *War of shadows: The struggle for Utopia in the Peruvian Amazon*. Berkeley: University of California Press.

Concha Contreras, J. d. D. (1975). Relación entre pastores y agricultores. *Allpanchis, 8*, 67–101.

Earle, T. (2000). Archaeology, property, and prehistory. *Annual Review of Anthropology, 29*, 39–60.

Eerkens, J. W., Vaughn, K. J., & Linares Grados, M. (2009). Pre-Inca mining in the southern Nasca region, Peru. *Antiquity, 83*, 738–750.

Espinoza Soriano, W. (1984). Los mitmas salineros de Tarma, Chinchaycocha y Pasco, siglos XX a.C. -XX d.C. *Allpanchis, 24*, 183–250.

Flores Ochoa, J. A., & MacQuarrie, K. (1994). Modern-Day Herders: An Andean legacy continues. In J. A. Flores Ochoa, J. Portús Pérez & K. MacQuarrie (Eds.), *Gold of the Andes: The llamas, alpacas, vicuñas and guanacos of South America* (Vol. 2, pp. 101–193). Barcelona: Patthey.

Gose, P. (2008). *Invaders as ancestors: On the intercultural making and unmaking of Spanish colonialism in the Andes*. Toronto: University of Toronto Press.

Inamura, T. (1981). Adaptación ambiental de los pastores altoandinos en el sur del Perú: Simbiosis económico-social con los agricultores. In S. Masuda (Ed.), *Estudios Etnográficos del Perú Meridonial* (pp. 65–84). Tokyo: Tokyo University Press.

Jennings, J. (2002). *Prehistoric imperialism and cultural development in the Cotahuasi valley, Peru* (Unpublished Ph.D. dissertation). University of California, Santa Barbara

Jennings, J. (2006). Core, peripheries, and regional realities in Middle Horizon Peru. *Journal of Anthropological Archaeology, 25*, 346–370.

Jennings, J., & Glascock, M. (2002). Description and method of exploitation of the Alca obsidian source, Peru. *Latin American Antiquity, 13*(1), 107–118.

Jennings, J., & Yépez Álvarez, W. (2008). The Inca conquest and consolidation of the Cotahuasi valley of southern Peru. *Ñawpa Pacha, 29*, 119–152.

Kumaki, H. (2011). Ethnoarchaeological research on the historical landscape management of the Salineras de Mara (Unpublished B.A. thesis). Harvard University.

Megaw, V., Morgan, G., & Stöllner, T. (2000). Ancient salt-mining in Austria. *Antiquity, 74*, 17–18.

Murra, J. (1972). El 'control vertical" de un máximo de pisos ecológicos en la economía de las sociedades Andinas. In I. Ortiz de Zúñiga (Ed.), *Visita de la Provincia de León de Huánuco*, (Vol. 2, pp. 429–468). Huánaco: Universidad Nacional Hermillo Valdizán.

Palomino Meneses, A. (1985). Las salineras de Maras: Organización y conflicto. *Allpanchis, 26*, 151–181.

Petersen, G. (1994). Pre-Hispanic metallurgy and mining in the Andes: Recent advances and future tasks. In A. K. Craig & R. C. West (Eds.), *Quest of mineral wealth: Aboriginal and colonial mining and metallurgy in Spanish America* (Geoscience and man, Vol. 33, pp. 37–73). Baton Rouge: Geoscience Publications, Department of Geography and Anthropology, Louisiana State University.

Petersen, G. (2010). *Mining and metallurgy in ancient Perú (trans: William E. Brooks), special paper*. Boulder: Geological Society of America.

Shimada, I. (1994). Pre-hispanic metallurgy and mining in the andes: Recent advances and future tasks. In A. K. Craig & R. C. West (Eds.), *Quest of mineral wealth: Aboriginal and colonial mining and metallurgy in Spanish America* (Geoscience and man, Vol. 33, pp. 37–73). Baton Rouge: Geoscience Publications, Deptartment of Geography and Anthropology, Louisiana State University.

Sonnenfeld, P. (1995). The color of rock salt—A review. *Sedimentary Geology, 94*(3–4), 267–276.

Stöllner, T. (2003a). The economy of Dürnberg-bei-Hallein: An iron age salt-mining centre in the Austrian Alps. *The Antiquarians Journal, 83*, 123–194.

Stöllner, T. (2003b). Mining and economy—A discussion of spatial organizations and structures of early raw material exploitation. In T. Stöllner, G. Körlin, G. Steffens, & J. Cierny (Eds.), *Man and mining: Studies in honour of Gerd Weisgerber* (Vol. Beiheft 16, pp. 415–446). Bochum: Der Anschnitt.

Stöllner, T. (2008a). Mining landscapes in early societies—Imprinting processes in pre- and protohistoric economies? In C. Bartels & C. Küpper-Eichas (Eds.), *Cultural heritage and landscapes in Europe: Proceedings of the international conference, Bochum, June 8–10, 2007* (pp. 65–92). Bochum: Veröffentlichungen aus dem Deutschen Bergbau-Museum.

Stöllner, T. (2008b). Montan-archaeology and research on old mining: Just a contribution to economic history? In U. Yancin (Ed.), *Der Anschnitt Zeitschrift für Kunst und Kultur im Bergbau, Beiheft 21* (pp. 1–30). Bochum: Veröffentlichungen aus dem Deutschen Bergbau-Museum.

Tibesar, A. S. (1950). The salt trade among the Montana Indians of the Tarma area of eastern Peru. *Primitive Man, 23*(4), 103–107.

Trawick, P. (2003). *The struggle for water in Peru: Comedy and tragedy in the Andean commons.* Stanford: Stanford University Press.
Verese, S. (2002). *Salt of the mountain: Campa Asháninka history and resistance in the Peruvian jungle (trans: Susan Gierbach Rascón).* Norman: University of Oklahoma Press.
Verese, S. (2006). *La sal de los cerros: Resistencia y utopía en la Amazonía Peruana.* Lima: Fondo Editorial del Congresso del Perú.
Weller, O. (2002). The earliest rock salt exploitation in Europe: A salt mountain in the Spanish Neolithic. *Antiquity, 76*, 317–318.

Chapter 7
Hunter–Gatherer–Fisher Mining During the Archaic Period in Coastal Northern Chile

Diego Salazar, Hernán Salinas, Jean Louis Guendon, Donald Jackson, and Valentina Figueroa

Introduction

Red and yellow colorants have been used throughout human history. Abundant and diverse archaeological and ethnographic contexts offer ample proof of this. They may be obtained both from organic and mineral sources, the latter constituted mainly by high-grade iron oxides or iron oxides mixed with clays, silicates, and other minerals (Popelka-Filcoff et al. 2007, Rapp 2009, Triat 2010).

In the Old World, iron oxides first appear in archaeological sites dated to the Lower Paleolithic, but show diverse uses from the Middle Paleolithic onwards (Beaumont 1973, Barham 2002, Henshilwood et al. 2011, Marean et al. 2007, Mason 1983, Petru 2006, Schmandt-Besserat 1980,Watts 2002, Zilhão et al. 2010). It seems the oldest evidence comes from the Wonderwork Cave, in South Africa, where iron oxide finds have been dated to nearly 800,000 years before present, or they could be from even earlier according to the reported use of hematite by early hominins (Weisgerber and Pernicka 1995). Regardless of what turns out to be the

D. Salazar(✉) • H. Salinas • D. Jackson
Departamento de Antropología, Facultad de Ciencias Sociales, Universidad de Chile, Casilla 10115, Santiago, Chile
e-mail: dsalazar@uchile.cl

J.L. Guendon
Université de Provence, UMR 6636 du CNRS (Centre National de la Recherche Scientifique), MMSH, 5 rue du Château de l'Horloge, BP 647, 13094, Aix-en-Provence cedex 2, France
e-mail: j.l.guendon@orange.fr

V. Figueroa
Université de Paris 1 Panthéon-Sorbonne, Paris, France
e-mail: Valentina.Figueroa-Larre@malix.univ-paris1.fr

N. Tripcevich and K.J. Vaughn (eds.), *Mining and Quarrying in the Ancient Andes*, Interdisciplinary Contributions to Archaeology, DOI 10.1007/978-1-4614-5200-3_7,

earliest evidence, the use (and therefore the extraction and processing) of this raw material predates the appearance of modern humans; it is an activity of great antiquity to the genus Homo. By the Upper Paleolithic, the systematic presence of red, black, and yellow pigments (hematite, manganese, and goethite) in various prehistoric contexts has been considered direct evidence of symbolic behavior by early *Homo sapiens* (Hovers et al. 2003).

At the same time, recent studies demonstrate that from the earliest periods these pigments were also used for a variety of functional purposes such as medicine, as a preserver of food and wood, an aid in working animal hides, as an adhesive, as insect repellent, or even for sun protection (Erlandson et al. 1999, Wadley et al. 2004, among many others). While some debate the dichotomy of functional versus symbolic uses of iron oxides in prehistory (MacDonald 2008), today it is well established that this raw material has been used in a wide variety of human activities from the Paleolithic through modern times.

The Americas are no exception in this regard. In the New World, iron oxides have also been reported in some of the earliest known Paleoindian sites from North America all the way down to Patagonia (Gillespie 2007; Lahren and Bonnichsen 1974; Massone 2004; Mazzia et al. 2005; Scalise and Di Prado 2006; Sepúlveda 2011). Furthermore, Stafford et al. (2003) have considered the early use of these pigments as a marker of cultural affiliation between Paleoindian groups and Upper Paleolithic groups from the Old World.

Despite the importance of iron oxides in human prehistory and its connection to symbolic activity of great antiquity, most archaeological evidence to date is primarily from consumption contexts. Recently, we have seen a great increase in provenience-based studies aiming at identifying the sources of the archaeological pigments through diverse archaeometric methods (d'Errico et al. 2010; Eiselt et al. 2011; Erlandson et al. 1999; Gialanella et al. 2011; Mooney et al. 2003; Popelka-Filcoff et al. 2007, among others), yet we still know very little about the extraction and processing of these ores (but see Henshilwood et al. 2011). We may infer that mining techniques were commonly required in order to extract iron oxides, but there are few archaeological mines known worldwide. The earliest iron oxide mining known to date is the Lion Cave site, in Swaziland, which has been dated to ca. 40,000 B.P. (Beaumont 1973). Slightly later dates have been reported for iron oxide mines in Australia (MacDonald 2008), Hungary, and Poland (Erlandson et al. 1999), as well as for the Island of Thassos (Koukouli-Chryssanthaki and Weisgerber 1996). But on the whole the evidence is lacking for iron oxide mining among hunter–gatherer groups, and many studies are still quite underpublished. There is mounting evidence from the Old World (see for example, Beaumont and Boshier 1974; Geniola et al. 2006; Goldenberg et al. 2003; Hammel et al. 2000; Hejl and Tippelt 2005; Robbins et al. 1998; Thackeray et al. 1983), but these are usually related to later Neolithic or Mesolithic populations.

In the Americas the evidence for ancient iron oxide procurement is far less abundant. We have little in the archaeological record of iron oxide extraction and/or processing sites (Popelka-Filcoff et al. 2007). To date the only secure data comes from Mina Primavera, a hematite mine in southern Peru dated to the Early

Intermediate Period and the Middle Horizon (Eerkens et al. 2007; Vaughn et al. 2007, 2011; Chap. 8). The Powars II site in Wyoming, central North America, has been interpreted as a Paleoindian "mine" (Stafford et al. 2003), but we remain unconvinced because there are no known dates for the site, and in particular because the published material lacks evidence to support ancient mining activities such as prehistoric mining tools. The high frequency of projectile points and the absolute absence of hammerstones are completely atypical for prehistoric mining operations worldwide, so we cannot rule out the possibility that a historical mining operation redeposited lithic materials from a Paleoindian site. Thus, to our knowledge there is only one prehistoric iron oxide mine known in the Americas besides the Mina Primavera in southern Perú: the San Ramón 15 site in coastal northern Chile (Salazar et al. 2010, 2011). Here we wish to present new data on this site and discuss its integration into local hunter–gatherer–fisher settlement systems during the Early and Late Holocene. We discuss the geology of the mine, the stratigraphic deposits found during excavations, and the lithic mining tools. The chronology of the trench mine will be inferred from stratigraphic deposits and the radiocarbon dates available. With all these data, we interpret the mining techniques and strategies used by Early Archaic (ca. 12,000–8,500 cal. BP) and Late Archaic (ca. 5,500–3,500 cal. BP) miners at San Ramón 15. Finally, we consider the role of this task-specific site in regional settlement systems of hunter–gatherer–fisher groups that inhabited the coast of Taltal between 12,000 and 3,500 cal. BP. It is worth noting that the data we discuss here complements, and in some regards modifies, our previous interpretations of the site (cf. Salazar et al. 2011).

Red Pigments in Coastal Northern Chile: The San Ramon 15 Mine

The use of iron oxides has been reported for the first human occupations of what is today northern Chile (Llagostera et al. 2000; Núñez et al. 2005; Santoro et al. 2011). In coastal northern Chile, iron oxides were in use by the Huentelauquén culture in Antofagasta, Chañaral, and Los Vilos (Cervellino 1996; Costa-Junqueira 2001; Jackson et al. 1999; Llagostera et al. 2000). This culture is considered the first maritime adaptation and the earliest inhabitants of this part of the Pacific coast of South America (Sandweiss 2008). During the Middle and Late Holocene, red pigments were intensely used for the elaboration of the Chinchorro Mummies, especially the so-called red-mummies (Arriaza 1994; Standen 1997). In other non-Chinchorro funerary contexts they have also been reported (Capdeville 1921; Capdeville 2010; Llagostera and Llagostera 2010; Mostny 1964). The use of this material was still documented during the nineteenth century in ritual and domestic contexts, so it can be said that iron oxides were part of the cultural tradition of the northern Chile's coastal populations for more than 10 millennia. Notwithstanding, we know very little if anything at all about how these pigments were extracted and processed by the early hunter–gatherer–fisher societies of the coastal Atacama Desert.

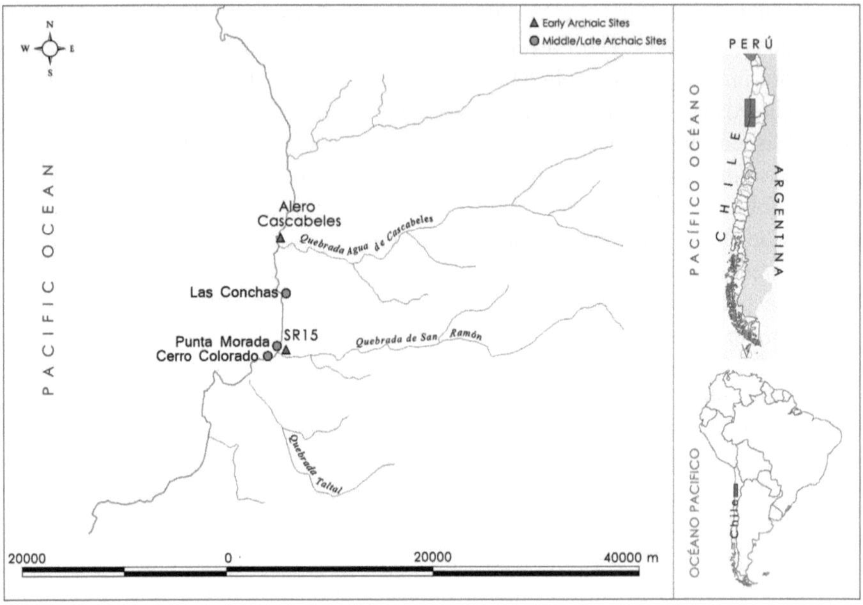

Fig. 7.1 Taltal in the context of coastal northern Chile, and the main sites discussed in the text. *Triangles* represent Early Archaic sites, while *circles* represent Middle and Late Archaic sites

San Ramón 15 (SR-15) provides new evidence that can help us to reconstruct, at least partially, these aspects of prehistoric technology, economy, and the associated funerary tradition.

The site is located in the western margin of the Coastal Cordillera, some 5 km northeast of the city of Taltal (25°23′02″S/70°26′30″W; see Fig. 7.1). Topographically the area is characterized by a marked climb up to 2,000 m above sea level and a thin (in some cases absent) coastal shelf. Presently, in spite of its hyper-arid weather with drinkable water only available in small and dispersed springs, the coast has a rich ecosystem as a consequence of the influence of the Humboldt Current (Bittmann 1986; Llagostera 1979). Paleoenvironmental records for northern Chile's highlands and the southern coast of Peru suggest more humid conditions in those areas during Late Pleistocene and Early Holocene periods (Betancourt et al. 2000; Moreno et al. 2009; Mächtle et al. 2010), so it is likely that more humid conditions prevailed in Taltal as well.

SR-15 was discovered in 2008 as a result of systematic archaeomineralurgical surveys carried out by the Fondecyt 1080666 Project (Chile) in an area near the coast of northern Chile (Salazar et al. 2010, 2009). The surveys focused on the coastal Cordillera region, where minerals exploited in prehistory are mainly concentrated. Ravines cross-cutting the Cordillera in an east to west direction were intensely surveyed, focusing on the identification of fresh water sources. It was assumed that in this arid environment, local fresh water sources would be a key factor in the organization of prehistoric mining production given the hyper-arid environment present

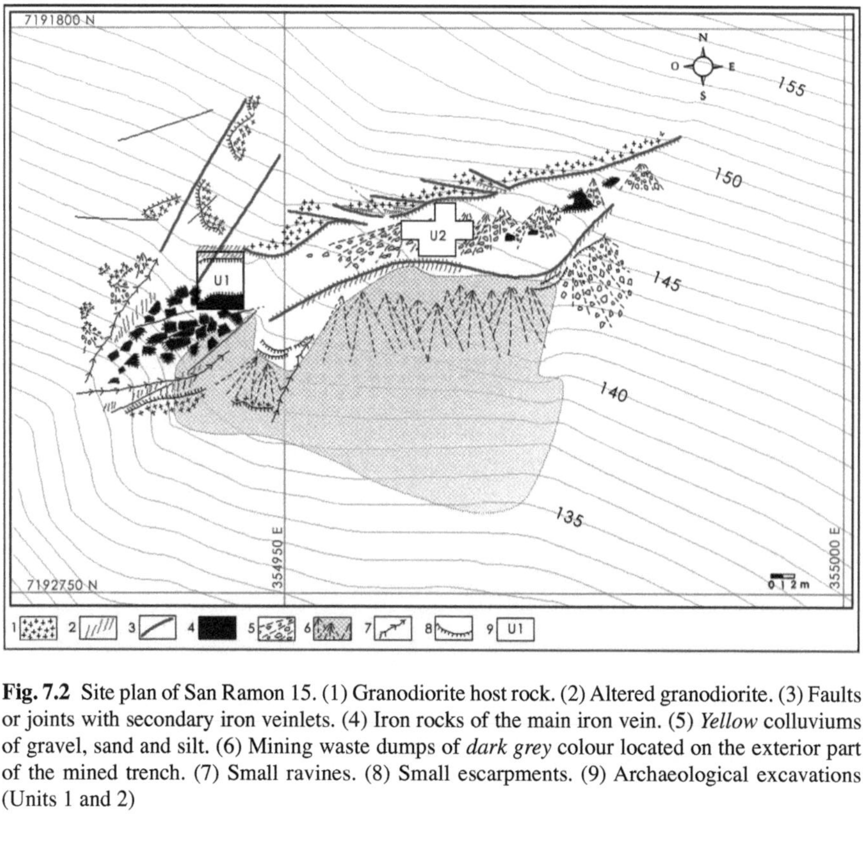

Fig. 7.2 Site plan of San Ramon 15. (1) Granodiorite host rock. (2) Altered granodiorite. (3) Faults or joints with secondary iron veinlets. (4) Iron rocks of the main iron vein. (5) *Yellow* colluviums of gravel, sand and silt. (6) Mining waste dumps of *dark grey* colour located on the exterior part of the mined trench. (7) Small ravines. (8) Small escarpments. (9) Archaeological excavations (Units 1 and 2)

in the area since the Middle Holocene (Grosjean et al. 2001). Hills and smaller gorges surrounding the water sources were then intensely surveyed looking for remains of human activity and landscape modification. SR-15 is located approximately 150 m uphill from the main fresh water source for the lower part of Quebrada San Ramón.

The site consists of a prehistoric mining trench dug into an iron vein striking N70°E hosted in Lower Cretaceous plutonic rocks of the Coastal Cordillera (Figs. 7.2 and 7.3). It is 40 m long and 2–6 m wide. Upon discovery its depth was less than 50 cm (Fig. 7.4), but our excavations have revealed an original depth of more than 8 m in some parts of the prehistoric trench (see below). Five intensive archaeological excavation field seasons have taken place between 2009 and 2011. These excavations focused on the central-eastern part and the western end of the prehistoric mining trench and followed natural as well as artificial layers (10 cm). In Unit 1, located in the western end of the trench, twelve 1 × 1 m square were dug. The excavation exposed the original mining trench reaching the bottom of the prehistoric exploitation at approximately 3 m from the present day surface. In Unit 2, located on the central-eastern part of the trench, twenty-four 1 × 1 m square were dug. Our excavation has reached more than 8 m below today's surface, but the

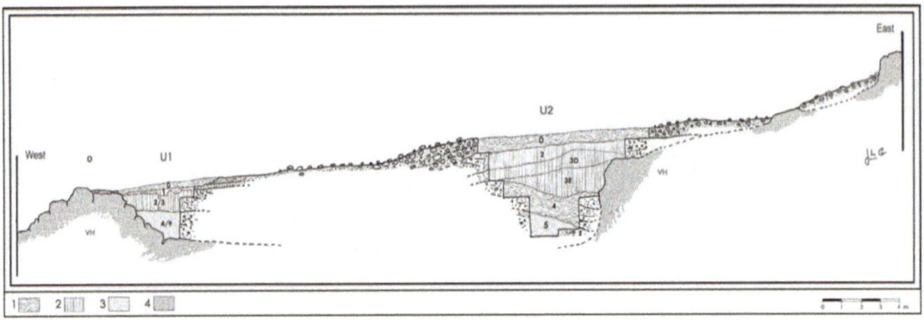

Fig. 7.3 Schematic East to West profile of the prehistoric mine of SR15, showing the two main units (U1 and U2) excavated so far and the relative position of the main stratigraphic units 0 through 5 identified

Fig. 7.4 West to east view of SR15 in 2009 during the first excavations. In the foreground, Unit 1

bottom of the prehistoric mine has not yet been found. Sediments excavated inside the prehistoric trench consist mainly of mining tailings and debris from indigenous exploitation and, to a lesser degree, colluvial deposits that covered the Prehispanic mining area presumably when the site was not in use.

The iron vein exploited in prehistory shows a filling pattern parallel to the walls, gangue minerals being dominated by quartz and fine and coarse calcite crystals, with iron oxide cementing irregular breccia bodies and veinlets (Salazar et al. 2011). Iron oxides are also present as red and yellow ochre filling small veinlets and as fine precipitate on the altered granodiorite. These ochre ores can be interpreted as secondary enrichment by remobilization and oxidation. X-ray diffraction analyses carried out on powdered (1–3 μm particle size) specimens of ochre pigments sampled from the prehistoric mine identified them as hematite and goethite.

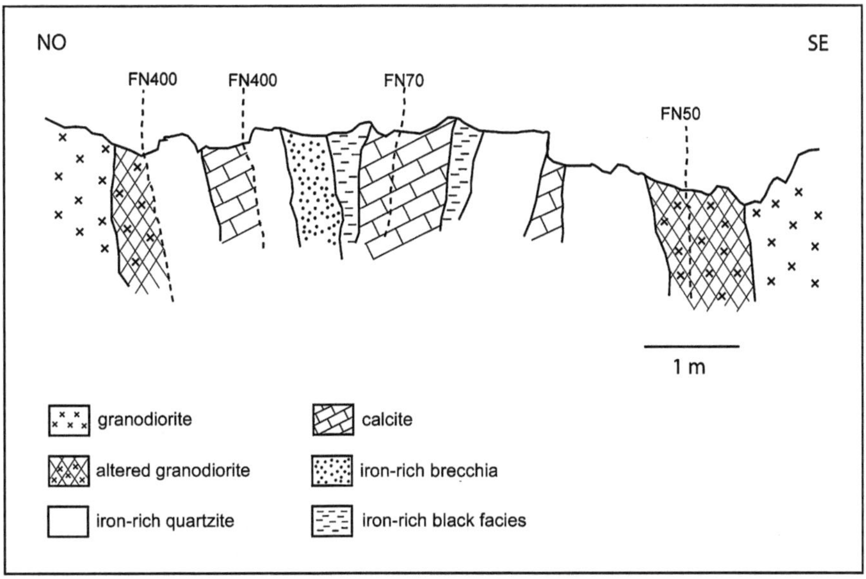

Fig. 7.5 Profile of the iron vein exploited at SR15. It is made of a series of distinct facies almost parallel to each other and to the host rock (granodiorite)

Ancient miners were mainly after two particular mineralization structures within the main iron vein (a) ca. 10–20 cm wide veinlets of pigment (goethite and hematite) that were formed between the main facies of the iron vein as a result of secondary oxidation processes and (b) iron-rich breccia-like facies within the iron vein with high concentration of pure hematite veinlets (Fig. 7.5).

Stratigraphical and Chronological Context of SR-15

Both units excavated in SR-15 have a similar stratigraphy. Most of the deposits that infilled the prehistoric trench are tailings, mining dumps, and debris. The only exceptions are the superficial layers in both units, a layer in Unit 2 (see below), and small layers in Unit 1 (see below) which are result of colluvial deposits due to natural factors.

An examination of sedimentological criteria and cultural material identified ten layers in Unit 1 (Salazar et al. 2011). Six of these correspond to the Early Archaic (Layers 4 through 9) and two to the Late Archaic (Layers 2 and 3), while the two superficial ones (Layers 0 and 1) are natural sediments deposited after the abandonment of the mine (Fig. 7.6). Using the same criteria, in Unit 2 twelve layers have been identified so far. All of them date to the Late Archaic, but they probably removed earlier debris both from Early and Middle Archaic exploitations.

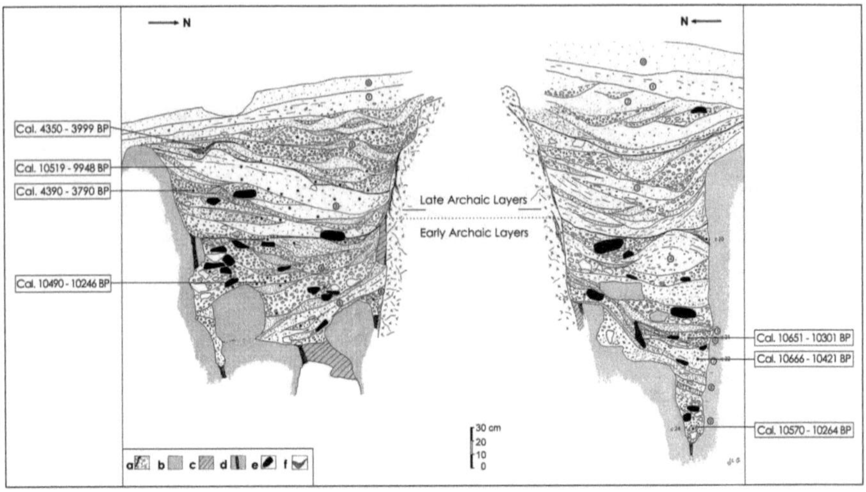

Fig. 7.6 West and East profiles of Unit 1, showing the stratigraphic location of the dated radiocarbon samples. (a) Altered granodiorite with thin iron oxide veinlets. (b) Main iron vein. (c) Massive iron facies with silica. (d) Calcite facies. (e) Iron oxide veinlets within the main iron vein. (f) hammerstones. (g) hearth

The superficial layer of this unit (Layer 0) is also a colluvial postabandonment stratum (Fig. 7.7).

All cultural layers in both units showed high frequencies of complete and broken hammerstones, flakes, and big rock fragments which were extracted from the iron vein and the host rock. The color of the sediment is dark gray due to the abundance of iron particles from the vein. The exposed mine walls show evidence of intense battering and are in direct contact with these sediments. The data thus supports our claim that most stratigraphic layers were the result of mining debris and tailings that gradually filled the prehistoric excavation as the operation moved either east to west or west to east. Besides this mining evidence, all layers produced remains of charcoal, fish and mammal bones, and mollusk shells, though in different frequencies (see Salazar et al. 2011).

The chronology of the stratigraphic sequence in both units was based on a total of 14 radiocarbon (AMS) dates (Table 7.1). In Unit 1, four of these dates consistently situate the lower layers in the Early Archaic, specifically between 10,200 and 10,700 cal B.P. (Table 7.1, Fig. 7.6). There is evidence that some early layers from Unit 1 may have been removed by Late Archaic miners, but layers 4 through 9 remain in situ since the Early Holocene (Fig. 7.8). Among the early deposits in Unit 1, layers 5, 6, and 8 have different sedimentological characteristics. They are mostly sands with graded stratification and a yellowish color due to the virtual absence of iron particles in the sediment. It seems these layers were naturally deposited, most likely during rainy episodes that occurred during the Early Holocene. Layers 4, 7, and 9, on the other hand, are in situ tailings from the earliest exploitation of the site.

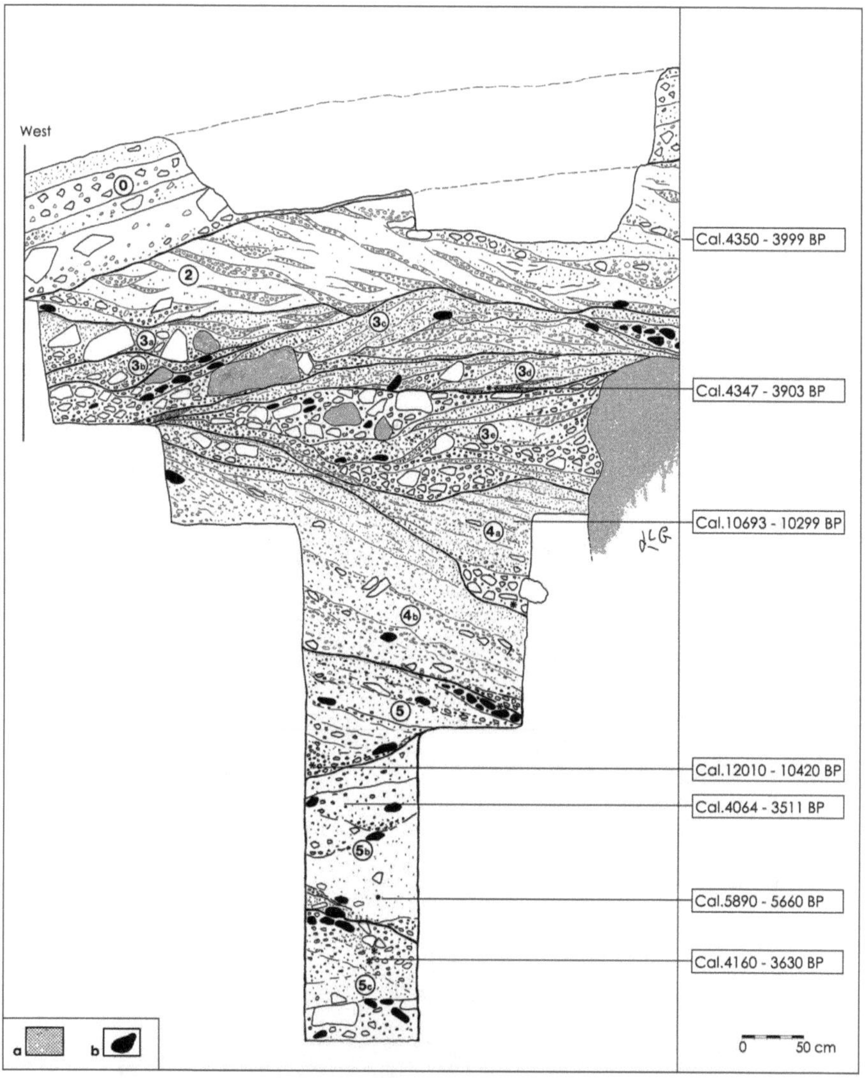

Fig. 7.7 North profile of Unit 2, showing the stratigraphic location of the dated radiocarbon samples. (a) Iron vein and rocks extracted from the Iron Vein; (b) Hammerstones

In contrast to Unit 1, radiocarbon dates indicate that in Unit 2 all layers were deposited during the Late Archaic, being roughly contemporary with Layers 2 and 3 of U1 (see Table 7.1 and Fig. 7.7). Considering the Early and Middle Archaic dates in the sequence of U2, it is reasonable to consider that Late Archaic miners removed and redeposited part of the earlier mining debris in order to enlarge the exploitation. It is also noteworthy that within this Late Archaic stratigraphic sequence, we identified an intermediate layer of almost 100 cm of natural sediments (Layer 3E).

Table 7.1 Radiocarbon dates from SR-15

Lab code	Context	Material	^{14}C year B.P.	Cal year B.P.
Beta 280992	U2 L5	Shell	10,110±280	12,010–10,420
UGAMS 5442	U1, L7	Charcoal	9,390±30	10,666–10,421
POZ 41243	U2, L4A	Charcoal	9,380±50	10,693–10,299
UGAMS 5441	U1, L6	Charcoal	9,360±30	10,651–10,301
POZ 32943	U1, L9	Charcoal	9,310±50	10,570–10,264
UGAMS 5440	U1, L4	Charcoal	9,250±30	10,490–10,246
Beta 255687	U1 L2-3	Charcoal	9,160±80	10,519–9,948
Beta 312875	U2, L5B	Charcoal	5,010±30	5,890–5,820 5,820–5,810 5,760–5,660
Beta 312876	U2, L5C	Shell	3,920±100	4,160–3,630
Beta 257858	U1, L3	Shell	4,074±100	4,397–3,791
Beta 300553	U2, L5B	Shell	3,777±100	4,064–3,511
UGAMS 5439	U1, L2	Charcoal	3,850±30	4,347–3,903
UGAMS 5443	U2, L3	Charcoal	3,850±30	4,347–3,903
Beta 261668	U2, L2	Charcoal	3,800±60	4,350–3,999

All dates have been calibrated with OxCal 4.01. The ShCal04 curve has been used for <10,000 BP radiocarbonic dates, and the IntCal09 curve has been used for >10,000 radiocarbonic dates. All dates correspond to 2σ (confidence interval of 94.5%). The results of the shell samples have been corrected for local reservoir effect following Ortlieb et al. (2011)

Fig. 7.8 Early Holocene mining trench identified in Unit 1. On the *left*, the granodiorite host rock. On the *right*, remains of the unexploited iron vein. Note the fill of the prehistoric trench with dark iron-rich sediment and fragments of hammerstones (East profile of Unit 1)

The color, organization, and absence of cultural material suggest that sediments from 3E correspond to episode(s) of natural processes that eroded the previous tailings and deposited colluvial material from outside the trench. The fact that some dates below and on top of this thick natural layer are contemporary (see Fig. 7.7) seems to indicate that it was a single event probably related to El Niño (ENSO).

In any case, this dense deposit of natural sediments is not present in Unit 1, so it may have been removed by the same Late Archaic miners. This interpretation is consistent with stratigraphic observations in Unit 1 which shows Late Holocene deposits directly over Early Holocene ones. Contact between both layers is almost completely horizontal, both in a north to south and an east to west sense (Fig. 7.6). This trait is quite distinct from the characteristic deposits of the other mining layers which always show significant inclinations. We have interpreted this stratigraphic contact as resulting from the excavation of earlier layers by Late Archaic miners in order to prepare a level working area and/or in order to reach the veinlets in the mine walls left unworked by previous miners. The Early Archaic radiocarbon date within this later layer gives additional support to our claim.

Layer 2 in both units is the final cultural mining debris deposited in SR-15 (only naturally deposited sediments were identified above). Two radiocarbon dates are currently available for this layer, one from U1 and one from U2. Both gave identical results, being synchronic to mining activity in Layer 3 and probably to the exploitation of Unit 2 as a whole. Since both samples from Layer 2 come from in situ hearths, the last moments of exploitation of San Ramón 15 seem to date to somewhere between 3,900 and 4,300 cal. B.P.

Thus, we may conclude that the trench in SR-15 is the outcome of mining activities during the Archaic Period. Two episodes of exploitation have been recorded by stratigraphic analysis and 14 radiocarbon dates (Table 7.1). The first activity probably occurred between 12,000 and 10,200 cal. B.P. and the second took place between 4,300 and 3,900 cal. B.P.

The fact that an in situ event of Early Archaic mining was preserved in Unit 1gives us the opportunity to compare the two periods of hunter–gatherer–fisher mining at San Ramón 15. It is interesting to consider the differences in mining strategy between both occupations, as we will see below. In Unit 1 it is evident that the Early Archaic exploitation of SR-15 excavated almost all the trench as we see it today, being thus the most intensive occupation of that area of the mine, while in Unit 2 Late Archaic mining was clearly more intense, removing and re-depositing all of the earlier dumps and extraction areas.

Mining Techniques and Strategies

The characteristics of the sediments deposited inside the mining trench in Unit 1 together with the radiocarbon dates indicate that subsurface mining at SR-15 began during the Pleistocene–Holocene transition (somewhere before 10,200 cal. BP). For reasons still unknown, mining stopped at the western end of the trench at around 10,200 BP, before all the available pigments were extracted.

In order to extract the main pigment veins and veinlets, it was necessary for the first miners of SR-15 to remove pigment-sterile rock from the main iron vein (mainly the quartzite and calcite-rich facies). These facies are very hard, so ancient miners probably fractured the rocks looking for the weaker angles, which generally coincide with the contact between the different facies of the iron vein. The tools used for this task will be discussed below. The size of the mining trench corresponding to the Early Archaic (roughly 1 m wide) indicates that these miners exclusively followed the higher grade pigment veinlets formed as result of secondary oxidation occurred between the main facies of the iron vein.

Exploitation during the Late Archaic, however, showed a different pattern. As we have already seen, during the second period of mining activities at the site, the entire 5 m wide iron vein was exploited, at least in the eastern part of the mine (Unit 2). The remainder of the iron vein left untapped by Early Archaic miners in U1 was worked by Late Archaic miners as well, but for some reason the exploitation was not completed and thus the deeper deposits of the Early Archaic remained stratigraphically undisturbed.

It is quite clear, at least in U1, that the Early Archaic miners pursued the main veinlets of iron oxide. On the other hand, the Late Archaic exploitation was targeted towards the extraction of hematite and goethite pigment from the breccia-like facies which is rich in iron oxides, as well as from small veinlets located in the altered granodiorite that hosts the main iron vein. That is the reason why almost all of the iron vein was exploited during this period, at least in Unit 2.

Unfortunately, we do not have in situ evidence for Middle Archaic mining (8,500–5,500 cal. BP). A single radiocarbon date obtained in U2 seems to support our claim that SR 15 was also being worked at that time, but there is no secure stratigraphic context to characterize this occupation. In any case, our stratigraphic interpretation poses a methodological problem in regard to the chronology of the artifacts found in layers 2 and 3 of U1 and throughout the stratigraphic sequence of U2, since it implies that Early (and Middle) Archaic artifacts entered Late Archaic layers being thus impossible to differentiate. In the following segment, we attempt to characterize the technological ensemble of the site and wherever possible we offer preliminary interpretations regarding the differences between Early and Late Archaic mining instruments at SR-15.

Mining Instruments of SR-15

The excavation of both Unit 1 and 2 in SR 15 has yielded approximately 2,500 lithic artifacts, generically known as hammerstones (Fig. 7.9). Nearly half of these are fractured (N=ca.1,600). Many flakes from these hammerstones have also been recovered, as well as a few artifacts made of shell, a fragmented bone instrument, a few siliceous flakes, and some food remains (mammal and fish bones, and mollusk shells) (see Salazar et al. 2011).

While nearly all the hammerstones were used manually (without a haft) we found that 0.5% of the analyzed artifacts showed evidence of hafting. We analyzed the raw material, dimensions, and morphology of the active edge and wear traces in the whole collection of hammerstones.

Fig. 7.9 A sample of lithic hammerstones from SR15

Results indicate that eight different raw materials were used for hammerstones by Archaic miners at SR-15: andesite, metandesite, granite, granodiorite, tuff, basalt, sandstone, and quarzite. Almost all these rocks are locally available, either from the coastline or the quebradas surrounding the site. It is likely that in both periods there were similar raw material provisioning systems and a good knowledge of the quality of different lithic raw materials to accomplish mining activities. However, during the Early Archaic large stones from the coastline were quite frequent, whereas they were rarely been found within Late Archaic layers. The higher frequency of beach stones in the Early Archaic suggests at least some of the artifacts were selected on the coast, while most Late Archaic hammerstones were selected in the ravine next to the site itself. This is interesting in the light of our recent yet unpublished bathimetric studies in the Taltal bay showing that during the Early Holocene the coastline was at least 1.5 km away from today's shore. Significantly, large coastal cobbles of up to 16 kg would have had to be carried more than 3 km in order to have been used at SR-15.

Regarding the dimensions of the hammerstones, it is likely that the main criteria for selecting the rocks was an appropriate size for manual handling and a shape permitting a degree of multifunctionality from the artifacts. Notwithstanding, the use of very big (more than 30 cm long) and very small (around 6 cm long) artifacts seems to indicate that in both periods rocks were chosen in order to accomplish specific functions within the mine. This suggests a rather well-developed knowledge of mining.

Preliminary use-wear studies indicate that Early Archaic hammerstones show more use-wear traces than Late Archaic ones, specifically regarding crushing, scarring, and abrasion marks (Fisher's exact=0.003). More abrasion in the early artifacts indicates they were used for a longer period and/or against harder surfaces. Early Archaic hammerstones show use-wear traces on all extreme edges and even on the face of the artifacts. This also points to a more prolonged and more physically demanding use of the artifacts. Thus, the hardest rocks available (coastal stones) were more frequently selected during the Early Archaic and all artifacts from this period show evidence of more use in terms both of time and/or intensity when compared to the Late Archaic mining tools. More research is needed to better understand these patterns.

San Ramón 15 and Archaic Hunter–Gatherer–Fisher Economies

Our data shows the beginning of mining activity at SR-15 during the Early Archaic Period and then a more intensive exploitation of the mine during the Late Archaic (in terms of metric tons removed).[1] Our current calculations indicate that more than ca. 3000 metric tons of rocks were removed from San Ramón 15, most of which was extracted during the Late Archaic as all the iron vein was battered for pigments. To accomplish this task in both periods the appropriate technological and logistical organization was required. This was surely different for both periods, but unfortunately available information remains scarce for characterizing hunter–gatherer–fisher settlement systems in the locality. We do know however, that for the Early Archaic Huentelauquén groups were highly mobile, relying on a diversified maritime subsistence economy (mollusks, fish, sea mammals, sea birds), and to a lesser degree on terrestrial fauna (Llagostera et al. 2000). In the coastal sites of the Huentelauquén populations siliceous rocks have been reported, which come from quarries more than 20 km inland from today's coastline (Castelleti 2007, Galarce 2008). Access to these quarries, however, seems to have been part of logistical systems with little or no investment in infrastructure and very short stays inland. Thus, iron oxide mining was probably not a part of this logistical mobility pattern of coastal hunter–gatherer–fishers looking for good-quality toolstone. Nor does it seem like it was part of an "embedded procurement strategy" (*sensu* Binford 1979) since the coastline was more than 3 km away from the site and almost all subsistence resources were available by the sea. It is thus more reasonable to suggest that iron oxide procurement demanded its own mobility systems and special-purpose strategies (Gould and Saggers 1985).

San Ramón 15 appears to be a task-specific site with very little evidence of domestic activities which suggests that the work force required for exploiting this mine came from base camps that were probably located at the coastline following other resource patterning such as availability of fresh water. While our surveys have not identified any of these base camps for the Early Archaic, the fact remains that

[1] Although the bottom of the prehistoric mine was not reached in U2, it is unlikely we will find evidence of in situ pre-Late Archaic layers. On the contrary, we believe earlier mining was more superficial (probably similar to the 3-m deep trench in U1), so its remains were completely altered as Late Archaic miners expanded the mine.

decisions to move to and from base camps must have reflected the need to procure iron oxides for the group. Given the known marine transgressions during the Early Holocene, it is likely that such a camp is now under coastal waters.

We know that mollusks, sea and terrestrial mammals, and fish from SR-15 are consistent with local Huentelauquén subsistence economy. This suggests that the social organization of these early groups allowed for a task force of miners working in SR-15 for some periods in the annual cycle to be provisioned by food from the coast.

While Huentelauquén populations seem to have been highly mobile and low-density groups, the situation was quite different for the Late Archaic, when SR-15 was also exploited by hunter–gatherer–fisher–miners. In the latter case, archaeological evidence in the arid coasts of the Antofagasta Region suggests higher population densities, semisedentary occupations including the use of architecture and burials under domestic floors, more extensive logistical mobility (especially inland), as well as probable cultural and genetic exchanges with other hunter–gatherer populations of what today is northern Chile (Bittmann 1986; Castelleti 2007; Cocilovo et al. 2005; Llagostera 1989; Núñez 1984).

In Taltal, Late Archaic is characterized by a local settlement system more complex and diversified than in previous periods (Castelleti 2007), including several campsites with architecture and burials (Capdeville 1921, Contreras et al. 2007, Llagostera 1989) as well as more ubiquitous presence of objects and resources from the highlands. At least six base camps with architecture and burials are today known in the surroundings of Taltal (Capdeville 1921; Contreras et al. 2007; Llagostera and Llagostera 2010). These are complemented by task-specific sites of different hierarchy and characteristics that are clustered at variable distances around these nuclear and semisedentary base camps (Castelleti 2007). Less than 2 km away from SR-15, two of these base camps have been known since early twentieth century: Morro Colorado and Punta Morada (Bird 1943; Capdeville 1921; see Fig. 7.1). Both show important occupation during the Middle Archaic, and then continuously until Late Archaic, when stone structures were built as dwellings with human interments in the floors.

As we previously noted, in Late Archaic base camps iron oxides are very common, especially associated with funerary rituals where the dead are placed under the floors of residential structures. The more permanent occupation at these sites, and the high levels of iron oxide consumption, seems consistent with the intensification of iron oxide extraction at SR-15. Since during the Late Archaic SR-15 was also a task-specific site with little evidence for domestic activities, it is likely that the miners came from base camps located near the mine. It is likely that Punta Morada and/or Morro Colorado were the sites from which the miners went daily to exploit SR-15.

Summary and Conclusions

The use of iron oxide pigments is amply documented in archaeological contexts of coastal northern Chile, from the Early Archaic until historical times, being especially used during the Late Archaic in funerary rituals. However, until recently we knew nothing about primary extraction sites and how these articulated with local hunter–gatherer–fisher economies. SR-15 is thus an important source of information for investigating the early pigment mining economy during the preceramic, including a better understanding of mining technology and techniques, and its transformation through time.

Our studies estimate that mining at SR-15 extracted around 3,000 metric tons of rock in order to obtain red and yellow pigments (hematite and goethite). Two periods of exploitation have been documented at the site. The first one during the Early Archaic (12,000-10,200 cal. B.P.) and the second one during the Late Archaic (4,300-3,900 cal. B.P.), being thus the earliest mine yet dated in the New World.

The first exploitation of the mine occurred during the Early Archaic by task groups that were part of the Huentelauquén Cultural Complex. The characteristics of the operation suggest these groups had an adequate mining skill and were thus probably the continuation of an earlier mining tradition that may have been developed elsewhere and then brought to the region with these first human populations.

Early Archaic Huntelauquén groups have been described as the first maritime adapted peoples of the Pacific Coast of northern Chile (Castelleti 2007; Llagostera et al. 2000; Núñez 1984; Sandweiss 2008). With evidence from SR-15 we can further characterize these groups as having had the technical knowledge for trench mining as well. We may therefore conclude that mining practices were also an important component in the lives of these early occupants, or at least some of its members, and that these groups articulated a more complex system of productive activities than hereto suspected.

Overall, we can now say that our data shows that the extraction of iron oxides was not an expedient or peripheral activity taking place within mobility systems driven by food requirements. On the contrary, Archaic settlement systems in the area were greatly influenced by the need to access iron oxide mines which would yield the necessary pigments to be later used in ceremonial and domestic contexts (Salazar et al. 2010). The mobility system associated with the exploitation of SR-15 seems closer to the foraging end of Binford's (1980) continuum, inasmuch as access to the site was made from base camps located on the shoreline, which in turn were presumably relocated following energetic and social requirements. It is interesting to notice that our data seems to be showing that these early foraging mobility systems were not only the result of the spatial structure of subsistence resource, but at least in part they were structured according to the spatial distribution and geological characteristics of other resources important for the social and symbolic reproduction of these coastal hunter–gatherer–fishers, notably iron oxides.

In sum, we are investigating a society of hunter–gatherer–fisher–miners that demands thorough study in order to better understand its social and economic organization. The evident continuities between the technology and mining techniques between the Early and Late Archaic episodes at SR-15 suggest a deep tradition in the region linking mineral procurement with social reproduction and ceremonial practices. That said, some important differences were detected at SR-15 between the Early and Late Archaic, and these call for further exploration.

Acknowledgments Our successive field seasons at SR-15 have been possible by two generous grants from the FondoNacional de Ciencia y Tecnología (ProyectosFondecyt 1080666 and 1110196). We are very appreciative of this support. We are also very grateful for our field and laboratory teams, which have been an important help for the success of this research. The paper benefited from the comments of an anonymous reviewer as well as by the comments and suggestions of the editors of this volume, even though we alone are responsible for its flaws. Additional thanks to Nico Tripcevich and Kevin Vaughn for their invitation to present our paper in this book. Last but not the least, our debt of gratitude to Mrs. Nevenka Conejeros for her logistic support and hospitality during our stays at Taltal.

References

Arriaza, B. (1994). Tipología de las momias Chinchorro y evolución de las prácticas de momificación. *Chungara, 26*, 11–48.

Barham, L. S. (2002). Systematic pigment use in the Middle Pleistocene of South-Central Africa. *Current Anthropology, 31*, 181–190.

Beaumont, P. B. (1973). The ancient pigment mines of Southern Africa. *South African Journal of Science, 69*, 140–146.

Beaumont, P., & Boshier, A. (1974). Report on test excavations in a prehistoric pigment mine near Postmasburg, Northern Cape. *South African Archaeological Bulletin, 29*(113/114), 41–59.

Betancourt, J. L., Latorre, C., Reich, J. A., Quade, J., & Rylander, K. A. (2000). A 22,000-year record of monsoonal precipitation from Northern Chile's Atacama desert. *Science, 289*, 1542–1546.

Binford, L. R. (1979). Organization and formation processes: looking at curated technologies. *Journal of Anthropological Research, 35*, 255–273.

Binford, L. R. (1980). Willow smoke and dogs' tails: hunter-gatherer settlement systems and archaeology. *American Antiquity, 45*, 4–20.

Bird, J. (1943). *Excavation in Northern Chile* (Anthropological Papers, Vol. XXVIII, Pt. IV). New York: American Museum of National History.

Bittmann, B. (1986). Recursos naturales renovables en la costa del norte de Chile: modos de obtención y uso. In S. Mazuda (Ed.), *Etnografía e historia del mundo andino: continuidad y cambio* (pp. 269–334). Tokio: University of Tokio Press.

Capdeville, A. (1921). Notas acerca de la arqueología de Taltal. II: Civilización Dolménica. Gentes de los círculos de piedra. *Boletín de la Academia Nacional de Historia, 2*(5), Quito.

Capdeville, A. (2010). Augusto Capdeville Rojas, notas. *Taltalia, 2*, 98–110.

Castelleti, J. (2007). *Patrón de Asentamiento y uso de recursos a través de la Secuencia ocupacional prehispana en la costa de Taltal* (Memoria para optar al grado de Magister en Antropología con mención en Arqueología). San Pedro de Atacama: Convenio Universidad Católica del Norte—Universidad de Tarapacá.

Cervellino, M. (1996). Breve análisis del desarrollo cultural prehispánico de la costa de la Región de Atacama, a la luz de viejas y nuevas evidencias. I[a] Parte. *Boletín del Museo Arqueológico de La Serena, 19*. doi:149.

Cocilovo, J., Varela, H., Costa-Junqueira, M. A., & Quevedo, S. (2005). Los pescadores arcaicos de la desembocadura del río Loa (Norte de Chile): El Sitio Caleta Huelen 42. *Chungara, 37*(1), 5–19.

Contreras, R., Cruz, J., Llagostera, A., Gracés, H., Núñez, P., Rodríguez, O., Gárate, H., & Becerra, G. (2007). *Los Bronces-1. Un Asentamiento de 5500 años en la Costa de Taltal*. Soloimagen, Taltal: Museo Augusto Capdeville.

Costa-Junqueira, M. A. (2001). Modalidades de enterramientos humanos arcaicos en el norte de Chile. *Chungara, 33*(1), 55–62.

D'Errico, F., Salomon, H., Vignaud, C., & Stringer, C. (2010). Pigments from the Middle Palaeolithic levels of Es-Skhul (Mount Carmel, Israel). *Journal of Archaeological Science, 37*, 3099–3110.

Eerkens J. W., Vaughn, K. J., Linares Grados, M., & Edwards, M. J. (2007) La Ballena: A mining base camp in the Southern Nasca region, Peru. *Antiquity Protect Gallery, 81*(314).

Eiselt, B. R., Popelka-Filcoff, J., & Darling, M. G. (2011). Hematite sources and archaeological ochres from Hohokam and O'odham sitesin central Arizona: an experiment in type identification and characterization. *Journal of Archaeological Science, 38*, 3019–3028.

Erlandson, J., Robertson, J. D., & Decantes, C. (1999). Geochemical analysis of eight red ochres from Western North America. *American Antiquity, 64*(3), 517–526.

Galarce, P. (2008). Aprendizaje y talla lítica en sociedades prehistóricas: contextos sociales y correlatos materiales. In D. Jackson, D. Salazar, & A. Troncoso (Eds.), *Puentes hacia el pasado. Reflexiones teóricas en arqueología* (pp. 93–110). Santiago: Monografías de la Sociedad Chilena de Arqueología.

Geniola, A., Larocca, F., & Vurro, F. (2006). *Approvvigionamento di risorse minerarie nella Grotta della Monaca (Sant'Agata di Esaro—Cosenza),* in AA. VV., *Materie prime e scambi nella Preistoria italiana*, Atti della XXXIX Riunione Scientifica dell'Istituto Italiano di Preistoria e Protostoria (Firenze, 25–27 novembre 2004), Vol. III (pp. 1349–1359). Firenze: IIPP.

Gialanella, S., Belli, R., Dalmeri, G., Lonardelli, I., Mattarelli, M., Montagna, M., & Toniutti, L. (2011). Artificial or natural origin of hematite-based red pigments in archaeological contexts: the case of Riparo Dalmeri (Trento, Italy). *Archaeometry, 53*(5), 950–962.

Gillespie, J. D. (2007). Enculturing an unknown world: caches and Clovis landscape ideology. *Canadian Journal of Archaeology, 31*, 171–189.

Goldenberg, G., Maass, A., Steffens, G., & Steuer, H. (2003). Hematite mining during the linear ceramics culture in the area of the Black Forest, South West Germany. In T. Stöllner, G. Körlin, G. Steffens, & J. Cierny (Eds.), *Man and mining-Mensch und Bergbau. Studies in honour of Gerd Weisgerber on occasion of his 65th birthday* (Vol. Der Anschnitt Beiheft 16, pp. 179–186). Deutsches Bergbau-Museum: Bochum.

Gould, R., & Saggers, S. (1985). Lithic procurement in central Australia: a closer look at Binford'sidea of embeddedness in archaeology. *American Antiquity, 50*, 117–136.

Grosjean, M., van Leeuwen, J. F. N., van der Knaap, W. O., Geyh, M. A., Ammann, B., Tanner, W., Messerli, B., Nuñez, L. A., Valero-Garcés, B. L., & Veit, H. (2001). A 22,000 ^{14}C year BP sediment and pollen record of climate change from Laguna Miscanti 23°S, northern Chile. *Global and Planetary Change, 28*, 35–51.

Hammel, A., White, C., Pfeiffer, S., & Miller, D. (2000). Pre-colonial mining in Southern Africa. *Journal of the South African Institute of Mining and Metallurgy, 100*, 49–56.

Hejl, E., & Tippelt, G. (2005). Prehistorical pigment mining on Santorini's neighbouring island Anafi Cyclades, Greece. *Austrian Journal of Earth Sciences, 98*, 22–33.

Henshilwood, C., d'Errico, F., van Niekerk, K. L., Coquinot, Y., Jacobs, Z., Lauritzen, S. E., Menu, M., & García-Moreno, R. (2011). A 100,000-year-old ochre—processing workshop at Blombos Cave, South Africa. *Science, 334*, 219–222.

Hovers, E., Ilani, S., Bar-Yosef, O., & Vandermeersch, B. (2003). An early case of color symbolism: ochre use by modern humans in Qafzeh cave. *Current Anthropology, 44*, 491–522.

Jackson, D., Seguel, R., Báez, P., & Prieto, X. (1999). Asentamientos y evidencias del Complejo Huentelauquén en la comuna de Los Vilos, Provincia del Choapa. *Anales del Museo de Historia Natural de Valparaíso, 24*, 5–28.

Koukouli-Chryssanthaki, C., & Weisgerber, G. (1996). Palaiolithiko orycheio ochras sti Thaso, *Archaiologia kai Technes, 60*, 82–89.

Lahren, R., & Bonnichsen, R. (1974). Bone foreshafts from a Clovis burial in southwestern Montana. *Science, 186*, 147–150.

Llagostera, A. (1979). Tres dimensiones en la conquista prehistórica del mar. Un aporte para el estudio de las formaciones pescadores de la costa sur andina. In, Actas del VIII Congreso de Arqueología Chilena, (pp. 217–245). Santiago: Ediciones Kultrún.

Llagostera, A. (1989). Caza y pesca marítima (9000 a 1000 a.C). In J. Hidalgo, V. Schiappacasse, H. Niemeyer, C. Aldunate, & I. Solimano (Eds.), *Prehistoria. Desde sus Orígenes hasta los Albores de la Conquista* (pp. 57–79). Santiago: Editorial Andrés Bello.

Llagostera, A., & Llagostera, V. (2010). Enterratorios del sitio arqueológico Los Bronces-1, Comuna de Taltal (Región de Antofagasta). *Taltalia, 3*, 7–20.

Llagostera, A., Weisner, R., Castillo, G., Cervellino, M., & Costa, M. A. (2000). El Complejo Huentelauquén bajo una perspectiva macroespacial y multidisciplinaria. *Contribución Arqueológica, 5*, 461–480.

MacDonald, B. L. (2008). *Ochre procurement and distribution on the central coast of British Columbia*. Hamilton, ON: Master of Arts, McMaster University Department of Anthropology.

Mächtle, B., Unkel, I., Eitel, B., Kromer, B., & Schiegl, S. (2010). Molluscs as evidence for a late Pleistocene and early Holocene humid period in the southern coastal desert of Peru (14.5°S). *Quaternary Research, 73*, 39–47.

Marean, C., Bar-Matthews, M., Bernatchez, J., Fisher, E., Goldberg, P., Herries, A., Jacobs, Z., Jerardino, A., Karkanas, P., Minichillo, T., Nilssen, P. J., Thompson, E., Watts, I., & Williams, H. (2007). Early human use of marine resources and pigment in South Africa during the Middle Pleistocene. *Nature, 449*, 905–908.

Mason, R. (1983). Prehistoric mining in South Africa and iron age copper mines in the Dwarsberg, Transvaal. *Journal of the South African Institute of Mining and Metallurgy, 82*(5), 134–142.

Massone, M. (2004). *Los Cazadores Después del Hielo. Colección de Antropología N°VII*. Santiago: Centro de Investigaciones Diego Barros Arana, Dirección de Bibliotecas, Archivos y Museos.

Mazzia, N., Flegenheimer, N., & Poiré, D. (2005). Not only flaked artifacts in early Pampean lithic assemblages (Argentina). *Current Research in the Pleistocene, 22*, 25–28.

Mooney, S., Geiss, C., & Smith, M. (2003). The use of mineral magnetic parameters to characterize archaeological ochres. *Journal of Archaeological Science, 30*(5), 511–523.

Moreno, A., Santoro, C., & Latorre, C. (2009). Climate change and human occupation in the northernmost Chilean Altiplano over the last ca. 11 500 cal. a BP. *Journal of Quaternary Science, 24*(4), 373–382.

Mostny, G. (1964). *Arqueología de Taltal. Epistolario de Augusto Capdeville con Max Uhle, Tomo I y II*. Santiago de Chile: Fondo Histórico y Bibliográfico J. T. Medina.

Núñez, L. (1984). *Tráfico de complementariedad de recursos entre las tierras altas y el Pacífico en el área Centro-Sur Andina* (Unpublished Ph.D. dissertation). Departamento de Antropología Cultural, Universidad de Tokio, Japón.

Núñez, L., Grosjean, M., & Cartajena, I. (2005). *Ocupaciones humanas y paleoambientes en la Puna de Atacama*. San Pedro de Atacama: Universidad Católica del Norte—Taraxacum.

Ortlieb, L., Vargas, G., & Saliége, J. F. (2011). Marine radiocarbon reservoir effect along the northern Chile–southern Peru coast (14–24°S) throughout the Holocene. *Quaternary Research, 75*, 91–103.

Petru, S. (2006). Red, black or white. The dawn of colour symbolism. *Documenta Praehistorica, XXXIII*, 203–208.

Popelka-Filcoff, R., David Robertson, J., Glascock, M. D., & Descantes, C. (2007). Trace element characterization of ochre from geological sources. *Journal of Radioanalytical and Nuclear Chemistry, 272*(1), 17–27.

Rapp, G. (2009). *Archaeomineralogy*. Alemania: Springer.

Robbins, L., Murphy, M., Campbell, A., & Brook, G. (1998). Intensive Mining of Specular hematite in the Kalahari ca. A.D. 800–1000. *Current Anthropology, 39*(1), 144–150.

Salazar, D., Castro, V., Michelow, J., Salinas, H., Figueroa, V., & Mille, B. (2010). Minería y metalurgia en la costa arrecia de la región de Antofagasta, norte de Chile. *Boletín del Museo Chileno de Arte Precolombino, 15*, 9–23.

Salazar, D., Castro, V., Salinas, H., & Varela, V. (2009). Nuevas investigaciones sobre la prehistoria y la antigua minería de Taltal. *Taltalia, 2*, 111–118.

Salazar, D., Jackson, D., Guendon, J. L., Salinas, H., Morata, D., Figueroa, V., Manríquez, G., & Castro, V. (2011). Early evidence (ca. 12,000 BP) for iron oxide mining on the Pacific coast of South America. *Current Anthropology, 52*, 463–475.

Sandweiss, D. (2008). Early fishing societies in Western South America. In H. Silverman & W. H. Isbell (Eds.), *Handbook of South American archaeology* (pp. 145–156). New York: Springer.

Santoro, C., Ugalde, P., Latorre, C., Salas, C., Osorio, D., Jackson, D., & Gayó, E. (2011). Ocupación humana pleistocénica en el desierto de Atacama: primeros resultados de la aplicación de un modelo predictivo de investigación interdisciplinaria. *Chungara, 43*(Número Especial), 353–366.

Scalise, R., & Di Prado, V. (2006). Early use of ochre in the Pampean region of Argentina. *Current Research in the Pleistocene, 23*, 66–68.

Schmandt-Besserat, D. (1980). Ochre in prehistory: 300,000 years of the use of iron ores as pigments. In T. A. Wertime & J. D. Muhly (Eds.), *The coming of the age of iron* (pp. 127–150). New Haven and London: Yale University Press.

Sepúlveda, M. (2011). Pinturas rupestres y tecnología del color en el extremo sur de Chile. *Magallania, 39*(1), 193–210.

Stafford, M., Frison, G., Stanford, D., & Ziemans, G. (2003). Digging for the color of life: Palaeoindian red ochre mining at the Powars II Site, Platte county, Wyoming, USA. *Geoarchaeology, 18*, 71–90.

Standen, V. (1997). Temprana Complejidad Funeraria en la Cultura Chinchorro: (Norte de Chile*)*. *Latin American Antiquity, 8*, 134–156.

Thackeray, A., Thackeray, J. F., & Beaumont, P. (1983). Excavations at the Blinkklipkop specularite mine near Postmasburg, Northern Cape. *South African Archaeological Bulletin, 38*(137), 17–25.

Triat, J. M. (2010). *Les Ocres*. París: CNRS Editions.

Vaughn, K., Linares Grados, M., Eerkens, J., & Edwards, M. (2007). Hematite mining in the ancient Americas: Mina Primavera, A 2,000 year old Peruvian mine. *Journal of Minerals, Metals and Material Society, 59*(12), 16–20.

Vaughn, K., Van Gijseghem, H., Linares M., & Eerkens J. (2011). Minería de hematita en la costa sur del Perú: Investigaciones arqueológicas en la Mina Primavera, *Chungara* (in press).

Wadley, L., Williamson, B., & Lombard, M. (2004). Ochre in hafting in middle stone age southern Africa: a practical role. *Antiquity, 78*(301), 661–675.

Watts, I. (2002). Ochre in the middle stone age of southern Africa: ritualized display or hide preservative? *South African Archaeological Bulletin, 57*(175), 1–14.

Weisgerber, G., & Pernicka, E. (1995). Ore mining in prehistoric Europe:An overview. In G. Morteani & J. P. Northover (Eds.), *Prehistoric gold in Europe* (pp. 159–182). Netherlands: Kluwer Academic Publishers.

Zilhão, J., Angelucci, D. E., Badal-García, E., d'Errico, F., Daniel, F., Dayet, L., Douka, K., Higham, T. F. G., Martínez-Sánchez, M. J., Montes-Bernárdez, R., Murcia-Mascarós, S., Pérez-Sirvent, C., Roldán-García, C., Vanhaeren, M., Villaverde, V., Woodand, R., & Zapata, J. (2010). Symbolic use of marine shells and mineral pigments by Iberian Neandertals. *Proceedings of the National Academy of Sciences United States of America, 107*, 1023–1028.

Chapter 8
The Organization of Mining in Nasca During the Early Intermediate Period: Recent Evidence from Mina Primavera

Kevin J. Vaughn, Hendrik Van Gijseghem, Verity H. Whalen, Jelmer W. Eerkens, and Moises Linares Grados

Introduction

As some have noted (Shimada 2007), each stage in the broader activity of craft production, whether it is the extraction of raw materials, the processing of material, production, or consumption should be evaluated as inseparable parts of the same basic process. At least one of these stages, the extraction of raw materials—whether it is through mining or quarrying—is still much more poorly understood in comparison with other aspects of craft production. Indeed, mining and quarrying, the first stages in the activities related to the manufacture of crafts including metals, ceramics, and stone is key to understanding the development of complex social and political organizations in antiquity. In the Andes, evidence for ancient mining has been relatively scarce, most likely because modern (and even later Prehispanic) mining has destroyed evidence for early mining (Cantarutti Chap. 9; Eerkens et al. 2009; Núñez 1999). Despite this, ancient mining in the Andes is clearly understood as one important part of the production of significant crafts as the papers in this volume demonstrate (see also, Aldunate et al. 2008; Bird 1979; Burger and Matos

K.J. Vaughn (✉) • V.H. Whalen
Department of Anthropology, Purdue University,
700 West State Street, West Lafayette, IN 47907, USA
e-mail: kjvaughn@purdue.edu; vwhalen@purdue.edu

H. Van Gijseghem
Universitё de Montreal, Montreal, Quebec, Canada
e-mail: h.van.gijseghem@gmail.com

J.W. Eerkens
University of California-Davis, Davis, CA, USA
e-mail: jweerkens@ucdavis.edu

M.L. Grados
Arqueocare, Lima, Peru
e-mail: arqueocare@gmail.com

N. Tripcevich and K.J. Vaughn (eds.), *Mining and Quarrying in the Ancient Andes*,
Interdisciplinary Contributions to Archaeology, DOI 10.1007/978-1-4614-5200-3_8,

Table 8.1 Chronology for the Southern Nasca Region (SNR) and the Central Nasca Region (CNR)

Horizons/periods	Local period	Culture	Approximate date in the SNR (Schreiber and Lancho Rojas 2003)	Approximate date in the CNR (Reindel 2009; Unkel et al. 2007)
Late Horizon	*–Inca–*	Inca	AD 1476–1532	AD 1400–1532
Late Intermediate	*–Tiza–*	Tiza	AD 1000–1476	AD 1000–1400
Middle Horizon	*–Loro–*	Loro, Wari	AD 750–1000	AD 620–1000
Early Intermediate Period	*–Nasca–*	Late Nasca	AD 550–750	AD 440–620
		Middle Nasca	AD 450–550	AD 325–440
		Early Nasca	AD 1–450	AD 90–325
		Proto (Initial) Nasca	100 BC–AD 1	200 BC–AD 90
Early Horizon	*–Formative–*	Paracas	800–100 BC	800–200 BC

Mendieta 2002; Lechtman 1976; Núñez 1999, 2006; Petersen 1970; Salazar et al. 2010; Shimada 1985, 1994, 1998; Stöllner 2009; Vaughn et al. 2007).

In this paper, we consider mining in Nasca from the perspective of Mina Primavera, a 2,000-year-old hematite mine located in the Ingenio Valley of the South Coast of Peru. Our work at the site has revealed mining technology and activities within the mine including the extraction and processing of hematite and ritual as evidenced by the playing of music. Our work has also shown that while there may have been limited mining during the Early Horizon at Mina Primavera (Table 8.1), exploitation of the mine began in earnest during the first part of the Early Intermediate period during a phase known locally as Nasca 1, an important time in the development of Nasca society. Mining accelerated rapidly, however, during later stages of the Early Intermediate period, especially during Early Nasca (phases 2–4) as seen through artifactual evidence, radiocarbon dates, and evidence for the intensification of hematite processing within the mine. After the Early Intermediate period, extraction of hematite from Mina Primavera appears to have decreased until by the Middle Horizon, it was apparently abandoned either because high-quality hematite had been exhausted, or because there was no longer a demand for the raw material.

Archaeological Context

Mina Primavera is located in the Ingenio Valley, south coast of Peru (Fig. 8.1). The Ingenio Valley is part of the Central Nasca Region (CNR[1]) of the Nasca "heartland." Contemporary archaeological investigations in the CNR have included Silverman's

[1] This region is sometimes referred to as the northern Nasca region (see Chap. 14). We use the term CNR to distinguish the region from the Southern Nasca Region (SNR) and the Northern Nasca Region (NNR) as we have defined in previous work (see Vaughn 2009).

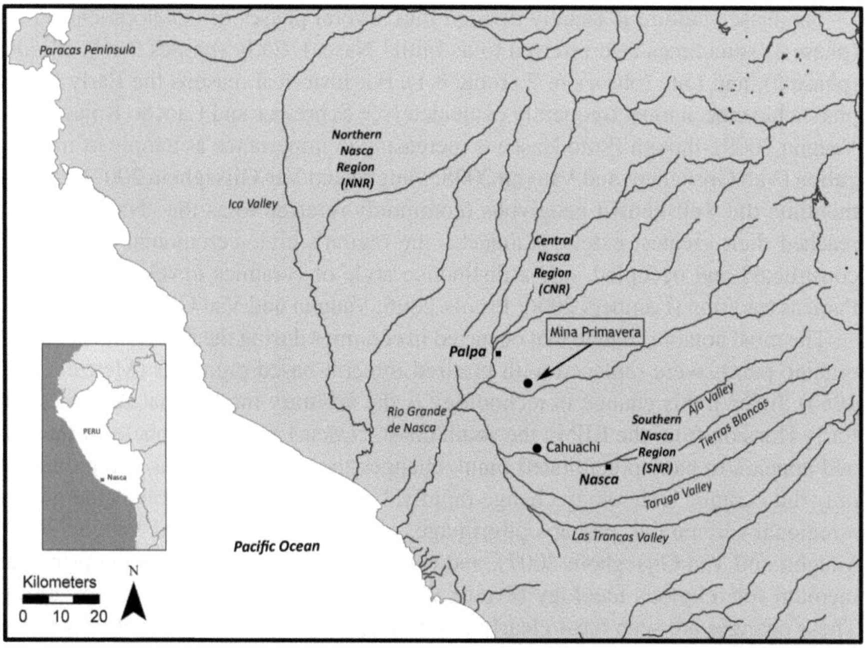

Fig. 8.1 The Nasca Region with the Ingenio Valley and the location of Mina Primavera highlighted. Note the Southern Nasca Region (SNR), Central Nasca Region (CNR), and the Northern Nasca Region (NNR). Redrawn from an original by Stefanie Bautista

(2002) settlement survey of the Ingenio Valley and Reindel, Isla and colleagues (Reindel 2009; Chap. 14) extensive investigations of the Palpa, Viscas and Rio Grande drainages.

The Nasca region is known for the development of several indigenous cultures including Paracas dating to the Early Horizon (ca. 800–200 BC[2]), and Nasca dating to the Early Intermediate period (ca. 200 BC–AD 620, hereafter EIP). Paracas is known for very elaborate textiles (Paul 1990) and mummy bundles excavated by Tello in the early twentieth century (Tello and Mejía Xesspe 1979), and excavations at several habitations around the Nasca region (Reindel and Isla Cuadrado 2006; Van Gijseghem 2006). Since its discovery, Nasca has been assumed to be directly related to the Paracas culture. Indeed, as Menzel and colleagues (Menzel et al. 1964: 251) suggested: "…the distinction between "Ocucaje" (meaning Paracas in Ica) and a later "Nasca" style at Ica is…arbitrary…Both are parts of a single tradition in which there is a strong element of continuity between any two successive phases."

[2] As this valley is part of the Central Nasca Region, we employ the extensively documented chronology used by Reindel (2009) and colleagues and that contrasts with the Southern Nasca Region chronology that we have used previously (see, for example, Vaughn 2009).

The Nasca culture is usually divided into several phases of development: Proto (phase 1, sometimes also referred to as Initial Nasca), Early (phases 2–4[3]), Middle (phase 5), and Late (phases 6, 7; Table 8.1). For historical reasons the Early Nasca phases have been most frequently evaluated (see Schreiber and Lancho Rojas 2003; Vaughn 2009), though Proto Nasca is increasing in importance as a topic of investigation (Van Gijseghem and Vaughn 2008; Vaughn and Van Gijseghem 2007). During this time the well-known geoglyphs (commonly referred to as the "Nasca Lines") reached their greatest extent, Cahuachi, the region's civic-ceremonial center, was constructed and occupied, and a distinctive style of ceramics developed from the Paracas tradition (Lambers 2006; Proulx 2006; Vaughn and Van Gijseghem 2007).

The most notable change that occurred in ceramics during the EIP is that postfired organic paints were replaced with prefired mineral-based pigments (Menzel et al. 1964). Indeed, this change in technology is the arbitrary marker that separates the Early Horizon from the EIP in the south coast. Nasca 1 is the first phase of the EIP and appears to have precipitated many changes throughout the Nasca region. Not only did ceramic technology change rapidly, Cahuachi appears to have emerged as a regional ceremonial, perhaps pilgrimage, center (see Kantner and Vaughn 2012; Vaughn and Van Gijseghem 2007), and ceramics replaced textiles as the principal medium for religious ideology (Proulx 2006; Vaughn and Van Gijseghem 2007). These changes are seen most clearly in reorganization of settlement patterns in the SNR (Schreiber and Lancho Rojas 2003), an emergence of a "Cahuachi cult" involving feasting, and new ceramic types (burnished blackwares and very thin, well-fired cream and red-slipped vessels), a percentage of which appear to have been produced at Cahuachi (Vaughn and Van Gijseghem 2007). These changes coincide with gradually improving climatic conditions and interannual predictability in climatic conditions (Kantner and Vaughn 2012). Furthermore, the emergence of the Cahuachi cult at this time appears to be the initial steps toward the Early Nasca pattern where polychromes were employed as the principal medium for ideology. Two of us (Vaughn and Van Gijseghem 2007) have argued that this most likely reflects the growing power of a resident elite at Cahuachi and an increasing effort to materialize a religious ideology.

While these have been proposed as testable hypotheses, the question remains from where did Nasca craft producers obtain the materials necessary to manufacture their goods? One of the goals of the Early Nasca Craft Economy project undertaken between 2002 and 2007 was to seek evidence for sources of raw materials for ceramic production (see Vaughn et al. 2006), and as part of this, a major discovery was recording a relatively undisturbed Prehispanic hematite mine called Mina Primavera (Vaughn et al. 2007). Though evidence for mining activities in the region

[3] Note that Reindel and colleagues typically group phase 4 with 5 into Middle Nasca. In the SNR, phase 4 is grouped with Early Nasca. While geographically, the Ingenio Valley and Mina Primavera are part of the CNR, we continue to group phase 4 into Early Nasca as we typically find Nasca 4 ceramics with earlier phases. Indeed, this is the pattern that we have found at Mina Primavera.

have been previously recorded (e.g., Eerkens et al. 2009), collaborative work at Mina Primavera reported here provides rare insight into the character of mining during the EIP in the Nasca region.

Mina Primavera

Located just to the north of the Ingenio Valley in the Portachuelo Formation, a formation composed of Cretaceous marine sedimentary deposits with intervals of intrusive volcanic pyroclastics, Mina Primavera was originally recorded in 2004 (Vaughn et al. 2007). It was found by working with local informants including itinerant miners as part of our larger project investigating the sources of raw materials used in ceramic production (e.g., see Vaughn and Neff 2004). Mina Primavera is approached on a very rough road out of Estudiante in the Ingenio Valley built by a mining company. Mining operated from the 1960s to the 1970s and focused on the extraction of barite ($BaSO_4$; used as a weighting agent in petroleum production). While the mining operation has since been abandoned, itinerant miners frequent the area in their prospecting for gold and copper.

The site itself is a small human-made cave in a cliffside of a narrow quebrada across from one such barite mine. We were initially suspicious of our informant's claim that it was Prehispanic because it clearly has modern construction consisting of a masonry wall reinforced by concrete complete with a doorframe (Fig. 8.2). While the door has been removed, informants stated that the wall was built by contemporary miners who were exploiting the modern barite mine across the canyon in order to securely store dynamite and tools within Mina Primavera. Indeed, initial exploration of the inside of the mine revealed evidence for the storage of explosives (e.g., instructions accompanying dynamite) and other contemporary debris such as fragments of newspapers from the twentieth century (Fig. 8.2).

The opening of the cave made by Prehispanic miners measures some 20 m across and is from 2 m in height on the eastern end of the entrance narrowing down to about 30 cm in height—enough for a single person to crawl through—on its western end. Entering the cave, one encounters a deep crimson hematite [or ochre, an iron oxide (Fe_2O_3) recognized in Quechua as "*tacu*" or "*taco*" (Petersen 1970)] seam with obvious ancient hammerstone marks and a floor littered in mining debris (Fig. 8.3). The cave is approximately 20×30 m with an average height of 2 m, and was constructed within a small stratigraphic pocket of hematite in the local sedimentary bedrock. The floor of the mine drops down several meters to the south and to the west to form three separate galleries.

Gallery 1, the mine's most accessible, is approximately 100 m^2 in area and slopes gently back to the south toward Gallery 2 and to the west toward Gallery 3. Gallery 2 is the southern-most gallery whose access is difficult because of a restricted entrance toward the back of Gallery 1. Once one passes through the narrow opening, Gallery 2 opens up considerably to form an approximately 150 m^2

Fig. 8.2 (**a**) The entrance to Mina Primavera. The height of the original eastern entrance to the left (now blocked by a doorframe) is approximately 2 m. (**b** and **c**) Fragments of historical materials found on the surface of the site within the mine. Note that the date of the newspaper is 1939

(10 × 15 m) space, though it is difficult to assess its exact size because discarded mining debris is covering the southernmost boundary of the mine. Gallery 3 is the western-most gallery. As in Gallery 2, the full extent of this gallery is unknown because discarded mining debris is covering the edge of the mine. In addition to the three galleries, a small 2 m^3 chamber is accessible via Gallery 3 (though the entrance is mostly covered by mining debris), and exploration of some areas of the mine hints at the existence of a number of other chambers yet to be cleared of debris. Because of accessibility, most excavations have been undertaken directly inside Gallery 1, with several test units at the entrances to the mine as well as in Galleries 2 and 3.

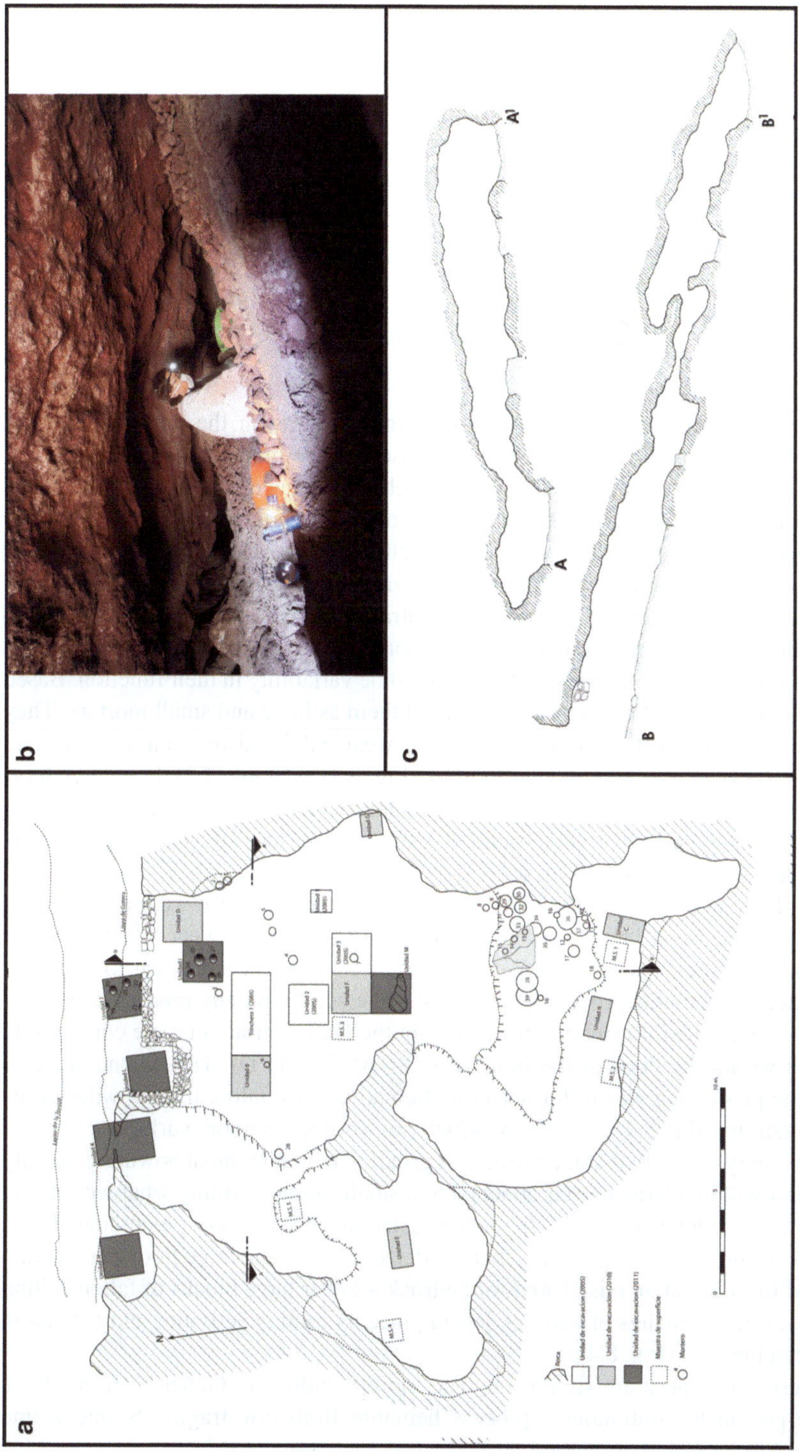

Fig. 8.3 **(a)** Map of Mina Primavera. The dark gray units were excavated in 2011. **(b)** Photo of Gallery 1. Note hammerstone marks. **(c)** Profiles of mine drawn in same scale as **(a)**

Table 8.2 Groundstone features in Mina Primavera

Groundstone feature	Gallery 1	Gallery 2
Bowl mortar	8	14
Conical mortar	2	0
Bottle mortar	1	0
Cup mortar	1	7
Total	12	21

Groundstone Features

In our work we have recorded 33 groundstone features in the floor of the mine (Table 8.2). Most of these were found exposed on the modern surface of the site, while several were buried under mining debris and encountered in excavations. Thus, the total number of these features within the site certainly exceeds the number recorded so far. These features are found exclusively in Galleries 1 and 2 as well as just outside the entrance to Gallery 1. No groundstone features have been documented either in Gallery 3 or outside the entrance to Gallery 3 despite test excavations there. The features were recorded as mortars used for hematite processing, but further inspection has shown that there was some variability in their function. Based on their size distribution, we have classified them as large and small mortars. They appear to covary further by the amount of wear exhibited on their surfaces and whether they are attached to other mortars (Fig. 8.4 and Table 8.2). Five large mortars in Gallery 1 have what we refer to as a satellite mortar—a small mortar directly attached to a large mortar via a rounded, well-worn surface suggesting the mortars were used simultaneously (Fig. 8.4A^2).

The large and small mortars also differ in the amount of use-wear exhibited on their surfaces. Most large mortars exhibit little to moderate wear while all small mortars exhibit extensive wear. We interpret these differences in wear to represent two stages of hematite processing. The first stage was to initially process large fragments of hematite into smaller pieces within the large mortars that are consistently located within the floor of the mine (Fig. 8.4A^1, B, and D). The second stage of hematite processing was to further grind hematite down into a finer powder in the small mortars (Fig. 8.4A^2, C), all of which are worn to a smooth surface.

In Gallery 1, this two-stage process of grinding hematite into a powder generally occurred within a large mortar attached to a small mortar forming what we refer to as a mortar station (Fig. 8.4A). In Gallery 2, however, secondary processing almost always occurred in small mortars that are detached from the large floor mortars and located in isolation on raised areas of bedrock—either large blocks of fallen ceiling or long, thin overhangs of bedrock left in place by mining activities and extending out from the wall (Fig. 8.4D).

These differences are significant. The mortar stations in Gallery 1 allow for a single person to continuously process hematite from raw fragments into a fine

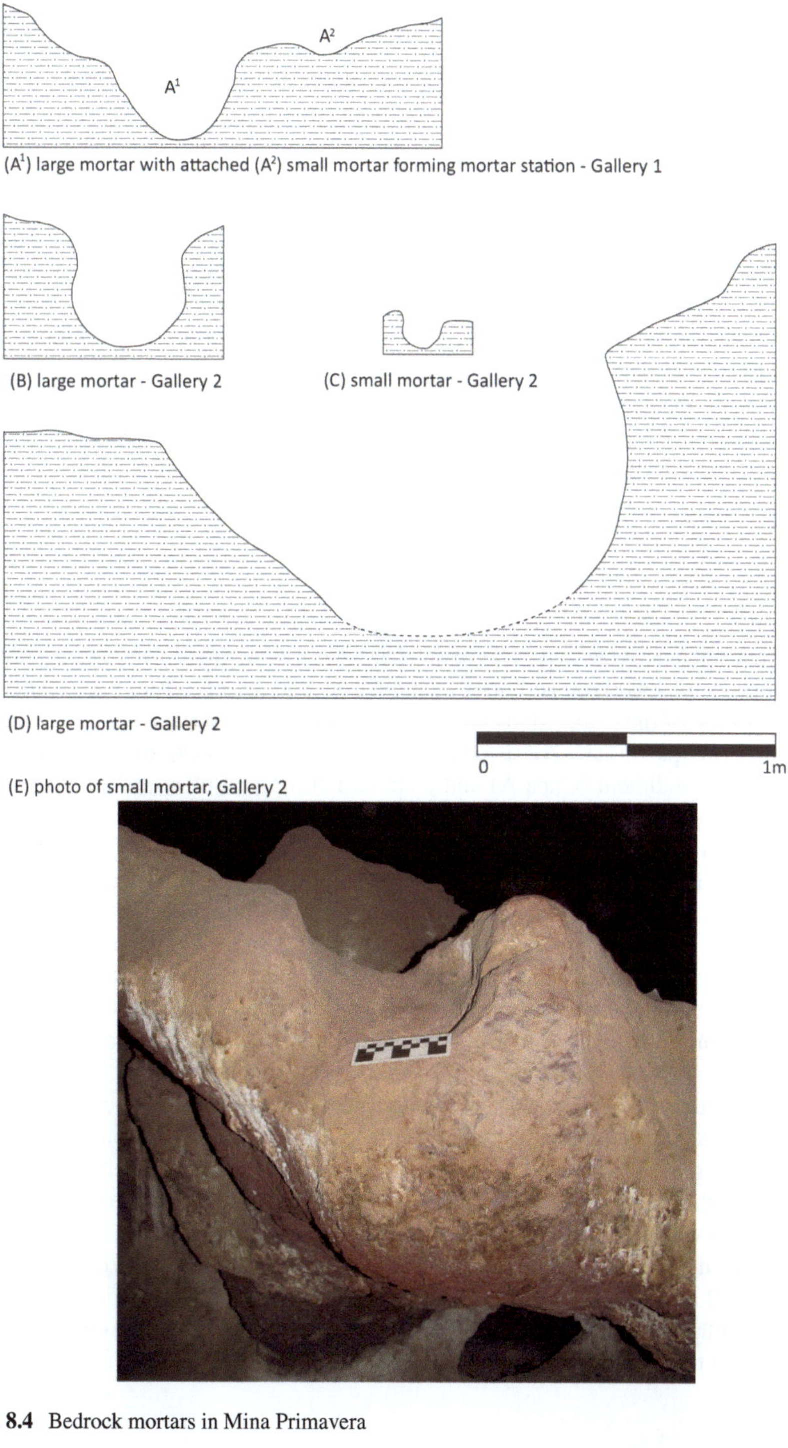

Fig. 8.4 Bedrock mortars in Mina Primavera

powder—first working the initial fragments in a large mortar, and then transferring the material to the adjacent small mortar to grind the hematite further. The layout of these stations allowed hematite to be moved directly from the large mortars into the attached small mortars for further processing; the worn surfaces between the two attest to this repeated act. The pattern is very different in Gallery 2 where compact mortar stations are absent. Instead, the mortars in Gallery 2 comprise what we refer to as a "mortar complex" consisting of multiple large mortars (e.g., #s 29–33 in Fig. 8.3) that are separated from multiple small mortars used for further processing (such as #s 13–15). Instead of processing centers for a single person, the scale of Gallery 2 complexes suggest multiple people could have worked simultaneously—some breaking up raw hematite fragments in the large mortars, while others were further processing them in adjacent mortars, and some working the hematite into a fine powder in the series of small mortars.

Excavations

A total of 60 m^2 excavated within Mina Primavera over the course of several field seasons from 2004 to 2011 have revealed that activities within the mine included not only mining and processing of hematite, but what we interpret to be camping within the mine as well as ritual activities including the playing of music. We turn here to a description of the excavations undertaken at Mina Primavera.

Two principal strata were revealed in excavations: a postexploitation stratum of wind-blown sediment (Capa A) and a stratum composed of mixed mining debris representing the principal exploitation of the mine (Capa B). In several excavation contexts, an additional stratum was identified (Capa C) composed of mixed mining debris and artifacts that appear to represent an earlier epoch of mining. The principal mining strata (Capas B and C) were usually separated by a shallow lens of wind-blown silt indicating a break in mining (Fig. 8.5). Artifacts include diagnostic ceramics, textiles, botanical and faunal remains, stone tools and other special-use artifacts including drilled *Spondylus princeps* pendants, fragments of *Spondylus princeps*, and musical instruments. With the exception of the musical instruments, all of these artifacts were found in a disturbed mining context.

Diagnostic ceramics found in excavations included Formative through LIP sherds (Fig. 8.6). A single ceramic fragment from the Early Horizon was found in Gallery 2, and a single LIP sherd was found in Capa A in Gallery 1 suggesting LIP mining or prospecting that postdated the principal exploitation in Capas B and C. We also recovered evidence for a Wari presence within the mine in Gallery 1, Capa B. This included a nose from a Wari face-neck jar and a handful of other sherds—including a rim sherd from a Chakipampa lyre cup—that clearly date to the Middle Horizon. Several of these vessel types, especially the face-neck jar and the lyre cup have been interpreted to be elite artifacts (see Cook 1984–1985; Edwards 2010: 341).

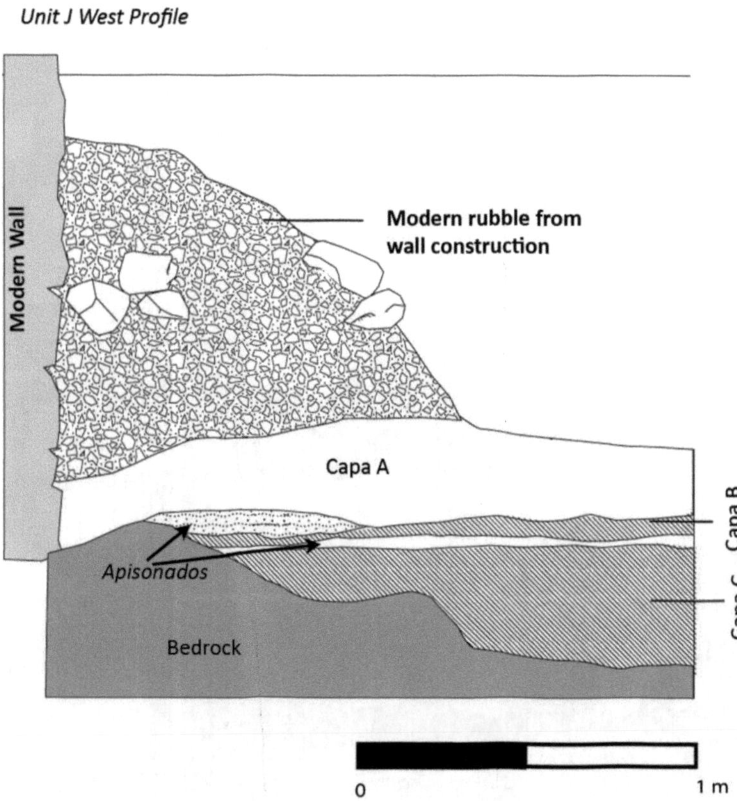

Fig. 8.5 Profile of Unit J, west wall. Capas B and C are separated by an *apisonado*, a shallow accumulation of wind-blown silt indicating a hiatus in mining activities for a period of time

Despite this range of diagnostic ceramics, the majority of pottery dates to the EIP as Nasca ceramics were found in all three galleries. In particular, an abundance of Nasca 1 (Proto Nasca) and Nasca 2–4 (Early Nasca) sherds are found within the mine. While Early Nasca sherds were found in all three galleries within Capa B, Nasca 1 sherds were restricted mostly to units near the contemporary entrances to Galleries 1 and 3 of the mine and in Capas B and C, while a few were found near the back of Gallery 1 in Capa B, but in an eroded condition. We have yet to recover any clear Middle (phase 5) or Late (phases 6, 7) Nasca sherds. Overall, the quantity of Proto and Early Nasca ceramics suggests that exploitation during this part of the EIP was the most persistent in ancient times.

Seven radiocarbon assays range from approximately 50 BC–AD 1400 (cal 2σ). Capa A dates range from AD 900–1400 while Capa B dates cluster in the early part of the first millennium AD (note that Capa C was identified in a few units just recently and we have yet to run radiocarbon assays from these contexts; Table 8.3).

Fig. 8.6 Diagnostic ceramics from Mina Primavera separated by galleries. While Early Nasca (Nasca 2-4) sherds were found in all three galleries of the mine, Nasca 1 sherds were restricted to units near the entrance to the mine

Table 8.3 Radiocarbon dates from Mina Primavera calibrated using INTCAL 09 (Reimer et al. 2009)

Sample #	Gallery	Unit	Stratum	^{14}C date	Calendar years (cal. 2σ)
AA72022	1	2	A	505 ± 43	AD 1316–1455
Beta 195717	1	1	A	990±70	AD 896–1210
AA72023	1	2	B	1901±44	AD 17–230
AA72024	1	3	B (A2)	1951±35	38 BC–AD 126
AA72021	1	T1	B	1961±43	48 BC–AD 128
AA94038	3	E	B (D)	1832±35	AD 83–254
AA94039	2	C	B	1722±35	AD 241–403
Panpipe	1	F	B	1480±55	AD 463–763

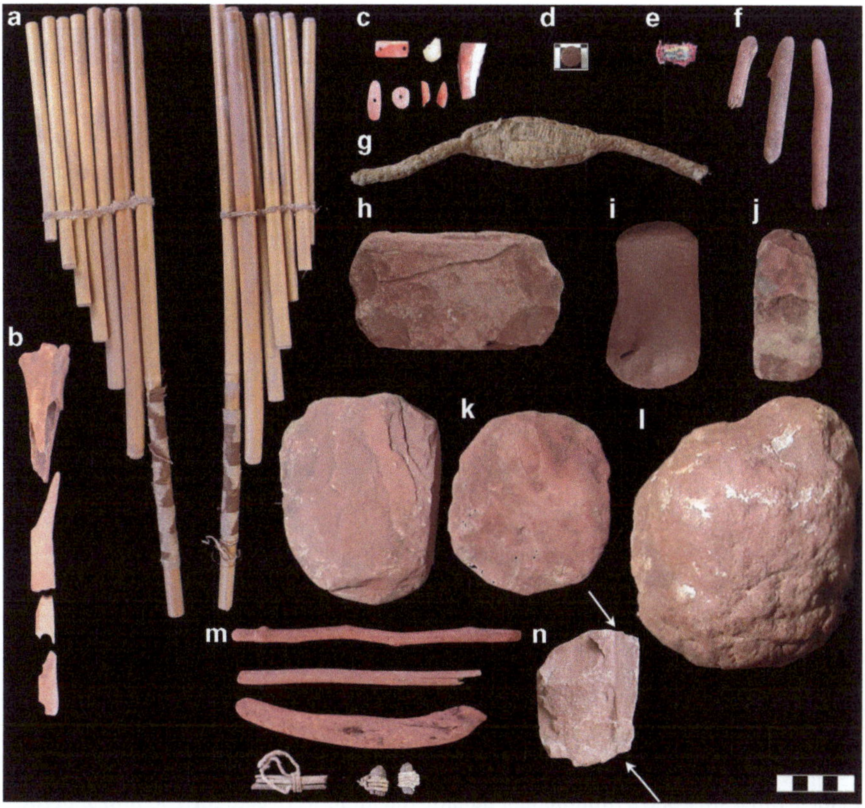

Fig. 8.7 Other artifacts from Mina Primavera. (**a**) Pair of cane panpipes. (**b**) Bone flute. (**c**) Spondylus including unworked fragments and finished ornaments. (**d**) Panpipe plug made of bottle gourd. (**e**) Fragment of dyed wool textile. (**f**) Wooden pries. (**g**) Sling. (**h**) Stone wedge. (**i**) Polisher. (**j**) Stone pry. (**k**) Axes. (**l**) Hand-held hammerstone. (**m**) Wooden handles. (**n**) Axe with evidence of hafting highlighted

The radiocarbon dates from Capa B further support the proposition that much of the exploitation taking place at the mine was done during Proto and Early Nasca.

Other artifacts recovered include fragments of well-woven cotton bags in the simple, plain warp-predominant weave typical of backstrap loom produced textiles in the Andes from ca. 800 BC. The majority of the fragments were stained red with signs of tearing. We interpret this to indicate that textiles were used to transport mined material to an outside location for further processing, though other possible functions include that some textiles were used to protect the hands of miners while using hammerstones to extract hematite. Clumps of string, a wool tab from an Early Nasca cloth, and a nondiagnostic sling were also recovered in excavations (Fig. 8.7). The sling was found in Gallery 1, Capa A, so is not associated with the principal exploitation of the mine. Botanical remains included an abundance of maize and

Table 8.4 Lithics from Unit M, Capas A–C

Unit	Tool	Capa A #	Capa A kg[a]	Capa B #	Capa B kg[a]	Capa C #	Capa C kg[a]
Finished tools							
M	Spades	11	2	15	2.5	61	9.0
M	Chisels	10	1.5	5	0.5	70	12.0
M	Cobble hammerstones	10	4.5	13	2.5	74	38.5
M	Axes	26	5.5	27	6.5	232	56.0
M	Picks	10	2	18	3.5	53	15.5
M	Polishers	3	0.1391	35	1.5	78	14.5
Lithic debris							
M	Cores	5	1.5	5	0.5	37	7.5
M	Modified flakes	28	1	58	1.5	92	3.5
M	Flakes	252	5.5	1,153	23.0	1,272	27.0
M	Mining debris/shatter[b]	249	17	788	51.5	1,555	123.0
M	Hematite chunks[c]	14	2.5	50	9.5	81	30.5

[a]Weights were calculated in the field using a hand scale with 0.5 kg accuracy. Anything less than 0.5 kg was weighed using a precision balance accurate to 0.1 g
[b]Unmodified shatter from mining. Raw material siltstone
[c]Fragments of discarded hematite

bottle gourd as well as a small amount of *pacay*, *lúcuma*, peanuts, and *achupalla* (probably used as a bast fiber).

Excavations revealed numerous tools related to the mining of hematite. Tools made from a siltstone available just outside the mine were categorized into six types: (1) flat spade-like wedges, or spades (Fig. 8.7h), (2) columnar-shaped wedges/chisels (Fig. 8.7j), (3) hand-held cobble hammerstones (Fig. 8.7l) , (4) semi-circular axes (Fig. 8.7k), some with evidence of hafting (Fig. 8.7n) to wooden handles (also found in abundance; Fig. 8.7m), (5) hand-held picks, and (6) smooth polishing stones (Fig. 8.7i). Also recorded were a number of wooden tools most likely used as pries (Fig. 8.7f). In some areas of the mine, virtually the entire excavated matrix consisted of remnants of stone tools. For example, in Unit M, a 2×2 unit placed near the back of Gallery 1 that consists of approximately 50 % bedrock (thus the unit is effectively a 1×2), 259 axes weighing on average 240 g each were found in Capas B and C (Table 8.4).

The mining tools bear a strong resemblance to those found with the well-known "Copper Man" from the Chuquicamata mine, Chile (Bird 1979; see also Craddock et al. 2003). This strong resemblance of mining tools from sites thousands of kilometers away has been noted in other Prehispanic mining contexts (see, for example, Salazar et al. 2011, Chap. 7; Shimada 1994) suggesting that this toolkit was common throughout the Andes in antiquity.

Faunal remains were relatively few and were found mostly in Capa A and associated with postexploitation natural activity by small mammals. There were some seashell remains including *Spondylus* (Fig. 8.7c), well-known as a symbol of water and fertility in the Andes, found in Capa B, Gallery 1. Some of the *Spondylus* remains had drilled holes for use as adornment, while others appeared not to be worked. Also used as adornment were wood and bone beads that were likely part of necklaces as well.

A pair of intact cane panpipes were also found in Gallery 1 (Unit F, Capa B; Fig. 8.7a) as well as fragments of a bone flute (Fig. 8.7b). The panpipes are replicas of each other, each with seven well-worked cane tubes of diminishing size tied together with cotton string. The longest tube of each pair is wrapped in brown cotton string with a nondiagnostic design. The center five tubes of each pair have bottle gourd discs plugging the base of the tube so that it could be played. Because the fragile panpipes were found in a well-preserved condition (they were neither decayed, nor broken), they appear to have been carefully deposited and found in a primary context (the only primary context recovered thus far in excavations). A single gourd disc from a panpipe was submitted for AMS dating with results of AD 463–763 (calibrated, 2σ) placing them in the late EIP or early MH. The presence of these panpipes does not seem to constitute a unique anomaly, as a number of gourd disc plugs were encountered in excavation, suggesting that in other instances, panpipes were brought to the mine and played.

Discussion

Our findings reveal important evidence regarding the organization and evolution of mining in the Nasca region. Artifacts and features in Mina Primavera suggest that by at least 2,000 years ago miners were bringing food, textiles for transport, wood for tools, as well as offerings and items for ritual activity with them to the mine. The domestic refuse in excavations suggests that miners probably camped in the mine itself. In survey, a number of temporary Prehispanic encampments have been recorded consisting of small artificial terraces with hammerstones, flaked stone, and broken pots within a short walk of Mina Primavera (Eerkens et al. 2009). Though these encampments could have been used to stage mining activities in the region, they are very small compared to other temporary mining camps investigated in nearby valleys. Thus, we believe that most staging activities took place within Mina Primavera. Whether this pattern is limited to Mina Primavera is unknown since the site is unique in the region in terms of its size and preservation.

While in the mine, miners extracted hematite using tools made onsite with locally available siltstone and also with wood brought to the site, processed the extracted mineral into smaller fragments in large mortars on the mine floor, further processed the smaller fragments of hematite into a powder in the small mortars and then transported the ground mineral in plain textile bags to some external location, where it was likely processed further to be used as a pigment. Other findings at the mine shed light on our understanding of the chronology and evolution of mining within Mina

Primavera, ritual that took place within the mine, and the organization of mining throughout the mine's history. We discuss each of these below.

The Evolution of Mining at Mina Primavera

Even though mines are difficult to date since they represent the remains of an extractive industry (see Stöllner 2003) the sequence and chronology of exploitation at Mina Primavera is fairly clear. The earliest mining recorded thus far seems to have been during the Early Horizon as indicated by a ceramic fragment found in the back of Gallery 2. This is not surprising given that painted ceramics and textiles were important for Paracas, and hematite in the form of a pigment would have been a valuable resource (Fester and Cruellas 1934: 156; Petersen 1970). This ceramic fragment, however, was clearly not found in a primary context, and its location is representative of the movement of materials within the mine given that the earliest ceramic fragment was found near the back of the mine where we would have expected evidence for the latest exploitation. Even so, only one sherd can be securely dated to the Early Horizon and none of the radiocarbon dates indicate exploitation during this time period. Until more sherds dating to the Early Horizon are found, we are not confident in suggesting that significant mining took place before the EIP.

What is clear, however, is evidence for exploitation of the mine during Nasca 1 (Proto Nasca). Most Nasca 1 sherds that have been recovered have been found in two units (K and L) located just outside the entrance to the mine into Gallery 3 and toward the bottom of Capa B or in Capa C. While not in a primary context, the Nasca 1 sherds appear to have been quickly buried by subsequent mining activities during Early Nasca as Capa B in these units tends to contain evidence for Nasca 2, 3, and 4. Furthermore, the sherds found in the lowest stratum tend to exhibit relatively sharp breaks. In contrast, the few Nasca 1 sherds found inside the mine (that is, not at its entrance) are mixed with other debris in Capa B and are fairly eroded. This suggests they have been subjected to considerable postdepositional movement and are perhaps far removed from their original context. Nasca 1 sherds were not found in the units in the immediate vicinity of the entrance to Gallery 1 (Units D, I, and J) and the artifacts found in these units are limited to Early Nasca sherds and a single Wari fragment from a face-neck jar (in Unit D, Capa B).

This suggests that the earliest concerted effort to exploit the mine occurred during Nasca 1 and this took place in the western portion of the mine's entrance leading into Gallery 3 and possibly the western edges of Gallery 1. In contrast, Early Nasca miners appear to have intensified their exploitation and mined persistently in all three galleries. Gallery 2 appears to represent the latest exploitation of the mine. This is due to its location (one had to extract the hematite from Gallery 1 prior to Gallery 2), and the single radiocarbon date from Gallery 2 provides the latest date from Capa B with a range of AD 241–403. Most of Mina Primavera appears to have been mined over the course of several centuries.

As stated, there is no evidence for Middle or Late Nasca exploitation of Mina Primavera. It is possible that the majority of the high-quality hematite had been removed by this time, or that other sources of hematite were found and exploited. But it is also possible that broader changes in settlement affected the exploitation of Primavera. For example, Reindel and colleagues (Reindel 2009; Chap. 14) have shown that extensive desiccation in the region led to a resettlement during Late Nasca times into the upper portions of the CNR valleys. Much of the middle and lower valley settlements were abandoned during the latter part of the EIP and perhaps this played a part in the abandonment of Mina Primavera as well.

While mining at Mina Primavera might have stopped by Middle Nasca, the site does not seem to have been forgotten. Indeed, there is very clear evidence for a Wari presence in the mine during the Middle Horizon in the form of ceramics and probably panpipes (see below). This Wari presence does not necessarily indicate Wari exploitation, of course. In fact, we do not have clear evidence for Wari mining as the Wari materials are found toward the top of Capa B and the radiocarbon dates from Capa B materials (with the exception of the panpipes) fall within a relatively narrow range within the EIP. The question, of course, is what was Wari doing in the mine, especially when there is no evidence for Middle Horizon occupation in this part of the Central Nasca Region (see Reindel and Stöllner Chap.14)? We explore this question below.

The Symbolic Importance of Mina Primavera

Numerous studies have shown that mines and many quarries were sacred spaces in the Andes (Chap. 9; Dean 2010; Nuñez 1999, 2006; Chap. 3; Chap. 12; Shimada 1994; Chap. 13; Chap. 1; Vaughn et al. i.p.). Though we do not wish to draw a direct analogy between the ethnographic present, the Inca, and the pre-Incaic Andes, we offer two lines of artifactual evidence to suggest that Mina Primavera was symbolically important to those who exploited it: the presence of *Spondylus* and musical instruments in the mine.

Several fragments of *Spondylus* (most likely *Spondylus princeps*) broken in the mining debris were found in Gallery 1, Capa B, and several pieces of worked *Spondylus* that were parts of necklaces were also found in Gallery 1, Capa B. *Spondylus princeps*, of course, is found off the coast of Ecuador and was frequently used in rituals in the Andes including those related to water and fertility (see Cordy-Collins 2001; Glowacki 2005). The very presence of a symbolically charged item (especially the unworked *Spondylus*) in such a remote location demonstrates the symbolic and ritual importance of the mine (see, for example, Chap. 12; Shimada 1994). While the *Spondylus* remains have not been directly dated, they have generally been found in Capa B and associated with Early Nasca materials.

Additional archaeological evidence for the symbolic importance of Mina Primavera includes panpipes and fragments of bone flutes, both of which demonstrate that activities taking place included the playing of music. At least in the contemporary context, rituals within mines were often accompanied by the playing of music (Nash 1979; see also Chap. 1; Vaughn et al. i.p.). Again, we hesitate to make a direct historical analogy, but given the rich and extensive ethnographic and ethnohistoric evidence that demonstrates the symbolic importance of mines in the Andean landscape, we interpret the instruments to have been used in some kind of generalized ritual, perhaps related to propitiation, within the mine.

The radiocarbon date (AD 463–763, cal. 2σ) of the panpipes places them in the latter part of the EIP (during the Late Nasca phases) or the early part of the Middle Horizon within the CNR chronology. Several lines of evidence, however, suggest that the panpipes date to the Middle Horizon rather than the EIP. First, there is a complete lack of evidence for Late Nasca exploitation or even Late Nasca presence within the mine. Second, to our knowledge there has not been an example of cane panpipes found that date to the EIP. All Nasca panpipes found have been ceramic, most of these made by slip casting (Dawson 1964; Gruszczynska-Ziolkowska and Prusik 2000). Haeberli (1979) has suggested that panpipes in Nasca were ceramic until the latter part of the EIP and into the Middle Horizon when coastal people began to make cane panpipes. Obviously, this lack of evidence does not mean that Late Nasca people did not make cane panpipes, but so far the evidence simply does not indicate that they did. Third, while we lack evidence for Late Nasca exploitation or presence within the mine, we do have evidence for Middle Horizon presence in the form of Wari sherds. These are found in Gallery 1, always in Capa B, though usually near the top of Capa B, and they seem to be mostly elite artifacts. As a final line of evidence, there is indication in patterns of ceramic breakage and wear that the deposit below which the panpipes were found was regularly moved and disturbed throughout the mine's history. The panpipes, however, were found in a context that was undisturbed from the position in which they were laid indicating that they were deposited after the principal periods of activity in the mine.

Beyond artifacts related just to the activity of extracting and processing hematite within the site, the artifacts recovered from excavations at Mina Primavera clearly indicate the symbolic importance of the mine. Ritual was practiced within the mine from at least Early Nasca times, though it is impossible to reconstruct exactly what these rituals might have been. The presence of spondylus artifacts that most likely date to the EIP and musical instruments that appear to date to the Middle Horizon suggests that the mine retained symbolic importance through its exploitation and possible abandonment.

Wari presence may have been part of efforts to seek raw materials for ceramic production. It is well-known that Wari ceramic technology borrowed much from Nasca ceramic technology (Knobloch 1976, 2005; Menzel 1964), and this could have included pigment sources and processing. However, since it appears that Wari did not undertake much exploitation of the mine, perhaps they were there for other reasons. While empires co-opt resources, they also co-opt rituals, sacred landscapes, and symbolic capital (Schreiber 2005a) and Wari's interest in Mina Primavera may not have been just the hematite of the mine, but the mine itself as part of the sacred landscape of Nasca (see Schreiber 2005b).

The Organization of Mining in Nasca

The size of Mina Primavera suggests extensive Prehispanic mining operations. Estimates of the amount of hematite extracted from the mine (nearly 700 m^3,[4] or approximately 3700 metric tons at 5.3 g/cc; Vaughn et al. 2007) to create the cave make the mine one of the largest Prehispanic hematite mines in the New World, and equivalent in size to the San Ramon hematite mine in northern Chile (Salazar et al. 2011; Chap. 7). Because radiocarbon dates suggest that most exploitation occurred during the EIP, and within this period primarily during Proto and Early Nasca, most of the mine appears to have been exploited over the course of a few centuries (rather than what we had initially suggested of over 1,400 years). Thus, the previous estimate (Vaughn et al. 2007) of an average of ½ m^3 (2.65 metric tons) of hematite removed *per annum* grossly underestimates the actual rate of removal during the early part of the EIP.

Given the chronology of the site reconstructed here, what might this suggest about the organization of mining at Mina Primavera? We have previously suggested that the mining undertaken in Mina Primavera during Nasca times could have been done on an informal, part-time basis, perhaps by itinerant miners (Eerkens et al. 2009; Vaughn et al. 2007). Full exploitation of the mine over several centuries would not have required full-time mining; however, there appears to have been significant changes in organization in mining between Proto Nasca and Early Nasca. We draw on several lines of evidence to make this argument.

First, while there is substantial evidence for the exploitation of the mine during the Proto Nasca phase, this exploitation appears to be limited to the upper portions of Gallery 3 (at its entrance and its immediate interior), conforming with expectations of the presumed history of the mine, which would have moved inwards from an initial exterior exploitation. If there was any exploitation of Gallery 1 during this phase, it was probably limited to its western boundaries. There is no additional evidence for Nasca 1 exploitation elsewhere in the mine. In contrast, the artifactual and radiocarbon evidence suggests that almost the entire mine was exploited during Early Nasca.

Second, a closer analysis of the groundstone mortars preserved in the mine provides another line of evidence suggesting a change in the organization and intensity of mining at Mina Primavera. Thus far, there is no ceramic evidence to confirm Nasca 1 use of groundstone mortars, however, several mortars in or near Gallery 1 contained Capa C material (mining debris and lithic fragments) suggesting that they were in use during early exploitation of the mine, or at least that they were no longer used while mining was still being practiced. Even if hematite was processed during Proto Nasca, the evidence for it is fairly meager. There is no evidence for the processing of hematite in mortars in Gallery 3 within the mine, nor have any mortars outside the narrow entrance to Gallery 3 been recorded where there is the most

[4] Note that the mine is actually larger than this as the full extent of the mine's boundaries have yet to be determined.

extensive evidence for Nasca 1 exploitation. If mortars existed in this area of the site some would have been identified since large areas of bedrock suitable for making mortars are exposed. Thus, our working hypothesis is that whatever hematite was extracted during Nasca 1 was only minimally processed at the mine itself. The fact that there was minimal processing of hematite associated with the earliest exploitation of Mina Primavera is analogous to lack of evidence for processing in the San Ramon hematite mine in Chile documented by Salazar et al. (2011; Chap. 7).

In contrast, where there is evidence for Early Nasca exploitation, groundstone mortars are also present. This includes outside the entrance to Gallery 1, whose excavated unit produced only Early Nasca pottery and several groundstone mortars. This also includes the interiors of Galleries 1 and 2 where the ceramic assemblage is predominantly Early Nasca with some evidence for activity during the Middle Horizon.

In Gallery 1 the processing of hematite is small-scale, and appears to have been accomplished by individual miners. The same compact mortar station was used for the initial processing of large fragments of hematite as well as the more refined processing of hematite into powder. In contrast, the groundstone mortars in Gallery 2, the last gallery to be exploited, indicate a significant change in the *chaîne opératoire* leading to processed hematite. Specifically, the quantity of mortars, their size, and their spatial arrangement in Gallery 2 suggest that processing occurred on a larger scale and was undertaken as a group rather than by individuals. Here, the initial processing of large fragments of hematite occurred in a 5 m-wide complex of large mortars suitable for multiple individuals and located within the floor. Secondary refining took place in a separate location apart from this large mortar complex in small mortars located on large blocks of collapsed ceiling nearby. In at least one case, processing occurred within a three mortar complex where hematite was transformed from raw fragments to a fine powder in adjacent stations that could have been operated by multiple individuals at the same time. This evidence suggests a transition from "simple linear" to "complex simultaneous" task performance, implying profound transformations in the social organization of, and attitudes towards, labor (Wilk and Rathje 1982).

While the activities in the mine seem to have intensified during Early Nasca, there is no direct evidence that the mining was supervised in any way, though the change we have documented does imply some level of coordination for groups of miners to work together on a single task. Ethnohistoric sources are clear that the Inca supervised many mining operations (see Chap. 1). Full-time supervision would require permanent facilities to support administrators and supervisors (e.g., Chap. 9; Salazar et al. 2010, Chap. 12), yet there is no evidence for these kinds of permanent facilities at Mina Primavera.

Even so, evidence indicates an intensification of hematite processing during Early Nasca. Increased processing occurred during a time when activities within the mine began to include ritual as evidenced by the *Spondylus* remains encountered. Additionally, it is not trivial that this intensification occurred near the back of the mine where it was hot, lacked light and was probably full of dust that must have made for a rather unpleasant, if not dangerous (e.g., Murr 2009), experience.

Why was there a change in the way that hematite was processed at Mina Primavera? If this question is evaluated from the perspective of the archaeology of the region, intensification occurred during an "apogee" of Nasca when Cahuachi reached its peak of activity as a major regional ceremonial center. Activities there during this time included, among other things, ceramic production requiring quality hematite in addition to other mineral pigments. Among the most common colors on polychrome ceramics, black and various shades of red were made using forms of iron oxide including hematite (see Eerkens et al. n.d.; Vaughn et al. 2005). Part of the nature of Cahuachi's rise was its resident "theocratic authority" (Orefici 2006: 184) who are hypothesized to have been successful—during Early Nasca at least—in their rise because of their association with water and fertility concepts in a desiccated landscape (Kantner and Vaughn 2012). Furthermore, this association was most clearly materialized on polychrome ceramics (Kantner and Vaughn 2012; Vaughn 2009).

It is clear from our findings that Mina Primavera was not only an important economic resource, it was also symbolically important in the landscape of Nasca. The extraction of hematite from this sacred place must have been symbolically and ritually charged. Using the material from the mine to produce equally symbolically charged artifacts may have increased the need for the material laden with symbolism. Thus, one possibility explaining the change in processing was that there was an increased demand for hematite especially at the beginning of Early Nasca once potters began to incorporate colors requiring minerals such as hematite into their ceramic repertoire.

There is direct archaeological evidence for hematite use at Cahuachi. For example, Alfred Kroeber recovered several small quantities of ground hematite that he described as "paint" during his Captain Marshall Field Second Archaeological Expedition to Peru, 1926 and curated at the Field Museum in Chicago (Kroeber and Collier 1998). One particular sample analyzed (Eerkens et al. n.d.) is clearly hematite and comes from a grave (Aj-10; Kroeber and Collier 1998: 79) that contains, among other items, Nasca 2 and Nasca 3 ceramics. Stable iron isotope ratios in this pigment sample match those from hematite we have collected and analyzed from Mina Primavera. At the same time, hematite samples we have collected analyzed from other modern mines in the Nasca region display different isotopic signatures. Although our analyses with iron isotopes are on-going, this supports the notion that at least some of the hematite from Mina Primavera was transported to ceremonial centers such as Cahuachi.

Other evidence for hematite use at Cahuachi comes from extensive excavations at the site by Orefici and colleagues. Several cases of pigments including "red ochre" (hematite) from ceramic production contexts (Orefici and Drusini 2003: 70, 144) and funerary contexts (Orefici and Drusini 2003: 118) have been published by the excavators. Unpublished examples of hematite along with ceramic production paraphernalia (brushes, spatulas, pigment mortars, etc.) are also on display at the Antonini Museum in Nasca (Vaughn personal observation).

Although it clearly was used for ceramic production in Nasca (Petersen 2010: 9; Vaughn et al. 2005; Yacovleff and Muelle 1934), hematite also had other uses. Hematite was used in the Andes as part of offerings to the sea (Murúa 1925), as a flux in copper smelting (Shimada and Merkel 1991), as body paint (Llagostera et al. 2000; Salazar et al. 2011; Scalise and Di Prado 2006), and as a pigment for textiles, wooden objects, cosmetics, and murals (Bonavia 1959, 1985; Phipps 1989; Yacovleff and Muelle 1934). Indeed, at Cahuachi, hematite has also been found as part of ritual caches (Bachir Bacha and Llanos 2006: 66), rubbed on to trophy heads (Bachir Bacha and Llanos 2006: 68), deposited within painted polychrome vessels (Bachir Bacha and Llanos 2006: 68; Petersen 2010: 9), and painted on adobe walls (Orefici and Drusini 2003: 78; Rios Valladares 2007). However, among the many uses that hematite could have provided, painting ceramics was certainly important in Nasca.

If the acceleration of ceremonial activities and ceramic production in Nasca were even in part cause for the intensification of extraction and processing at Mina Primavera, it would not be the only documented case of this. Indeed, Bloxam (2011: 153) has suggested that "political and ideological change at key periods in history may provoke intensive production of a specific resource due to its symbolic association with an emerging religious cult." Following Bloxam we might have expected an intensification of activities at Mina Primavera, not to mention at other mines that provided raw materials for key Nasca ritual paraphernalia in the region during the height of the Nasca culture.

Conclusions

While work is continuing, excavations at Mina Primavera suggest a significant change in the way the mine was exploited over the course of its history from initial prospection to abandonment. We have documented here a shift from relatively small-scale exploitation to much more intensive mining with the advent of Early Nasca, and have shown that this shift in mining was contemporaneous with major societal changes occurring on the South Coast of Peru. Other documented activities that took place in the mine have included some unexpected activities such as ritual and the playing of music. Furthermore, evidence suggests that ritual activity within the mine may have continued even after exploitation of the mine slowed, or perhaps stopped, alluding to the importance of the mine not just as an economic resource, but as a sacred place on the landscape.

Of course, these conclusions are proposed as hypotheses suitable for testing in future work. Far more work needs to be done through excavations at the mine, evaluating the mineralogy and geochemistry of the hematite from the mine, and linking this hematite to other sites and artifacts in Nasca. Beyond these obvious directions for future research, what this work makes clear is that mining was an integral part of the economic, sociopolitical, and ritual landscape of Nasca. Furthermore, Mina Primavera must have been but one of many Prehispanic mines in the region. There were surely other hematite mines, but also mines of other important raw materials such as manganese, copper ores, and gold, not to mention clays and locally available stone such as chert. Surveys in the region by our team (Eerkens et al. 2009) and the team in Palpa (Reindel and Stöllner this volume; Stöllner 2009) have revealed a number of mines and quarries, but none as large or as well preserved as Mina Primavera. Unfortunately, we may never know the full story when it comes to mining in Nasca or more generally in the Andes. We need to accept that there must have been many mines that have been lost due to subsequent mining and other contemporary disturbances, making Mina Primavera a unique and important site.

Acknowlegments We thank the National Geographic Society for funding to Vaughn and the Social Sciences and Humanities Research Council of Canada for funding to Van Gijseghem to complete the fieldwork described in this paper. Many people have contributed to our fieldwork at Mina Primavera including Stefanie Bautista, Jessica Kaplan, Meghan Tierney, Jose Sereveleón, Luis Manuel González LaRosa, Craig Smith, John Kantner, Cynthia Hotujec, and Yoshiko Tsuji. We thank them for their efforts and for making the daily (and arduous) trek up to Mina Primavera! We also thank Rubén Garcia of the Ica INC and Susana Arcé of the Ica Museum for their assistance throughout this project. We greatly appreciate the comments of an anonymous reviewer who made helpful suggestions on an earlier draft of this paper.

References

Aldunate, S. C. d., Victoria, C. R., & Varinia, V. G. (2008). San Bartolo y Cobija: Testimonios de un modo de vida minero en las tierras altas y la costa de Atacama. *Estudios Atacameños, 35*, 97–118.

Bachir Bacha, A., & Llanos, O. (2006). El Gran Templo del centro ceremonial de Cahuachi, Nazca, Perú. *Dimensión Antropológica, 38*, 49–86.

Bird, J. B. (1979). The "Copper Man": A prehistoric miner and his tools from northern Chile. In E. P. Benson (Ed.), *Pre-Columbian metallurgy of South America* (pp. 105–132). Washington D.C: Dumbarton Oaks.

Bloxam, E. (2011). Ancient quarries in mind: pathways to a more accessible significance. *World Archaeology, 43*(2), 149–166.

Bonavia, D. (1959). *Una pintura mural de Pañamarca, Valle de Nepeña* (Arqueológicas, Vol. 5). Lima: Publicación del Instituto de Investigaciones Antropológicas, Museo Nascional de Antropología y Arqueología.

Bonavia, D. (1985). *Mural painting in ancient Peru*. Bloomington: University of Indiana Press.

Burger, R. L., & Matos Mendieta, R. (2002). Atalla: a center on the periphery of the Chavín Horizon. *Latin American Antiquity, 13*(2), 153–177.

Cook, A. (1984–1985). The Middle Horizon ceramic offerings from Conchopata. *Ñawpa Pacha, 22–23*, 49–90.

Cordy-Collins, A. (2001). Blood and moon priestesses: Spondylus shells in Moche ceremony. In E. Benson & A. Cook (Eds.), *Ritual human sacrifice in ancient Peru*. Austin: University of Texas Press.

Craddock, B. R., Cartwright, C. R., Craddock, P. T., & Wray, W. B. (2003). Hafted stone mining hammer from Chuquicamata, Chile. In P. Craddock & J. Lang (Eds.), *Mining and metal production through the ages* (pp. 52–68). London: The British Museum Press.

Dawson, L. E. (1964). Slip casting: a ceramic technique invented in ancient Peru. *Ñawpa Pacha, 2*, 107–112.

Dean, C. (2010). *A culture of stone: Inka perspectives on rock*. Durham: Duke University Press.

Edwards, M. J. (2010). *Archaeological investigations at Pataraya: A Wari outpost in the Nasca valley of Southern Peru*. Santa Barbara: University of California.

Eerkens, J. W., Vaughn, K. J., Barfod, G. H., Williams, P. R., & Lesher, C. E. (n.d.). *Tracing the geographic origin of iron oxides using stable iron isotope analysis: An exploration of pottery pigments from the Nasca region, Peru* (manuscript in authors' possession).

Eerkens, J. W., Vaughn, K. J., & Linares Grados, M. (2009). Prehispanic mining in the Southern Nasca region, Peru. *Antiquity, 83*, 738–750.

Fester, G. D., & Cruellas, J. (1934). Colorantes de Paracas. *Revista del Museo Nacional, 3*(1–2), 154–156.
Glowacki, M. (2005). Food of the Gods or mere mortals? Hallucinogenic spondylus and its interpretive implications for early Andean Society. *Antiquity, 79*, 257–268.
Gruszczynska-Ziolkowska, A., & Prusik, T. (2000). Tocanado los números. Las antaras nasquenses desde una perspectiva acústica. *Estudios Latinoamericanos, 20*, 99–110.
Haeberli, J. (1979). Twelve Nasca panpipes: a study. *Ethnomusicology, 23*(1), 57–74.
Kantner, J., & Vaughn, K. J. (2012). Pilgrimage as costly signal: religiously motivated cooperation in Chaco and Nasca. *Journal of Anthropological Archaeology, 31*, 66–82.
Knobloch, P. J. (1976). *A study of the Huarpa ceramic style of the Andean early intermediate period*. Binghamton: SUNY.
Knobloch, P. J. (2005). Monkey saw, monkey did: a stylization model for correlating Nasca and Wari chronology. *Andean Past, 7*, 111–134.
Kroeber, A. L., & Collier, D. (1998). *The Archaeology and Pottery of Nazca, Peru: Alfred L. Kroeber's 1926 Expedition*, edited by P. Carmichael. Walnut Creek, CA: AltaMira Press.
Lambers, K. (2006). *The geoglyphs of Palpa, Peru: Documentation, analysis, and interpretation*. Bonn: Forschungen zur Archäologie Aussereuropäischer Kulturen.
Lechtman, H. N. (1976). A metallurgical site survey in the Peruvian Andes. *Journal of Field Archaeology, 3*, 1–42.
Llagostera, A., Weisner, R., Castillo, G., Cervellino, M., & Antonieta Costa, M. (2000). El Complejo Huentelauquén bajo una perspectiva macroespacial y multidisciplinaria. *Contribución Arqueológica, 5*, 461–480.
Menzel, D. (1964). Style and time in the Middle Horizon. *Ñawpa Pacha, 2*, 1–106.
Menzel, D., Rowe, J. H., & Dawson, L. E. (1964). *The Paracas pottery of Ica: A study of style and time*. Berkeley: University of California Press.
Murr, L. E. (2009). Nanoparticulate materials in antiquity: the good, the bad and the ugly. *Materials Characterization, 60*(4), 261–270.
Murúa, M. d. (1925). *Historia de los Incas, reyes de Perú: Colección de libros y documentos referentes a la historia del Perú, Tomo V*. Sanmartí, Lima.
Nash, J. C. (1979). *We eat the mines and the mines eat us: Dependency and exploitation in Bolivian tin mines*. New York: Columbia University Press.
Núñez, L. (1999). Valoración minero-metalúrgica circumpuneña: menas y mineros para el Inka rey. *Estudios Atacameños, 18*, 177–221.
Núñez, L. (2006). La orientación minero-metalúrgica de la producción atacameña y sus relaciones fronterizas. In H. Lechtman (Ed.), *Esferas de interacción prehistóricas y fronteras nacionales modernas: los Andes surcentrales* (pp. 205–260). Lima: IEP-IAR.
Orefici, G. (2006). Reconstructing Nasca domestic and ceremonial life: A view from Cahuachi and Pueblo Viejo. In I. Shimada, H. Baba, K. Shinoda, & O. Masahiro (Eds.), *Nasca, wonder of the world. Messages etched on the desert floor. TBS, Tokyo* (pp. 176–184). Tokyo: TBS.
Orefici, G., & Drusini, A. (2003). *Nasca: Hipótesis y evidencias de su desarrollo cultural*. Brescia, Italy: Centro Italiano Studi E Ricerche Archeologiche Precolombiane.
Paul, A. (1990). *Paracas ritual attire: Symbols of authority in ancient Peru*. Norman: University of Oklahoma Press.
Petersen, G. (1970). Minería y metalurgia en el antiguo Perú. *Arqueológicas, 12*, 1–140.
Petersen, G. (2010). *Mining and metallurgy in ancient Perú (W. E. Brooks, Trans., Special Paper 467)*. Boulder, CO: The Geological Society of America.
Phipps, E. J. S. (1989). *Cahuachi textiles in the W. D. Strong collection: Cultural transition in the Nasca valley, Peru* (Unpublished Ph.D. dissertation). Department of Art History, Columbia University.
Proulx, D. A. (2006). *A sourcebook of Nasca ceramic iconography: Reading a culture through its art*. Iowa City: Iowa University Press.
Reimer, P. J., Baillie, M. G. L., Bard, e., Bayliss, A., Beck, J. W., Blackwell, P. G., et al. (2009). INTCAL09 and MARINE09 radiocarbon age calibration curves, 0-50,000 years cal bp, Radiocarbon, *51*, 1111–1150.

Reindel, M. (2009). Life at the edge of the desert – archaeological reconstruction of the settlement history in the valleys of Palpa, Peru. In M. Reindel & G. A. Wagner (Eds.), *New technologies for archaeology: Multidisciplinary investigations in Palpa and Nasca, Peru* (pp. 439–461). Berlin: Springer.

Reindel, M., & Isla Cuadrado, J. (2006). Culturas tempranas de la costa sur del Perú: sus evidencias en los valles de Palpa. *Boletín de Arqueología PUCP, 10*, 237–284.

Rios Valladares, P. D. (2007). *El Sector CAH 04, Y8—EXP 108: Las funciones de la Plaza al este de la Gran Pirámide en los complejos ceremoniales de Cahuachi (Nazca), Período Intermedio temprano*. Lima: Pontificia Universidad Católica del Perú.

Salazar, D., Castro, V., Michelow, J., Salinas, H., Figueroa, V., & Mille, B. (2010). Minería y metalurgia en la costa arrecia de la región de Antofagasta, norte de Chile. *Boletín del Museo Chileno de Arte Precolombino, 15*, 9–23.

Salazar, D., Jackson, D., Guendon, J. L., Salinas, H., Morata, D., Figueroa, V., et al. (2011). Early evidence (ca. 12,000 BP) for iron oxide mining on the Pacific coast of South America. *Current Anthropology, 52*(3), 63–475.

Scalise, R., & Di Prado, V. (2006). Early use of ochre in the Pampean region of Argentina. *Current Research in the Pleistocene, 23*, 66–68.

Schreiber, K. (2005a). Imperial agendas and local agency: Wari colonial strategies. In G. J. Stein (Ed.), *The archaeology of colonial encounters: Comparative perspectives* (pp. 237–262). Santa Fe: School of American Research Press.

Schreiber, K. (2005b). Sacred landscapes and imperial ideologies: The Wari empire in Sondondo, Peru. In K. J. Vaughn, D. E. Ogburn, & C. A. Conlee (Eds.), *The foundations of power in the prehispanic Andes* (pp. 131–150). Washington D.C: Archeological Papers of the American Anthropological Association.

Schreiber, K. J., & Lancho Rojas, J. (2003). *Irrigation and society in the Peruvian desert: The puquios of Nasca*. Lanham, MD: Lexington Books.

Shimada, I. (1985). Introduction. In Y. Masuda, I. Shimada, & C. Morris (Eds.), *Andean ecology and civilization: An interdisciplinary perspective on Andean ecological complementarity: Papers from Wenner-Gren Foundation for Anthropological Research Symposium no. 91* (pp. xi–xxxii). Tokyo: University of Tokyo Press.

Shimada, I. (1994). Pre-Hispanic metallurgy and mining in the Andes: Recent advances and future tasks. In A. K. Craig & R. C. West (Eds.), *In quest of mineral wealth: Aboriginal and colonial mining and metallurgy in Spanish America* (Louisiana State University Series: Geoscience and Man, Vol. 33, pp. 37–73). Baton Rouge: Geoscience Publications.

Shimada, I. (1998). Sicán metallurgy and its cross-craft relationships. *Boletín del Museo de Oro, 41*, 27–61.

Shimada, I. (2007). Introduction. In I. Shimada (Ed.), *Craft production in complex societies: Multicraft and producer perspectives* (pp. 1–21). Salt Lake City: University of Utah Press.

Shimada, I., & Merkel, J. F. (1991). Copper alloy metallurgy in ancient Peru. *Scientific American, 265*(80–86).

Silverman, H. (2002). Nasca Settlement and Society on the Hundreth Anniversary of Uhle's Discovery of the Nasca Style. In H. Silverman & W. H. Isbell (Eds.), *Andean archaeology I: Variations in sociopolitical organization* (pp. 121–157). New York: Kluwer Academic/Plenum Publishers.

Stöllner, T. (2003). Mining and economy: a discussion of spatial organisations and structures of early raw material exploitation. In T. Stöllner, G. Körlin, G. Steffens, & J. Cierny (Eds.), *Man and mining-Mensch und Bergbau. Studies in honour of Gerd Weisgerber on occasion of his 65th birthday*. Bochum: Der Anschnitt Beiheft 16.

Stöllner, T. (2009). Gold in southern Peru? Perspectives of research into mining archaeology. In M. Reindel & G. A. Wagner (Eds.), *New technologies for archaeology: Multidisciplinary investigations in Palpa and Nasca, Peru* (pp. 393–407). Berlin: Springer.

Tello, J. C., & Mejía Xesspe, T. (1979). *Paracas, segunda parte*. Lima: Universidad National Mayor de San Marcos and the Institute of Andean Research de Nueva York.

Unkel, I., Kromer, B., Reindel, M., Wacker, L., Wagner, G. A. (2007). A chronology of the pre-Columbian Paracas- and Nasca-culture in South Peru based on AMS-14C-dating, *Radiocarbon, 49*, 551–564.

Van Gijseghem, H. (2006). A frontier perspective on Paracas society and Nasca ethnogenesis. *Latin American Antiquity, 17*(4), 419–444.

Van Gijseghem, H., & Vaughn, K. J. (2008). Regional integration and the built environment in middle-range societies: Paracas and early Nasca houses and communities. *Journal of Anthropological Archaeology, 27*(1), 111–130.

Vaughn, K. J. (2009). *The ancient Andean village: Marcaya in prehispanic Nasca*. Tucson: University of Arizona Press.

Vaughn, K. J., Van Gijseghem, H., Linares Grados, M., & Eerkens, J. W. (still in in press). Minería de hematita en la costa sur del Perú: Investigaciones arqueológicas en la Mina Primavera. *Chungará, Revista de Antropología Chilena.*

Vaughn, K. J., Conlee, C. A., Neff, H., & Schreiber, K. (2006). Ceramic production in ancient Nasca: provenance analysis of pottery from the Early Nasca and Tiza cultures through INAA. *Journal of Archaeological Science, 33*, 681–689.

Vaughn, K. J., Eerkens, J. W., Linares Grados, M., & Edwards, M. J. (2007). Hematite mining in the ancient Americas: Mina Primavera, a 2,000 year old Peruvian mine. *JOM: The Journal of The Minerals, Metals & Materials Society, 59*(December), 21–25.

Vaughn, K. J., & Neff, H. (2004). Tracing the clay source of Nasca polychrome pottery: results from a preliminary raw material survey. *Journal of Archaeological Science, 31*(11), 1577–1586.

Vaughn, K. J., Neff, H., Conlee, C. A., & Schreiber, K. J. (2005). A compositional analysis of Nasca polychrome pigments: Implications for craft production on the prehispanic south coast of Peru. In R. J. Speakman & H. Neff (Eds.), *Laser ablation ICP-MS: A new frontier in archaeological characterization studies* (pp. 138–154). Albuquerque: University of New Mexico Press.

Vaughn, K. J., & Van Gijseghem, H. (2007). A compositional perspective on the origins of the Nasca cult. *Journal of Archaeological Science, 34*, 814–822.

Wilk, R., & Rathje, W. (1982). Household archaeology. *American Behavioral Scientist, 25*(6).

Yacovleff, E., & Muelle, J. C. (1934). Notas al trabajo colorantes de Paracas. *Revista del Museo Nacional, 3*(1–2), 154–168.

Part III
Metals

Chapter 9
Mining Under Inca Rule in North-Central Chile: The Los Infieles Mining Complex

Gabriel E. Cantarutti

Introduction

This chapter is based on the results of ongoing archaeological research conducted at the Inca Period (ca. AD 1450–1541) mining complex of Los Infieles, in north-central Chile. Through an archaeological survey, this research has documented more than 50 Inca Period sites, which form part of a large mining complex focused on the extraction of opaline silica and chrysocolla. Both resources were probably used as ornamental stones in the production of adornments, religious paraphernalia, and other prestige goods. Through the analysis of the data collected so far, this chapter presents an overview of the main sites and the organization of mining activities in the Los Infieles area during the Inca Period. Additionally, this research discusses the level of involvement of the Inca state in those mining operations.

By the early sixteenth century, the ruling elites of the Inca state (ca. AD 1400–1532) demanded (among other resources) the continuous procurement of rare minerals and metals across the Andes to support the various religious, political, and social institutions on which the organization of the empire relied (Carcedo and Vetter 2004; D'Altroy 2002; Earle 1994). The Incas generally used stone and metal objects to reward subject local leaders for political services and to integrate them into the imperial political structure (D'Altroy and Earle 1985). At the same time, metal and stone luxuries played not only a central role in materializing Inca state ideology (DeMarrais et al. 1996; Lechtman 2007) but were also critical to the practice of local religious institutions, and for enhancing the prestige of provincial elites (González 2004; González and Tarragó 2004). Goods made of gold, silver, and tin-bronze (copper–tin alloy) served as personal status markers, ritual paraphernalia, and as decorations in special buildings. Moreover, minerals—such as turquoise,

G.E. Cantarutti (✉)
Department of Anthropology, University of Illinois at Chicago,
MC 027, 1007 W. Harrison Street, Chicago, IL 60607, USA
e-mail: gcanta@uic.edu

N. Tripcevich and K.J. Vaughn (eds.), *Mining and Quarrying in the Ancient Andes*, Interdisciplinary Contributions to Archaeology, DOI 10.1007/978-1-4614-5200-3_9,

chrysocolla, or sodalite—were extensively used in the manufacture of personal adornments and inlays for other prestige goods (Carcedo and Vetter 2004; Núñez 1987), or deposited as offerings at sacred places (Berenguer 2004; Nielsen 1997; Pimentel 2009).

Mineral and metal wealth production in the Inca political economy has been largely studied through the lens of metal craft-production (Angiorama 2004; Earle 1994; González 2004; Howe and Petersenc 1994; Lechtman 1993, 2007; Rutledge and Gordon 1987). Based on the theoretical principles of "attached production" (Brumfiel and Earle 1987), scholars have suggested that elite control of the labor of skilled craft-workers was the most efficient way to control the production and circulation of luxury items, especially in the southern domains of the Inca Empire (Costin 1996; Earle 1994, 1997). Regardless of the efficiency of this form of wealth production control, its implementation does not exclude the utilization of additional forms of state control at earlier stages of wealth production, particularly during the initial extraction of critical ores and minerals (i.e., mining). In fact, documentary and archaeological evidence attests to the existence of Inca tributary control of gold and silver mining operations in the highlands of southern Peru and Bolivia (Berthelot 1986; Van Buren and Presta 2010).

The scarcity of systematic archaeological studies of Inca mining sites impedes a comparative understanding of how mining activities were organized, and the degree to which the Inca state was involved in the extractive operations of different mineral resources. However, the growing interest of scholars in studying mining centers across the Inca Empire is bringing new perspectives on the specific technologies, organizational means, and religious practices by which the Incas managed the extraction of minerals and metals for imperial ends (Bouysse-Cassagne 2005, 2008; Cruz 2009; Cruz and Absi 2008; Salazar 2002, 2005, 2008, Chap. 12; Van Buren and Mills 2005; Van Buren 2009; Van Buren and Presta 2010).

Mining Under Inca Rule in the Southern Quarter of the Empire

The prevailing views of Inca mining organization rely mostly on historical data from the gold mines of Carabaya and Chuquiabo (Berthelot 1986), and the famous silver mines of Potosí and Porco (Bouysse-Cassagne 2008; Cruz and Absi 2008; Platt et al. 2006; Van Buren and Presta 2010). According to historical sources (Berthelot 1986; Bouysse-Cassagne 2008), the Incas laid claim to all mineral resources, but took only the richest and most concentrated ore deposits for themselves, while leaving the more scattered and lower grade sources to the local communities and their elites. Following policies similar to those used in the annexation of cultivated fields and other resources, mineral deposits were claimed for the support of Inca state institutions, and for the coffers of individual ruling Incas. Local people, both men and women, pertaining to nearby groups were required to work in the mines on a rotating basis (*mita*), although resettled groups (*mitimaes*) could also be involved in the mining activities (Cruz and Absi 2008). The historical sources also indicate that while some mines were worked at the expense of the Incas, others were exploited at the expense of and for the benefit of local lords (Berthelot 1986).

Several differences have been observed in terms of workforce coordination, control, and extractive techniques involved in the exploitation of the Inca's and local lords' mines (Berthelot 1986). In the Inca gold mines of Carabaya, for example, the state controlled the number of workers to be employed on each occasion, managing the volume of production. The extractive tasks in these state mines also involved more complex mining techniques (excavation of galleries and channels) that required central coordination of the workforce. Moreover, the presence and work of the miners at the Carabaya state mines was overseen by guards, and state supervisors collected and recorded the amount of extracted minerals on a daily basis (Cieza de León 1996 [1554: Chap. 18]; de Polo Ondegardo 1940 [1561]; Sancho de la Hoz 1919 [1535]). In contrast, production at the local lords' gold mines operated at much smaller scales of production. In these cases, extractive techniques were simpler (e.g., gathering, troughs, and sieves), and the labor was conducted by a limited number of workers (Berthelot 1986).

While existing research has provided evidence of direct forms of Inca control of the exploitation of rich gold (Berthelot 1986) and most recently silver (Van Buren and Presta 2010) deposits, imperial involvement in the extraction of other mineral resources (e.g., copper ores, tin) is far less understood. Studies in the Calchaquí Valley of northwestern Argentina have suggested that the Inca state was not directly involved in controlling the mining and primary smelting of metallic copper ores (Earle 1994). According to this model, the initial mining and processing activities were left in the hands of local households, which produced copper ingots as tribute to the state. It has been argued that instead of investing in the coordination and supervision of mining activities, the Incas sought to centralize control of the production of tin-copper alloys and the manufacture of metal sumptuary items at state administrative settlements (Earle 1994). Although insightful, it is important to note that the Calachaquí copper production model is based on evidence recovered from an imperial administrative facility and a local residential site, but it does not include the study of Inca mining sites.

Investigations at the mining complexes of San José del Abra and Conchi Viejo in Atacama, Chile, support the idea that the Inca state was closely involved in the extraction of turquoise, chrysocolla, and pseudomalachite (Núñez 1999; Salazar 2008). When pre-Inca and Inca Period operations are compared, a significant growth in the scale of production is apparent at both mining complexes. Moreover, it is clear that the number of workers at the mines increased during Inca rule, and that they spent more time in mining-related activities at the complexes. Although most of the constructions and material remains at the sites reflect local styles, the presence and control of the state is observed in a number of critical aspects (Salazar 2008 and Chap. 12). These include the use of Inca style pottery in the preparation and serving of celebratory foods; the introduction of state religious institutions through the construction of ceremonial platforms near the mining sites; the positioning and use of storerooms with Inca style features; and the development of a transportation infrastructure of roads and facilities for the effective functioning of the mining complexes.

Fig. 9.1 Location of the Los Infieles area in relation to the Elqui Valley

The above antecedents illustrate that there were important and identifiable variations in how the Inca state was involved in mining activities across the empire. Nevertheless, further research is needed to determine whether the strategies, technologies, and organizational means inferred in these studies can be generalized to the extraction of specific mineral resources (e.g., gold, silver, copper, tin, precious stones) and/or to mining localities located in other regions of the empire.

Geographic and Prehistoric Background of the Elqui Valley

The Los Infieles mining complex is located in the Coquimbo region, in north-central Chile, approximately 40 aerial km northeast of the city of La Serena (Fig. 9.1). This area is characterized by a semiarid climate with annual mean temperatures close to 15°C. Annual precipitation is around 100 mm and rains are limited to the winter months. Rain periods alternate with drought intervals that intensify under the influence of the "El Niño/La Niña" (ENSO) event (Romero et al. 1988). The nearest permanent watercourse to the Los Infieles area is the east–west running Elqui River. This river irrigates a fertile valley dominated by agricultural activities, currently centered on fruit production. The region to the north of the Elqui Valley, where the

Los Infieles area is located, comprises unirrigated lands of rugged topography between 700 and 1,900 masl, where people principally work on goat herding, horticulture, and mining activities (Castillo 2003). In this area, the local vegetation is mostly composed of scattered low-shrub and cacti communities.

At the time of the Inca conquest, the Elqui Valley was occupied by groups that formed part of the Chilean Diaguita Culture (ca. AD 900–1541). Diaguita villages and hamlets were scattered across the valley bottoms and along the coastline, where groups practiced a mixed economy based on agricultural and marine resources (Ampuero 1989; Rosado and Vernacchio 1998). There is no evidence at pre-Inca Diaguita sites of social stratification or of a regional political hierarchy. In fact, current research suggests that the pre-Inca Diaguitas had a segmentary social structure comparable to that of tribal societies (Hidalgo 1989; Troncoso 1998). It is also accepted that the Elqui region was largely autonomous during pre-Inca times in terms of its cultural development (Ampuero and Hidalgo 1975; Ampuero 2007). Small-scale exchange and ideological interaction certainly occurred during this period, but cultural contacts appear to have been largely limited to groups located in north-central Chile and northwest Argentina. The development of the Diaguita culture is divided into three cultural phases that cover pre-Inca (Diaguita I and II) to Inca times (Diaguita III), and each phase is associated with diagnostic ceramic assemblages (Ampuero 1978, 1994; Cantarutti and Solervicens 2005). At the beginning of the Spanish conquest (ca. 1540), the population of the Elqui Valley was probably around 6,000 people (Hidalgo 1972).

Compared with earlier cultural traditions of the Elqui Valley, the Diaguita culture shows an increase in the quantity and variety of copper or bronze objects (Ampuero 1994). Besides the presence of metal adornments, the Diaguita sites provide evidence for utilitarian metal tools, including fishing hooks and chisels. The distinctive local Diaguita style of metal adornments and the recovery of utilitarian metal tools, both from graves and domestic contexts, support the idea that most of these artifacts are the products of a local, but small scale mining-metallurgical industry (Ampuero 1994). Adornments such as beads and pendants made of copper ores (possibly malachite, turquoise, and chrysocolla) are also relatively common in Diaguita tombs.

The Elqui Valley was incorporated into the Inca Empire during the late 1400s, most probably during the reign of Tupac Inca Yupanqui (Silva 1985). The influence of the Inca state in the region has been well documented in locally produced Inca-style pottery and in local-style ceramics that display mixtures of Diaguita and Inca features (Ampuero 1989, 1994; Cantarutti 2002, Cantarutti and Mera 2005; Cornely 1956; Rowe 1950; González 1995). Furthermore, a small number of facilities with Inca-style architecture and ceramics have been recorded in the region (Iribarren 1978; Stehberg 1995). These include at least four mountain shrines where Inca-style metal figurines, along with other offerings, have been recovered (Iribarren 1962; Krahl and González 1966; Tierney 2001). The most elaborate Inca Period burials are concentrated in the lower course of the Elqui Valley, in sites such as Altovalsol (Cornely 1956) and Fundo Coquimbo (Ampuero 1969). In fact, based on the materials recovered from these sites and information contained in sixteenth century chronicles, scholars have proposed that Inca representatives may have resided at them (Ampuero and Hidalgo 1975).

The Spanish captain Mariño de Lobera (1865 [1595]: 78) specifically notes that an Inca governor ruled north-central Chile from a settlement located in the lower Elqui Valley. Importantly, Mariño de Lovera also states that the Incas had a gold smelting facility at this site, where they also oversaw the production of "turquoise" and "crystal" (quartz?) craftworks. Mariño and other chroniclers indicate that the Inca provinces of Chile periodically sent a vast amount of tribute to Cuzco in the form of round gold ingots and nuggets (de Bibar 1966 [1558]: 76; de Góngora Marmolejo 1862 [1575]: 3; Mariño de Lovera 1865 [1595]: 21). Other early colonial writers support these observations and emphasize that the southern domains of the Inca Empire played an important role in providing the state with minerals and metals (de Betanzos 1996 [1557]; Ramírez de Velazco 1587, in Angiorama 2004; Sarmiento de Gamboa 2007 [1572]).

Although it is widely accepted that one of the primary motivations for the Inca conquest of Chile's southernmost territories was to establish control over mineral-rich territories (D'Altroy 2002; Llagostera 1976; Raffino 1981; Stehberg 1995), research on undisturbed Inca Period mining contexts in this region has been scarce. Nevertheless, since Iribarren's (1962) first description of the Los Infieles sites, it has been speculated that this particular area represented an important mining district during Inca times (Iribarren 1978; Ampuero 1989). This view was supported by Castillo (2007), who through a reconnaissance work established that the Los Infieles area contained many more mining sites than previously documented. Most recently, I completed a full coverage archaeological survey in this area. The data collected through this research indicate that the Los Infieles area was occupied by an Inca Period mining community involved in extractive, administrative, domestic, and ceremonial activities, which gave life to a large mining complex.

The Los Infieles Mining Complex

The Los Infieles area is largely defined by the drainage basin of a narrow dry ravine known as Quebrada Las Pircas. The entire basin is enclosed by two parallel rocky ranges of hills from which numerous steep ridges and other small ravines descend, creating a very rugged landscape (Fig. 9.2). In 2010, a full coverage archaeological survey of 50 km^2 was conducted in the Los Infieles area and 198 sites were documented. Surface ceramics and architecture were used to date 56 of the sites to the Inca Period, most of which correspond to mines and groups of stone structures.

Despite the fact that the survey confirmed the existence of pre-Inca sites in the Los Infieles area, none of these settlements was clearly related to mines or included surface remains that could be associated with mineral processing activities (e.g., evidence of mineral crushing and sorting, lithic debris resulting from different stages of lapidary work). Although the existence of pre-Inca occupations at the mining sites cannot be ruled out completely without excavation, the overwhelming presence of Inca Period materials at the mining sites indicates that the Los Infieles area only became the focus of a large and intense mining occupation during Inca times. In

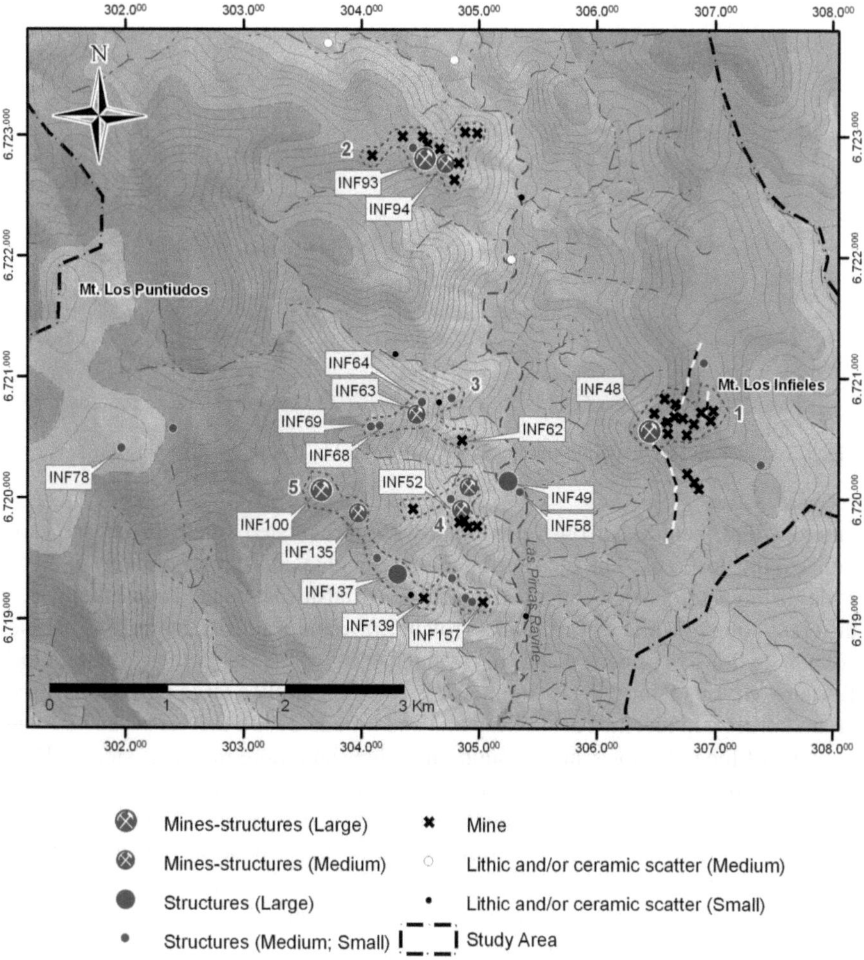

Fig. 9.2 Map of the study area showing the five mining clusters contained in the Los Infieles mining complex and sites mentioned in the text

contrast with other rich mineral localities exploited in the southernmost Inca territories (Iribarren 1973; González and Westfall 2008), the sites within the Los Infieles area remain in a good state of preservation. Although the sites have been affected by modern goat herding and some limited looting has occurred, they remain largely intact and unaltered by industrial mining operations.

The Mines

In accordance with the geologic configuration of the area, the distribution of the mines is associated with the presence of dioritic, granodioritic, and monzodioritic

Fig. 9.3 Mines and spoil mounds on the western slope of Mt. Los Infieles (mining cluster 1)

outcrops that include narrow veins filled with precipitated silicas in the cracks or faults of the host rock. The Inca Period miners sought veins (approximately 5–30 cm wide) in the outcrops that contained opaque and translucent banded opaline silicas of different colors (white, green, pink, brown, gray, and reddish), as well as chrysocolla, a hydrated copper silicate of an intense light blue to greenish-blue color[1]. Because Mt. Los Puntiudos (1,745 masl) is almost entirely a large mass of intrusive granodioritic and dioritic rocks (Emparan and Pineda 2000), the veins can be found in many of the outcrops scattered along its ridges and more rarely its slopes. Thus, the distribution of mines in Mt. Los Puntiudos follows a pattern of scattered clusters of sites that are found within a limited range of altitude, between 900 and 1,470 masl. In contrast, the veins in Mt. Los Infieles can be only found at higher elevations (between 1,200 and 1,350 masl), where intrusive monzodioritic rocks outcropped (Emparan and Pineda 1999). Accordingly, the mines and their associated constructions at Mt. Los Infieles follow a nucleated pattern.

The mines at Mt. Los Infieles (site INF48) were exploited as open-cut trenches with widths ranging between 3 and 6 m. Lengths are highly variable and difficult to measure because many of the open pits, particularly those that are closer to the hill summit, form discontinuous rows up to 130 m long (Fig. 9.3). Although the borders of the trenches are considerably eroded and filled with granular materials from the meteorized monzodioritic host rock, it is evident that the mines correspond to superficial workings that only occasionally reach a depth of four meters. In general, the presence of minerals on the surface of the mines and their periphery is scarce. However, most of what was found was bluish chrysocolla, and occasionally opaline

[1] In a strict sense, both chrysocolla ($(Cu,Al)_2H_2Si_2O_5(OH)_4 \cdot nH_2O$) and opaline silica ($SiO_2 \cdot nH_2O$) are not minerals, since they do not have a regular arrangement of atoms and a distinct crystal structure. Although it would be more appropriate to call them "mineraloids," I have opted to use the term mineral, which I define here as a solid aggregate that has been mined and from which something of value has been extracted.

Fig. 9.4 View of mine INF52e (mining cluster 4) from above. Open-cut trench excavated on an outcrop to exploit a chrysocolla vein

silica, phyrophyllite (aluminum silicate hydroxide), quartz, hematite (iron oxide), and very small amounts of malachite (hydrous copper carbonate).

The mines in the clusters of sites at Mt. Los Puntiudos also adopt the form of open-cut trenches or open-pits of alveolar shape. Despite the fact that chrysocolla was extracted in the mines of Mt. Los Puntiudos, the proportion of opaline silica exploited was higher than that at Mt. Los Infieles. Some of the trench-type mines consist of a long axis crossed by smaller perpendicular trenches, adopting the form of a series of "T"-shaped channels (e.g., mine INF93, and in part mines INF63, INF94, and INF100). Other mines take the form of large discontinuous "cracks" composed by two or more open-cut trenches (e.g., INF62, INF139). The mine INF52e is particularly notable because it was exploited by breaking the rocks both horizontally and vertically to take advantage of the steep slope where it is located. This open-cut trench is approximately 1–1.8 m wide, by 25 m long, reaching a maximum depth of 10 m (Fig. 9.4). The only Prehispanic mine in the entire study area that was worked through a nearly vertical shaft is the one included at the site INF135. The shaft is narrow (ca. 0.80 m) and at least six meters long until a point at which it is blocked by the influx of sediments and rocks. Apparently, it is connected with another shaft that is shorter and inclined, but currently blocked.

Fig. 9.5 Andesite stone mauls found on the spoil mound of mine INF52e (mining cluster 4)

Almost the only artifacts observed on the pits and spoil mounds, both in Mt. Los Infieles and Mt. Los Puntiudos, were complete, nearly complete, or broken andesite and granodiorite stone mauls, as well as flakes resulting from the use and breakage of these instruments (Fig. 9.5). Both andesite and granodiorites stones are materials available in the Qda La Pircas Basin. Some of the mauls show little modifications and others present shallow encircling grooves for hafting. The mineral debris of opaline silica and chrysocolla at the spoil mounds suggests that primary crushing and selection of mineral took place in different spots, right beside the pits. The mines are included within five clusters of Inca Period sites across the study area (Fig. 9.2).

Mining Cluster 1

The first cluster of mines and structures defined here is in fact a large single site (INF48) that extends along the western slope and summit area of Mt. Los Infieles (Fig. 9.3). This site was called "Los Infieles" by Iribarren (1962), who established that it dated to the Inca Period and provided a general description of its structures, mines and artifacts. Three decades later, Stehberg (1995) produced a more accurate description and map of the larger structures. Stehberg also excavated a room (3 m × 2 m) inside the largest building of the site (S-2). The excavation revealed an Inca Period component with no evidence of earlier occupations. Among the artifacts recovered from the excavation were several sherds of local Inca style *aríbalos* (long-necked jars), shallow plates, large local decorated storage vessels, and other fragments of Diaguita III (Diaguita-Inca phase) style pottery. In addition, a small amount of lithic debris, a projectile point, camelid bones, fish bones, and a probable musical instrument made from a sea lion bone (*Otaria* sp.) (Stehberg 1995; Stehberg and Cabeza 1991) were recovered.

The archaeological survey conducted in 2010 revealed that the site INF48 includes 27 ruined stone structures and two open flat spaces cleared of rocks located on the hill summit (Fig. 9.6). As it has been described by Stehberg (1995), the two largest structures of the site (S-1 and S-2) reflect provincial Inca style planning (Figs. 9.7 and 9.8). The structure S-1 is composed of a series of rectangular rooms that might have been used for storage functions. During our survey, fragments of large wide-mouth storage vessels of local Diaguita style were recorded inside these rectangular rooms. Both inside and outside the structure S-2, there were considerable quantities of ceramic sherds from serving, cooking, and storage vessels of local Inca and Diaguita III style, along with a small percentage of foreign style fragments (Fig. 9.9). Around the building we also observed grinding stones, pestles, and small flakes associated with the production and/or sharpening of bifacial instruments. The surface materials, as well as the artifacts recovered by Stehberg, suggest that the structure S-2 was used for residential and domestic activities.

In addition to the larger structures, there are 16 smaller constructions, most of which are located close to the mine pits and trenches. Most of these dry stone structures are subcircular in shape and with total areas that are less than 50 m^2. Mineral debris observed inside of them indicates that they were clearly related to mineral processing activities. For example, two of the rooms in the middle-size structure S-7 contained significant amounts of sorted chrysocolla cobbles (70–130 mm) and pebbles (<16 mm), which suggests that the rooms might have been used for secondary steps of crushing, selecting, and stocking[2] minerals.

On the very summit of Mt. Los Infieles there is a compound of features and structures that might have had a ceremonial function. According to oral accounts collected by Castillo (2007), a camelid metal figurine was supposedly found buried on a quadrangular structure (platform?) in this area of the site. Although it is not possible to verify this account, one of the structures recorded in the summit area was a significant accumulation of rocks (S-27, diameter of 4 m approximately) that could have served as a sort of landmark and/or as a feature with religious significance. Individual large heaps of stones with diameters ranging between 4 and 10 m were also recorded in other four mining sites of the Los Infieles area (INF93, INF63, INF52, and INF100). To a certain extent, these heaps of rocks resemble the well-known *apachetas*, that is, sacred cairns that are found along roads and mountain passes, where people prayed and asked for protection during their travels (Galdames 1990; Pimentel 2009; Vitry 2002). In the sites of the Los Infieles area the cairns are always built on top of outcrops and close to the mines, suggesting a possible practice of ceremonial activities related to the mining operations. Probably most of the rocks in the piles were part of the wastes discarded from the mines. Nevertheless, instead of been dumped downslope to the spoil mounds the rocks were placed upslope on top of an outcrop.

[2] Based on Pigott (1998), I use the concept of "stocking" to describe the action of piling and temporarily storing sorted minerals on open areas or in storage units near the mines or at mining camps, before the selected raw minerals are transported to smelting facilities, lapidary workshops, or other sites.

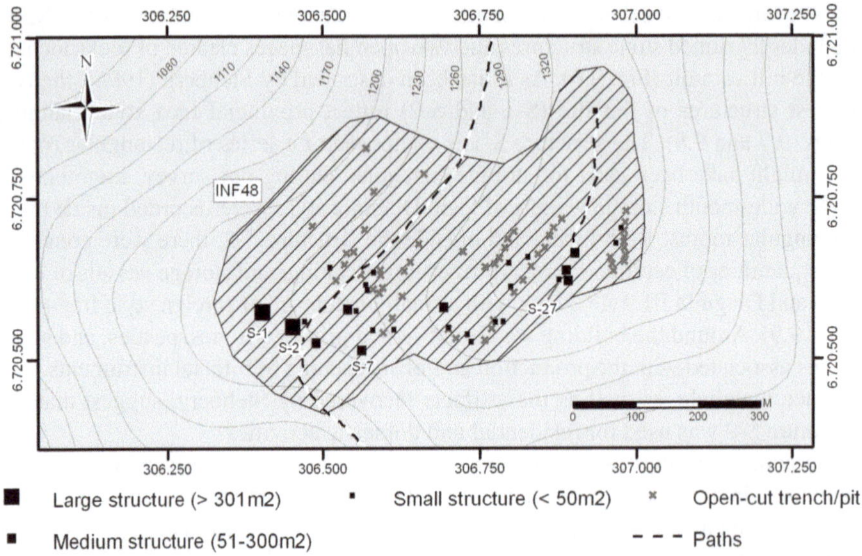

Fig. 9.6 Mines, structures, and paths in the mining cluster 1 on Mt. Los Infieles

Fig. 9.7 Provincial Inca style buildings (S-1 and S-2) at site INF48, west slope of Mt. Los Infieles

Mining Cluster 2

The second mining cluster includes 12 sites that are located on a ridge in the north-eastern extreme of Mt. Los Puntiudos, approximately 200 m above the stream bed of Quebrada Las Pircas (Fig. 9.10). The site INF94 includes three structures, two of which are ruined rectangular structures (S-2 and S-3), together with a smaller and simpler semicircular construction (S-1). The structure S-2 includes a well

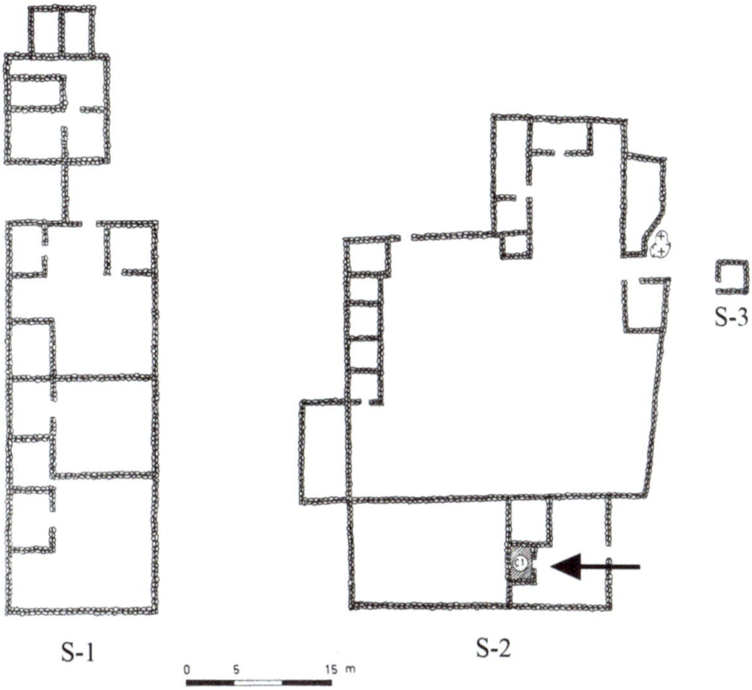

Fig. 9.8 Plan of Provincial Inca style buildings S-1 and S-2, at site INF48 on Mt. Los Infieles (according to Stehberg 1995: 130). The *arrow* points to the room excavated by Stehberg

Fig. 9.9 Local Inca and Diaguita style sherds found inside building S-2, at site INF48 (mining cluster 1). *1* and *6* fragments of arı́balos. *2*, *3* and *5* fragments of shallow plates. *4* possible fragment of "duck-shape" jar

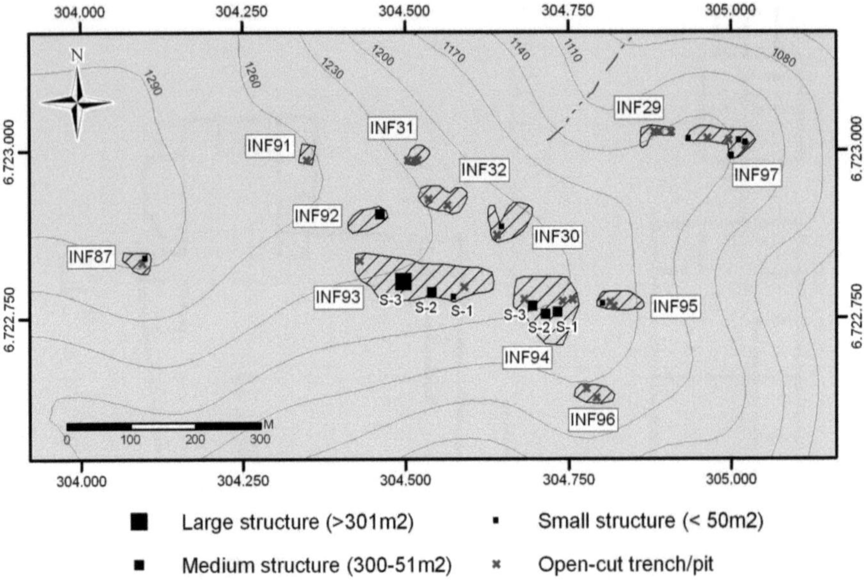

Fig. 9.10 Mines and structures in the mining cluster 2 on the northeastern slopes of Mt. Los Puntiudos

constructed but unfortunately looted circular room, which was full of small chrysocolla fragments. The room probably served as a storage unit for collecting sorted mineral from the different nearby mines.

Not far from the site INF94 is the site INF93, which includes three structures. The largest is a trapezoidal enclosure of ruined walls (S-3; ca. 35 m×25 m), in which several pestles and flat mortars (crushing slabs), along with small chrysocolla and opaline silica fragments were recorded. Within the same site, the structure S-1 corresponds to a large heap of granodioritic rocks placed on top of a natural boulder. As in the case of the heap of stones documented in the site INF48, the function of this structure is uncertain, but it may have served as a landmark of religious significance.

The higher frequencies of pottery were recorded at the sites INF92 and INF93, although scattered sherds were also found at sites INF94, INF95, INF31, and INF97. The diagnostic fragments correspond to local Inca style *aríbalos* or restricted vessels used for storing or serving liquids. Fragments of Inca style shallow plates, Diaguita III bowls, and cooking pots were also observed, but in lower frequencies.

The nature of the larger structures and the distribution of the surface materials associated with them suggest that while INF93 could have been involved in mineral crushing and sorting activities, INF92 and INF94 might have served as residential and storage areas. These three sites seem to constitute a mining camp associated with the maintenance of the people who worked at the nearby mines, which also included specific constructions that served in the final stages of mineral selection (S-3, INF93) and storage (S-2, INF94).

Mining Cluster 3

The third mining cluster is composed of seven sites distributed along a steep ridge, in the central-eastern section of Mt. Los Puntiudos (Fig. 9.2). Only two of the sites include mines (INF62 and INF63), of which the largest is contained at INF63. About 40 m downslope the large open-cut trench (ca. 100 m long) of this site, there is a large heap of rocks (S-1) of unknown function. Not far from this feature there is a medium-sized elliptical structure (S-2) where we found scattered ceramic fragments of food serving vessels (local Inca style shallow plates) and restricted containers. Upslope from the mine, we recorded a series of small terraces and structures that are probably related to the hand sorting and stocking of selected minerals. Within this last area we also found pottery fragments belonging to local Inca style *aríbalos*, large wide-mouth storage vessels of local Diaguita style, and smaller restricted containers. Near the northern extreme of the mining trench, we found another group of small structures that form the site INF64. Judging from the significant amount of mineral debris spread across this site, I believe that it was probably used for the secondary crushing, sorting and stocking of opaline silica and chrysocolla.

Another two sites, INF68 and INF69, were recorded upslope of the large mine. Judging from the location of these sites, more than 100 m upslope of the mine and its working areas, it is difficult to think that they could have served as residential camps for the miners. Among other alternatives or complementary functions, the sites could have served as checkpoints for overseeing the circulation of workers and the access to a shrine (INF78) that could be reached by heading upwards through this ridge.

Mining Cluster 4

This mining cluster is organized around the large site INF52, where we recorded several constructions and the impressive open-cut mine INF52e (Fig. 9.11). In the northeastern extreme of the site there is a possible ceremonial cairn (S-2) on top of an outcrop, which is between two medium-sized structures (S-1 and S-3), each with a front terrace facing to the east. Structure S-1 is a subrectangular building divided into a row of quadrangular rooms (Fig. 9.12). Its northern room showed evidence of what appears to be a midden deposit with burned bones and ceramic remains in an exposed matrix of ashy sediments. In addition, a small subrectangular room attached to the southeastern corner of S-1was filled with small fragments of opaline silica and chrysocolla, suggesting that it served as a mineral storage room (similar in function to that observed in the structure S-2 of site INF94, at mining cluster 2).

The front terraces of S-1 and S-3 contained numerous ceramic fragments belonging to local Inca style *aríbalos*, shallow plates, and large Diaguita storage vessels, together with flat mortars, pestles, and fine crushed minerals. Certainly, different activities took place around structures S-1, S-2, and S-3, such as storage, cooking, and serving of food for domestic and possibly also ceremonial activities. At the same time, the terraces were probably used to crush, sort, and stock minerals. Other structures within the site INF52 showed evidence of secondary crushing and mineral sorting

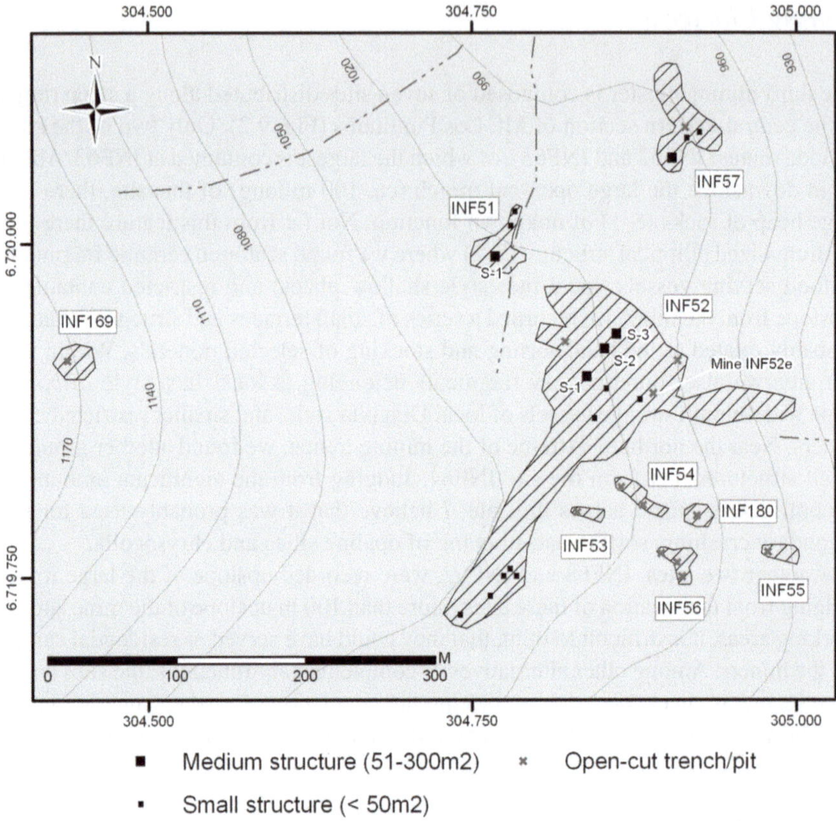

Fig. 9.11 Mines and structures in mining cluster 4 on the central-eastern slopes of Mt. Los Puntiudos

Fig. 9.12 Front terrace and linear multiroomed structure (S-1) at site INF52 (mining cluster 4)

activities (i.e., small size mineral debris associated with handheld hammerstones). These constructions were probably also related to the exploitation of the other five mines (INF53 to INF56, and INF180[3]) located near the site INF52.

The largest structure within this mining cluster was recorded at the site INF51. It is a subrectangular building (S-1) (20 m×8 m) where we found a few stone mauls, mineral debris, numerous ceramic sherds of local Inca style *aríbalos*, as well as fragments of Diaguita III style cooking pots, restricted containers, and bowls. This structure might have served as a camp for the miners that worked in the mining cluster. Alternatively, the site could have served as a logistical facility in which food and tool supplies were readily available for the nearby sites, or even as a checkpoint to control the circulation of workers and minerals to the mines of this and the fifth mining cluster. Just like in other sites of the Los Infieles area, excavations are needed to determine the function of this structure.

Mining Cluster 5

The fifth mining cluster is composed of ten sites, most of which are distributed along a steep ridge that descends from Mt. Los Puntiudos to Quebrada Las Pircas (Fig. 9.2). Four mines are included in this cluster (INF100, INF135, INF139, and INF157), the largest of which is related to a series of structures at the site INF100.

The mine of the site INF100 is located on top of a rough and conspicuous hump, where the miners excavated a long open-cut pit (ca. 50 m×1.8 m). On the very summit of the rocky hump there is also a large accumulation of stones that may have served a ceremonial purpose and a couple of small subrectangular structures, which seem to have been used for crushing and sorting minerals (chrysocolla and opaline silica). Additionally, on a relatively flat area to the west of the mine there is a series of larger subrectangular structures, including a quadrangular enclosure of ruined walls (ca. 22 m×18 m) (similar to that recorded in the second mining cluster; INF93, S-3). Probably, this compound provided temporary residential buildings and storage rooms, in addition to areas of mineral selection activities that complemented those conducted near the mine.

Compared to the other sites of this fifth mining cluster, the site INF137 stands out as a relatively large camp. The site is defined by a large elliptical structure (ca. 35 m×25 m) with collapsed walls, which includes a couple of small rooms and a charcoal kiln built by modern herders. Inside and outside of the elliptical enclosure, we found ceramic fragments that belonged to local Inca style *aríbalos*, shallow plates, large Diaguita storage vessels and other restricted containers. We also recovered a minor proportion of lithic debitage of andesitic and siliceous materials. This site was probably an important temporary residential camp and a source of supplies, especially for the miners who worked at the mines of the sites INF135 and INF139.

[3] Two late twentieth century mining test pits and a modern inclined mine shaft were detected within the limits of the site INF52, but they are not shown in the map of the fourth mining cluster.

Other Sites Recorded in the Los Infieles Area

In addition to the sites contained within the five mining clusters described above, there are three other sites, INF49, INF58 and INF78, which are relevant to understanding mining activities in the Los Infieles area during the Inca Period.

The sites INF49 and INF58 are located on the margins of Quebrada Las Pircas (Fig. 9.2). Both sites define an area that contains a mixture of historic and ruined provincial Inca buildings. Inca Period ceramic sherds were found at the two sites, but they were far more numerous along the site INF49. The ceramic sherds at INF49 included fragments of local Inca *aríbalos*, shallow plates with zoomorphic handles ("duck-shape"), and large Diaguita storage containers. Interestingly, one of the modified structures of the site contained an unusual concentration of foreign (and thus far unidentified) style fragments of restricted containers. Chrysocolla and opaline silica fragments were also recorded at the site but in small quantities. We also found andesitic lithic debris, some of which might be associated with the manufacture and repairing of stone mauls. The bed of Quebrada Las Pircas is a good source of andesite cobbles (Emparan and Pineda 1999) and therefore, the site INF49, as well as other small Inca Period sites located upstream could be related to the procurement and manufacture of andesite stone mauls. In any case, the location of the site INF49, on a flat plain between the mining clusters on Mt. Los Puntiudos and Mt. Los Infieles, as well as the evidence recorded at site, suggest that it could have served as an important mining camp for supporting and coordinating extractive activities, especially those conducted in the nearer mining clusters (third and fourth).

The site INF78 is known in the existing literature as the shrine of "Los Puntiudos" (Beorchia 1985; Stehberg 1995). It is located on the lowest of the three summits (1720 masl) of Mt. Los Puntiudos (Fig. 9.2) and includes two structures. One of these constructions is a large and low platform (ca.13 m × 6.5 m) that unfortunately is severely looted. No Prehispanic artifacts were found around these structures during our archaeological survey. Nevertheless, in the mid-twentieth century, local herders found a large cache composed of metal and *Spondylus* Inca style figurines buried in the platform. At present, a small portion of the offerings (54 objects) are housed at the Archaeological Museum of La Serena. The oral accounts collected by Iribarren (1962) and Castillo (2007) emphasize that these offerings were found in tombs. The probable association of these Inca style offerings with human bodies raises the possibility that a *capacocha* ceremony (Duviols 1976; Reinhard and Ceruti 2010) may have occurred at Mt. Los Puntiudos[4]. What is certain is that during the Inca Period, the site INF78 constituted an important shrine where state religious ceremonies were conducted, possibly to worship deities related to the mining

[4] In a similar way, it has been argued that the *capacocha* context found at Cerro Esmeralda (near Iquique, Chile) was probably related to the richness and sacred nature of the silver Huantajaya mines, near Iquique, Chile (Checura 1977). In contrast to the platform of Los Puntiudos and other Inca shrines located at elevations on top or very close to the Prehispanic mines (see Cruz 2009; Salazar, Chap. 12 in this volume), Cerro Esmeralda is approximately 5 km away from the silver mines of Huantajaya.

activities. Other Inca shrines associated with the cult of mine deities have been documented on San José del Abra (Chap. 12), Potosí, Porco (Cruz 2009), and Cerro Esmeralda (Checura 1977). Most of the time, the shrines correspond to ceremonial platforms built on the summit of the mined mountains or on high slopes overlooking the mining areas. All these cases illustrate the religious significance that the mining operations gained in the southern provinces of the empire, under the sponsorship and control of the Inca state.

Discussion

The archaeological survey conducted in the Los Infieles area reveals that during the Inca Period, this region became an important pole of activity for the mining industry, where more than 35 mines (including individual and discontinuous trenches) were exploited. Preliminary assessment based on the surface materials registered during the archaeological survey suggests that each of the five mining clusters defined included at least one large site, in which similar operational sequences of mining activities were conducted. These major sites are: INF48 (in the first mining cluster); INF 93 and INF94 (in the second mining cluster); INF63 (in the third mining cluster); INF52 and INF57 (in the fourth mining cluster); and INF136 and INF100 (in the fifth mining cluster). In general terms, the mining operational sequence identified included the following stages: (1) the extraction of minerals from one or more nearby mines; (2) the primary crushing and sorting of minerals, which seems to have occurred at the entrance and on the spoil mounds of the open-cut trenches; (3) the secondary crushing and sorting of high-quality minerals that took place inside specific structures at the major sites; and (4) the stock-piling of high-quality pebble-sized fragments of chrysocolla and/or silica rocks in storage rooms. These storage rooms form part of structures located at the major sites[5]. Further, most of these larger sites either include structures that could have served as temporary residential facilities (e.g., INF48, INF94, INF63, and INF100), or are located near other sites that could have provided lodging for the miners (e.g., INF66; INF51; INF49; INF137).

The similarities detected in terms of mining operational sequences between the larger sites suggest that each of the mining clusters could have developed its extractive activities with a high degree of independence from the other clusters. The highly similar organization could support the notion of state oversight over the different clusters or a management on grand scale, but further research is needed to

[5] I prefer to refer to these mineral storage units simply as "storage rooms" rather than describing them as "colcas." The concept of "colca" is traditionally used to describe rows of circular or rectangular stone structures included within or located near Inca style buildings that form part of state administrative facilities, such as "tambos" or provincial administrative centers (de Huaycochea 1994; Levine 1992). The proposed storage units mentioned in this chapter correspond to small rooms, usually a single circular or rectangular unit that forms part of a larger stone structure.

examine these possibilities. The relative contemporaneity among the mining clusters raises the question of whether they were ever exploited simultaneously in Inca times. Because of the short span of time encompassed by the Inca Period in north-central Chile (possibly ca. 100 years at most), it is unlikely that absolute dates would help to clarify a temporal sequence of extractive operations during Inca times. Nevertheless, a detailed study aimed at assessing which mines were exhausted and the character of site abandonment behaviors (e.g., sudden or programmed abandonment; O'Brien 1994; Stevenson 1982) could shed light on sites that might have been active until the Spanish conquest.

The pottery fragments and certain lithic instruments such as projectile points suggest that most of the people involved in the mining activities belonged to local Diaguita groups. The nearest Diaguita habitational settlements that could have provided a portion of labor force to the mining sites in the Los Infieles area are located roughly 25 km west, on the coast, or 35 km southwest, in the lower course of the Elqui Valley. Although the archaeological survey revealed the existence of several sites that certainly provided habitation areas for the people involved in the mining operations, none of these sites has either the size or the quantity and variety of artifactual remains expected for a habitational Diaguita settlement. Moreover, the maintenance of a relatively large and permanent population in the area would have required the local and continuous production of foods. The fact is that although certain small areas of the Quebrada Las Pircas basin were suitable for maintaining small cultivated fields, no agricultural facilities or sites that could be associated with these activities were registered. Considering the aridity of the area, the rather simple infrastructure of the sites, and the temporary basis of the mandatory tribute in labor (*mita*) to which the miners were generally subject across the Inca empire (Berthelot 1986; Murra 1980 [1956], 1982; Van Buren and Presta 2010), it is more likely that most of the Diaguita workers only resided in the area for certain periods of time. However, regardless of whether the workers resided in a permanent or temporary basis, it is apparent that this population must have been externally supplied with foods and provisions from other areas, especially from the well populated Elqui Valley. The evidence so far collected suggests that the Inca state was strongly involved in providing these services.

The involvement of the Inca state in the maintenance and management of mining activities in the Los Infieles area is first evidenced by the presence of buildings planned in Inca architectural canons at the sites INF48, INF49, INF52, INF58, INF94 and possibly INF100. Further, the structures S-1 and S-2 at the site INF48 constitute the largest provincial Inca style buildings documented across the entire area inhabited by Diaguita groups, between the rivers Huasco and Choapa (ca. 370 km; Stehberg 1995). Others sites in the Los Infieles area, like INF58, INF52 (S-1) and INF94 (S-2 and S-3), also include structures constructed in the simple orthogonal pattern that is characteristic of the local Inca style buildings in north-central Chile (Stehberg 1995). Worth noting is that, regardless of their architectural simplicity, these stone structures reflect Inca planning in a vast region where Diaguita groups built their huts with wattle and daub walls. Certainly, excavations are necessary to further understand the activities, functions, and status of the people that occupied the different local Inca buildings in the Los Infieles area. However, the size, form, and artifactual remains found in some of the rooms in these structures suggest that they probably served for storing foods and stocking minerals. This would entail the management of food supplies needed to sustain the workers and control over the final mineral production. As was pointed out by Stehberg (1995), it is possible that the compound of larger structures at the site INF48 may have played an important role in the administration, at least, of the mining operations at Mt. Los Infieles. Nevertheless, further research is required to better understand the activities that took place at the different sites and to compare more accurately their functions.

A second line of evidence for the involvement of the Inca state in the Los Infieles area is the significant amount of local Inca pottery at the sites. In all of the provincial Inca style buildings and in many of the other structures there were fragments of *aríbalos* and shallow plates in local Inca style. Both types of vessels are certainly among the most conspicuous symbols associated with the largesse of the Inca state as a generous provider of food and drinks in exchange for labor tribute and in contexts of public ceremonial feasting (Bray 2003; Morris 1995). If we assume that *aríbalos* were used as storage containers for *chicha* (maize beer), then we can infer that this drink was stored and regularly provided to the miners who worked in the Los Infieles area. Similarly, solid and semi-solid foods must have been served in shallow plates. The broad presence of these two types of vessels, traditionally used for food and drink consumption at Inca state facilities across the empire, strongly suggests that the mining activities in the Los Infieles area were to a large extent sponsored by the Inca state.

The ceremonial platform erected on the summit of Mt. Los Puntiudos arguably emphasizes the role of the Inca state as an important agent in promoting, through state religious practices, the success of the mining operations. The various anthropomorphic and camelid Inca style figurines deposited in the platform correspond to the characteristic state ritual paraphernalia that only state religious specialists could manipulate and offer to the *huacas* (shrines) acknowledged by the Inca state across the empire (Ceruti 2003; DeMarrais et al. 1996). Most probably, as were other mine shrines of the empire, the platform on Mt. Los Puntiudos was dedicated to a local deity and/or state gods who, according to wide spread Andean beliefs, favored and allowed the successful exploitation of the mines (Bouysse-Cassagne 2005, 2008; Cruz 2009; Chap. 12). Hence, the offerings would have been given to the gods in exchange for mineral extraction activities. Undoubtedly, the introduction of state religious institutions into a paramount working center such as the Los Infieles mining complex reinforced the role of ideology in the legitimization of Inca rule (Acuto 1999; Bauer 1996).

The mines at Mt. Los Infieles were first thought to be associated with the exploitation of metallic ores, specifically copper oxides, silver, and gold (Iribarren 1962). More recently, Cuadra and Arena (2001) suggested that the mines may have provided copper ores for both lapidary and metallurgical operations. However, the mineralogical examination of the mines in the entire study area by the mining engineer C. Canut de Bon (2008 Ms), and his microscopic observations of hundreds of mineral samples collected from the mines during our archaeological survey have made it clear that the extractive operations were concentrated on the exploitation of opaline silicas and chrysocolla. In contrast to copper oxides (cuprite and tenorite) and carbonates (malachite and azurite), chrysocolla contains a lower percentage of copper by weight and, as a silicate, is far more difficult to smelt. Unfortunately, very little is known about the circulation and contexts in which both chrysocolla and opaline silica were used across the empire (Carcedo and Vetter 2004). The conspicuous colors, as well as the brittle and laminar structure of the minerals extracted at the sites of the Los Infieles area suggests that they were exploited for ornamental purposes (e.g., the production of beads, pendants, incrustations, or small carved figures). Alternatively, the mineral could have played a ceremonial role, as part of offerings in which small raw pieces of it were deposited at ritual places or ceremonial structures. Nevertheless, this practice which is closely associated with llama caravan networks and mountain worship has rarely been documented in north-central Chile, and is far more common across the Atacama Desert and the southern Andean Altiplano (Berenguer 2004; Nielsen et al. 1999; Pimentel 2009).

The absence of lapidary workshop remains in the Los Infieles area, and the small size of the remaining sorted minerals at the sites suggest that the final products obtained from the mining operations were high-quality granule and pebble size minerals. Hence, this mineral product had to be transported to other places where it could be further worked, or incorporated into exchange networks as a raw material. It is worth noting that the Spanish captain Pedro Mariño de Lovera (1865 [1595]: 78) wrote that in the lower course of the Elqui Valley, specifically in the settlement where an Inca governor resided, the Incas kept large quantities of "turquoise" and "crystal." Could the Spanish captain have confused the chrysocolla with turquoise and white opaline silica with crystal? Chrysocolla, at least, is often mistaken for turquoise (hydrous phosphate of copper and aluminum; Petersen 2010), and opaline silica may look like a translucent crystal. Unfortunately, the habitational component of the site mentioned by Mariño de Lovera has not been investigated, and it is therefore not possible to assess at present this reasonable connection between that site and the Los Infieles area.

Certainly, much more research is needed to understand the economic and symbolic significance of chrysocolla and opaline silica within local and broader politico-economic contexts. However, the large number of mines and their associated facilities across the Los Infieles region suggest that these resources had much higher economic value than scholars tend to think during the Inca Period, at least in northern Chile. The mining complex of Los Infieles, like those of San Jose del Abra and Conchi Viejo (Salazar 2008 and Chap. 12), show that Inca control of mining activities was not only exercised on gold and silver mines, as it is described in early Colonial documents. As Lechtman (1976) has pointed out, the Colonial sources reflect more the Spanish fixation on these two precious metals rather than an accurate picture of indigenous practices and interests. The extent to which the Incas sought to centralize control over the exploitation of these valuable mineral deposits is currently under research in my ongoing investigation, but the evidence so far collected suggests that the Inca state was significantly involved in sponsoring and supporting the mining operations.

Acknowledgments I want to express my deepest gratitude to the members of the agrarian community "Olla de Caldera," especially to their council, for granting permission to conduct the fieldwork, and to our muleteers, Don Augusto and Rafael Robledo. The archaeological survey would have not been made possible without the help of my mountaineer friend Eduardo Olivier and the equipment kindly provided by the adventure travel agency "Sangotay Aventuras." Thanks to the crew that worked during the different survey seasons: Paulina Acuña, Macarena Crisóstomo, Italo Borlando, Pamela Orozco, Carlos Osorio, and Gabriela Vega. I am greatly indebted to Gastón Castillo for sharing information regarding his research in the Los Infieles area, and to the mining engineer Claudio Canut de Bon U. for his identification of mineral samples and the time he has dedicated to explain geological topics to me. I am grateful to Brian Bauer, Nicola Sharratt, and an anonymous reviewer for their insightful comments on drafts of this chapter. I would also like to thank Nicholas Tripcevich and Kevin Vaughn for their kind invitation to participate in this volume. The archaeological survey was funded by a dissertation fieldwork grant from the Wenner-Gren Foundation and by a research grant from the University of Illinois at Chicago.

References

Acuto, F. A. (1999). Paisaje y dominación: La constitución del espacio social en el Imperio Inka. In A. Zarankin & F. A. Acuto (Eds.), *Sed Non Satiata. Teoría social en la arqueología latinoamericana contemporánea* (pp. 33–75). Buenos Aires: Ediciones El Tridente.

Ampuero, G. (1969). Excavaciones arqueológicas en el Fundo Coquimbo, Departamento de La Serena. *Actas del V Congreso Nacional de Arqueología Chilena* (La Serena), pp. 153–166.

Ampuero, G. (1978). Notas para el estudio de la cultura Diaguita Chilena. *Publicaciones del Museo Arqueológico de La Serena, 16*, 111–124.

Ampuero, G. (1989). La Cultura Diaguita Chilena. En J. Hidalgo, V. Schiappcasse, H. Niemeyer, C. Aldunate, & I. Solimano (Eds.), *Culturas de Chile: Prehistoria, desde sus orígenes hasta los albores de la conquista* (pp. 277–287). Editorial Andrés Bello, Santiago.

Ampuero, G. (1994). *Cultura Diaguita. Serie Patrimonio Cultural Chileno—Colección Culturas Aborígenes*. Santiago: División de Extensión Cultural del Ministerio de Educación.

Ampuero, G. (2007). *Los Diaguitas en la perspectiva del siglo XXI*. Santiago: LOM.

Ampuero, G., & Hidalgo, J. (1975). Estructura y proceso en la prehistoria y protohistoria del Norte Chico de Chile. *Chungará, 5*, 87–124.

Angiorama, C. (2004). Acerca de Incas y metales en Humahuaca. *Producción metalúrgica en Los Amarillos en tiempos del Tawantinsuyu. Relaciones de la Sociedad Argentina de Antropología, 29*, 39–58.

Bauer, B. (1996). Legitimization of the state in Inca myth and ritual. *American Anthropologist, 98*(2), 327–337.

Beorchia, A. (1985). *El enigma de los santuarios indígenas de alta montaña*. Revista del CIADAM 5. Centro de Investigaciones Arqueológicas de Alta Montaña, San Juan.

Berenguer, J. (2004). *Tráfico de caravanas, interacción y cambio en el Desierto de Atacama*. Santiago: Sirawi Ediciones.

Berthelot, J. (1986). The extraction of precious metals at the time of the Inka. In J. V. Murra, N. Wachtel, & J. Revel (Eds.), *Anthropological history of Andean polities* (pp. 69–88). Cambridge, MA: Cambridge University Press.

Bouysse-Cassagne, T. (2005). Las minas del centro sur andino, los cultos prehispánicos y los cultos cristianos. *Bulletin IFEA* 34(3):443–462. 2008. Minas del sol, del Inka y de la gente. Potosí en el contexto de la minería prehispánica. In J. P. Cruz & J. Vacher (Eds.), *Mina y metalurgia en los Andes del Sur: Desde la época prehispánica hasta el siglo XVII (Actes e Mémoires de IFEA 23)* (pp. 303–348). Lima: Institut de Recherche pour le Developpement and IFEA.

Bouysse Cassagne, T. (2008). Minas del sol, del Inka y de la gente. Potosí en el contexto de la minería prehispánica. In *Mina y metalurgia en los Andes del Sur: Desde la época prehispánica*

hasta el siglo XVII, edited by J. P. Cruz and J.-J. Vacher, pp. 303–348. Institut de Recherche pour le Developpement e IFEA, Lima.
Bray, T. L. (2003). Inka pottery as culinary equipment: Food, feasting, and gender in imperial state design. *Latin American Antiquity, 14*(1), 3–28.
Brumfiel, E. M., & Earle, T. K. (1987). Specialization, exchange, and complex societies: and introduction. In E. M. Brumfiel & T. K. Earle (Eds.), *Specialization, exchange, and complex societies. New directions in archaeology* (pp. 1–9). Cambridge: Cambridge University Press, UK.
Cantarutti, G. E. (2002). *Estadio Fiscal de Ovalle: Redescubrimiento de un sitio diaguita-inca en el valle del Limarí*. Professional Title thesis. Universidad de Chile, Santiago.
Cantarutti, G. E., & Mera, C. R. (2005). Variabilidad estilística alfarera y reflexiones sobre la dominación incaica en el valle de Limarí, a la luz del sitio Estadio Fiscal de Ovalle. *Xama* 15–18, *51 Congreso Internacional de Americanistas* (Santiago 2003)—Actas del simposio Tawantinsuyu 2003: Avances recientes en arqueología y etnohistoria, pp. 199–215.
Cantarutti, G. E., & Solervicens, C. (2005). Cultura Diaguita preincaica en el valle del Limarí: Una aproximación a partir del estudio de colecciones cerámicas. En *Actas del XVI Congreso Nacional de Arqueología Chilena* (Tome, 2003) (pp. 147–156). Escaparate Ediciones, Concepción.
Carcedo, P., & Vetter, L. (2004). The use of minerals and precious metals according to the chronicles. In *The Incas: Art and Symbols* (pp. 169–217). Banco de Crédito del Perú, Lima.
Castillo, G. (2003). La Vuelta de los Años: Reseñas y Perspectivas sobre Comunidades, el Pastoreo y la Transhumancia en la Región Semiárida de Chile. In P. Livenais & X. Aranda (Eds.), *Dinámicas de los Sistemas Agrarios en Chile Árido: La Región de Coquimbo* (pp. 65–119). Santiago: LOM Ediciones Ltda.
Castillo, G. (2007). Los Puntiudos-Los Infieles: Bases para la contextualización de colecciones pertenecientes al Museo Arqueológico de La Serena. *Informes Fondo de Apoyo a la Investigación Patrimonial, 2006*, 117–127.
Ceruti, M. C. (2003). *Llullaillaco: Sacrificios y ofrendas en un santuario inca de alta montaña.* Universidad Católica de Salta, Salta.
Checura, J. (1977). Funebria incaica en el Cerro Esmeralda. *Estudios Atacameños, 5*, 127–144.
Cieza de León, P. (1996 [1554]). In F. Cantú (Ed.), *Crónica del Perú*. Segunda Parte. Pontificia Universidad Católica del Perú, Lima.
Cornely, F. (1956). *Cultura Diaguita Chilena y Cultura de El Molle*. Santiago: Editorial del Pacífico S.A.
Costin, C. L. (1996). Craft production and mobilization strategies in the Inca Empire. In B. Wailes (Ed.) *Craft specialization and social evolution: In memory of V. Gordon Childe* (pp. 211–228). University Museum Monograph 93. The University Museum of Archaeology and Anthropology. Philadelphia: University of Pennsylvania, PA.
Cruz, P. (2009). Huacas olvidadas y cerros santos. Apuntes metodológicos sobre la cartografía sagrada en los Andes del sur de Bolivia. *Estudios Atacameños* (38), 55–74.
Cruz, P., & Absi, P. (2008). Cerros ardientes y *huayras* calladas. Potosí antes y durante el contacto. In J. P. Cruz, & J.-J. Vacher (Eds.), *Mina y metalurgia en los Andes del Sur: Desde la época prehispánica hasta el siglo XVII* (Actes e Mémoires de IFEA 23) (pp. 91–120). Institut de Recherche pour le Developpement e IFEA, Lima.
Cuadra, W. & Arena, M. (2001). *El oro de Chile: Desde los tiempos prehispánicos (900 a.C.) hasta nuestra independencia (1810)*. LOM Ediciones, Santiago.
D'Altroy, T. N. (1992). *Provincial power in the Inka Empire*. Washington, DC: Smithsonian Institution Press.
D'Altroy, T. N. (2002). *The Incas*. United Kingdom: Blackwell Publishing.
D'Altroy, T. N., & Earle, T. K. (1985). Staple finance, wealth finance, and storage in the Inka political economy. *Current Anthropology, 26*(2), 187–197.
de Betanzos, J. (1996 [1557]). In R. Hamilton (Ed.), *Narrative of the Incas*. University of Texas Press, TX: Austin.
de Bibar, G. (1966 [1558]). In I. A. Leonard (Ed.), *Crónica y relación copiosa y verdadera de los reynos de Chile*. Fondo Histórico y Bibliográfico José Toribio Medina, Santiago.

Canut de Bon, C. (2008). *Informe sobre viaje a sector mina Los Infieles. Área sierra de Salapor – Cerro Los Puntiudos y Los Infieles. Provincia de Elqui-Comuna de La Serena-Región de Coquimbo.* Report preserved in the Museo Arqueológico de La Serena, Chile. Manuscript in posesión of the author.
de Góngora Marmolejo, A. (1862 [1575]). *Historia de Chile. Desde su descubrimiento hasta el año de 1575.* Imprenta del Ferrocarril, Santiago.
DeMarrais, E., Castillo, L. J., & Earle, T. K. (1996). Ideology, materialization, and power strategies. *Current Anthropology, 37*(1), 15–31.
Duviols, P. (1976). La Capacocha. Mecanismo y función del sacrificio humano, su proyección geométrica, su papel en la política integracionista y en la economía redistributiva del Tawantinsuyu. *Allpanchis, 9*, 11–57.
Earle, T. K. (1994). Wealth finance in the Inka empire: Evidence from the Calchaquí Valley, Argentina. *American Antiquity, 59*(3), 443–460.
Earle, T. K. (1997). *How chiefs come to power*. Stanford: Stanford University Press, CA.
Emparan, C., & Pineda, G. (1999). *Mapas Geológicos, No 12 Área Condoriaco—Rivadavia.* Santiago: Sernageomin.
Emparan, C., & Pineda, G. (2000). *Mapas Geológicos, No 18, Área La Serena—La Higuera.* Santiago: Sernageomin.
Galdames, L. (1990). Apacheta: La ofrenda de piedra. *Diálogo Andino, 9*, 9–25.
González, P. (1995). *Diseños cerámicos de la Fase Diaguita-Inca: Estructura, Simbolismo, Color y Relaciones Culturales.* Memoria para optar al Título de Arqueólogo. Santiago: Universidad de Chile.
González, L. R. (2004). *Bronces sin nombre. La metalurgia prehispánica en el noroeste argentino.* Ediciones Fundación CEPPA, Buenos Aires.
González, L. R., & Tarragó, M. (2004). Dominación, resistencia y tecnología: La ocupación incaica en el noroeste argentino. *Chungará, 36*(2), 393–406.
González, C., & Westfall, C. (2008). Atacameños en El Salvador: Nuevas apreciaciones sobre un fardo funerario del cementerio Las Turquesas. *Estudios Atacameños, 35*, 49–73.
Hidalgo, J. (1972). Culturas protohistóricas del Norte de Chile: *El testimonio de los cronistas.* Editorial Universitaria, Santiago.
Hidalgo, J. (1989). Diaguitas chilenos protohistóricos. In J. Hidalgo, V. Schiappcasse, H. Niemeyer, C. Aldunate, & I. Solimano (Eds.), *Culturas de Chile: Prehistoria, desde sus orígenes hasta los albores de la conquista* (pp. 289–293). Santiago: Editorial Andrés Bello.
Howe, E. G., & Petersenc, U. (1994). Silver and lead in the late prehistory of the Mantaro Valley, Peru. In D. A. Scott & P. Meyers (Eds.), *Archaeometry of Pre-Columbian sites and artifacts* (pp. 183–198). Los Angeles: Getty Conservation Institute, CA.
Huaycochea Núñez de la Torre, F de M. (1994). *Qolqas Bancos de Reserva Andinos.* Cusco: Editorial Universitaria, Universidad Nacional de San Antonio Abad del Cusco.
Iribarren, J. (1962). Minas de explotación por los incas y otros yacimientos arqueológicos en la zona de Almirante Latorre, Departamento de La Serena. *Publicaciones del Museo y la Sociedad Arqueológica de La Serena, 12*, 61–72.
Iribarren, J. (1973). *Una mina de explotación incaica: El Salvador-Provincia de Atacama* (pp. 229–266). N° Especial: Actas del VI Congreso Chileno de Arqueología. Boletín de Prehistoria.
Iribarren, J. (1978). Manifestations of Inca culture in two provinces of Chile. In D. L. Browman (Ed.), *Advances in Andean archaeology* (pp. 443–448). The Hague and Paris: Mouton Publishers.
Krahl, L., & González, O. (1966). Expediciones y hallazgos en la alta cordillera de la provincia de Coquimbo (cerros Las Tórtolas y Doña Ana) 1956–1958. In *Anales de Arqueología y Etnología 21* (pp. 101–125). Cuyo: Facultad de Filosofía y Letras. Universidad Nacional de Cuyo.
Lechtman, H. (1976). A metallurgical site survey in the Peruvian Andes. *Journal of Field Archaeology, 3*(1), 1–42.
Lechtman, H. (1993). Technologies of power: The Andean case. In J. S. Henderson & P. Netherly (Eds.), *Configurations of power: Holistic anthropology in theory and practice* (pp. 244–280). Ithaca: Cornell University Press, NY.

Lechtman, H. (2007). The Inka, and Andean metallurgical tradition. In R. L. Burger, C. Morris, & R. Matos (Eds.), *Variations in the expression of Inka power* (pp. 313–356). Washington, DC: Dumbarton Oaks Research Library and Collection.

Levine, T. Y. (Ed.). (1992). *Inka storage systems*. Norman: University of Oklahoma Press, OK.

Llagostera, A. (1976). Hipótesis sobre la expansión incaica en la vertiente occidental de los Andes Meridionales. In L. Núñez (Ed.), *Homenaje al Dr. Gustavo Le Paige, SJ* (pp. 203–218). Antofagasta: Universidad del Norte.

Mariño de Lobera, P. (1865 [1595]). *Crónica del Reino de Chile*. Reducida a nuevo método y estilo por el Padre Bartolomé de Escobar, de la Compañía de Jesús. Imprenta del Ferrocarril. Santiago.

Morris, C. (1995). Symbols to power. Styles and media in the Inka state. In C. Carr & J. E. Nietzel (Eds.), *Style, society and person* (pp. 419–433). New York: Planum Press, NY.

Murra, J. (1980 [1956]). *The economic organization of the Inca state*. Greenwich: JAI Press, CT.

Murra, J. (1982). The mit'a obligations of ethnic groups to the Inka state. In G. A. Collier, R. I. Rosaldo, & J. D. Wirth (Eds.), *The Inca and Aztec states 1400–1800: Anthropology and history* (pp. 237–262). New York: Academic Press, NY.

Nielsen, A. (1997). El tráfico caravanero visto desde la Jara. *Estudios Atacameños, 14*, 339–371.

Nielsen, A., Vázquez, M. M., Avalos, J. C., and Angiorama, C. I. (1999). Prospecciones arqueológicas en la reserva "Eduardo Avaroa" (Sud Lípez, Dpto. Potosí, Bolivia). *Relaciones de la Sociedad Argentina de Antropología 24*, 95–124.

Niemeyer, H., Castillo, G., & Cervellino, M. (1989). Los primeros ceramistas del Norte Chico: El Complejo El Molle (0–800 d.C.). In J. Hidalgo, V. Schiappcasse, H. Niemeyer, C. Aldunate, & I. Solimano (Eds.), *Culturas de Chile: Prehistoria, desde sus orígenes hasta los albores de la conquista* (pp. 227–263). Santiago: Editorial Andrés Bello.

Núñez, L. (1987). Tráfico de metales en el área centro-sur andina: Factos y expectativas. *Cuadernos Instituto Nacional de Antropología, 12*, 73–105.

Núñez, L. (1999). Valoración minero-metalúrgica circumpuneña: Menas y mineros para el Inka rey. *Estudios Atacameños, 18*, 177–221.

O'Brien, W. (1994). *Mount Gabriel: Bronze Age mining in Ireland*. Galway: Galway University Press.

Petersen, G. (2010). *Mining and metallurgy in ancient Perú*. (W. E. Brooks, Trans.). The Geological Society of America, Special Paper 467, CO: Boulder.

Pigott, V. (1998). Prehistoric copper mining in the context of emerging community craft specialization in northeast Thailand. In A. B. Knapp, V. Pigott, & E. W. Herbert (Eds.), *Social approaches to an industrial past. The archaeology and anthropology of mining* (pp. 205–225). New York: Routlegde, NY.

Pimentel, G. (2009). Las huacas del tráfico. *Arquitectura ceremonial en rutas prehispánicas del desierto de Atacama. Boletín del Museo Chileno de Arte Precolombino, 14*(2), 9–38.

Platt, T., Bouysse-Cassagne, T., & Harris, O. (2006). *Qaraqara-Charca. Malku, Inka y Rey en la Provincia de Charcas (Siglos XV-XVII)*. IFEA, Plural Editores. La Paz: University of St. Andrews, University of London, Fundación Interamericana, Fundación Cultural del Banco Central de Bolivia.

Polo de Ondegardo, J. (1940 [1561]). Informe del licenciado Juan Polo de Ondegardo al licenciado Briviesca de Muñatones sobre la perpetuidad de las encomiendas en el Perú. *Revista Histórica, 13*, 125–196.

Raffino, R. (1981). *Los Inkas del Kollasuyu*. La Plata: Ramos Americana Editora.

Reinhard, J., & Ceruti, M. C. (2010). *Inca rituals and sacred mountains: A study of the world's highest archaeological sites*. Los Angeles: Cotsen Institute of Archaeology Press, UCLA, CA.

Romero, H., Rovira, A., & Véliz, G. (1988). *Geografía*. Santiago: IV Región de Coquimbo. Instituto Geográfico Militar.

Rosado, M., & Vernacchio, J. (1998). Dental health and diet of two prehistoric populations from Chile's semiarid North. *Dental Anthropology, 13*(1), 2–8.

Rowe, J. H. (1950). Influencia incaica en la alfarería diaguita-chilena (carta del profesor J. H. Rowe). *Publicaciones de la Sociedad Arqueológica de La Serena, 5*, 28–29.

Rutledge, J. W., & Gordon, R. B. (1987). The work of metallurgical artificers at Machu Picchu, Peru. *American Antiquity, 3*(52), 578–594.

Salazar, D. (2002). *El complejo minero San José del Abra (1450–1536 d.C.), II Región: Una aproximación a la arqueología de la minería* (Unpublished Master's thesis). Facultad de Ciencias Sociales, Universidad de Chile.

Salazar, D. (2005). Investigaciones arqueológicas sobre la minería incaica en San José del Abra (II Región, Norte de Chile). *Xama*, 101–117.

Salazar, D. (2008). La producción minera en San José del Abra durante el Período Tardío atacameño. *Estudios Atacameños, 36*, 43–72.

Sancho de la Hoz, P. (1919 [1535]). Relación para S.M. de lo sucedido en la conquista y pacificación de estas provincias de la Nueva Castilla. In H. Ortega & C. A. Romero (Eds.), *Colección de libros y documentos referentes a la historia del Perú*, Vol. 5 (pp. 122–202). Lima: Imprenta y Librería Sanmarti y Ca.

Sarmiento de Gamboa, P. (2007 [1572]). Topa Inca Yupanki goes to conquer and put down the risen Collas. In B. S. Bauer & V. Smith (Eds.), *The history of the Incas* (pp. 160–162). Austin: University of Texas Press, TX.

Silva, O. (1985). La expansión incaica en Chile. Problemas y reflexiones. *Boletín del Museo Arqueológico de La Serena* 18 (Actas del IX Congreso Nacional de Arqueología Chilena, La Serena 1982), 2, 321–344.

Stehberg, R. (1995). *Instalaciones incaicas en el norte y centro semiárido de Chile*. Santiago: Dirección de Bibliotecas Archivos y Museos.

Stehberg, R., & Cabeza, A. (1991). Sistema vial incaico en al Chile Semiárido. *Comechingonia*, Año 9—N° especial, *El Imperio Inka, actualización y perspectivas por registros arqueológicos y etnohistóricos*, Vol. II, pp. 153–216.

Stevenson, M. G. (1982). Toward an understanding of site abandonment behaviour: Evidence form historic mining camps in the south west Yukon. *Journal of Anthropological Archaeology, 1*(3), 237–267.

Tierney, P. (2001). Pastos Salados. *Revista del Centro de Investigaciones Arqueológicas de Alta Montaña* (CIADAM) 6, 165.

Troncoso, A. (1998). La Cultura Diaguita en el valle de Illapel: Una perspectiva exploratoria. *Chungará, 30*(2), 125–142.

Van Buren, M. (2009). Proyecto Arqueológico Porco-Potosí: Inka and Spanish Colonial Silver Mining in the Bolivian Andes, Vol. 2009. Colorado State University. http://lamar.colostate.edu/~mvanbure/index.htm. Accessed 01.10.2009.

Van Buren, M., & Mills, B. H. (2005). Huayrachinas and Tocochimbos: Traditional smelting technology of the Southern Andes. *Latin American Antiquity, 16*(1), 3–25.

Van Buren, M., & Presta, A. M. (2010). The organization of Inka silver production in Porco, Bolivia. In M. A. Malpass & S. Alconini (Eds.), *Distant Provinces in the Inka Empire: Toward a deeper understanding of Inka imperialism* (pp. 171–192). Iowa City: University of Iowa Press, IA.

Vitry, C. (2002). Apachetas y mojones, marcadores espaciales del paisaje prehispánico. *Revista Escuela de Historia, 1*(1), 179–191.

Chapter 10
Amalgamation and Small-Scale Gold Mining in the Ancient Andes

William E. Brooks, Gabriela Schwörbel†, and Luis Enrique Castillo

Introduction

> Asimismo, es posible suponer que los peruanos de aquellas tierras conocían, ya en épocas remotas, el método de la amalgamación, empleando para ello el azogue que lograban obtener del cinabrio cuyas menas existían en Buldibuyo.
>
> Situación Actual de la Minería del Mercurio en el Perú, 1954
>
> Augusto Cabrera La Rosa

> It is possible that ancient Peruvians knew about the use of mercury amalgamation, and they used mercury obtained from the cinnabar ores found near Buldibuyo
>
> Overview of Mercury Mining in Peru, 1954
>
> Augusto Cabrera La Rosa

In 1532, Atahualpa's gold ransom (Hemming 1970: 48) was a physical confirmation of the abundant placer, vein, disseminated, and porphyry gold occurrences in the Andes (Cánepa 2005; Ministerio de Energía y Minas 2000; Noble and Vidal 1994; Petersen 2010: 49 [1970]; Petersen et al. 1990) as well as the successful small-scale gold mining methods used before European contact. Porphyry ores produced a copper–gold–silver alloy or *tumbaga*, which, through depletion gilding or other treatments, resulted in enhanced surficial gold (Petersen 2010: 57 [1970]). Nuggets and coarse gold would have been easily hand-picked from placer sources; however, recovering fine-grained disseminated, or vein gold, would have required gravity

W.E. Brooks (✉)
Geologist Reston, VA 20191, USA
e-mail: webgeology@aim.com

G. Schwörbel (deceased) • L.E. Castillo
Museo Nacional de Arqueología, Antropología e Historia, Lima, Peru
e-mail: gschworble@mcultura.gob.pe; lcastillo@mcultura.gob.pe

N. Tripcevich and K.J. Vaughn (eds.), *Mining and Quarrying in the Ancient Andes*, Interdisciplinary Contributions to Archaeology, DOI 10.1007/978-1-4614-5200-3_10,

separation, panning, or crushing, and then, as now, amalgamation of the gold with mercury. Placer gold was the main source of gold for humans in antiquity and provided more than two-thirds of the gold ever produced (Boyle 1979: 333).

Mines in Ancient Peru

Georg Petersen recognized the importance of mining in ancient Peru, and with encouragement from his archaeological colleagues in Lima, *Mineria y Metalurgia en el Antiguo Peru*, essentially a De Re Metallica (Agricola 1556/1912) for the New World, was published in 1970. His book includes information on gold, silver, copper, guano, salt, and other resources in the Andes. For example, one of these mines, Caylloma, near Arequipa, still produces silver and the rich guano resources of coastal Peru, which were exploited in ancient times, are still an important international commodity (Elton 2001). A translation of Petersen's book in 2010 includes an updated bibliography with papers on ancient placer gold mines (West 1952); a hematite mine (Vaughn et al. 2007; see also Chap. 8); copper mining sites (Shimada and Merkel 1991; Brooks et al. 2004); and cinnabar, the ore of mercury, which was mined from the Santa Barbara mine at Huancavelica (Brown 2001, Cooke et al. 2009; Robins 2011). Given the volume of metals produced in ancient Peru, the sites were likely numerous and only recently has attention been given to studies of the mines and mining technology in the ancient Andes.

Gold is surely the benchmark metal of the ancient Andes and there are numerous papers that include photographs and descriptions of the numerous gold artifacts from the region. However, some of the gold may have been sourced from copper porphyry ores, as *tumbaga,* or derived from placer (alluvial) sources. The placer gold mining sites are especially difficult to identify because the location of the placer occurrences will change due to factors linked to the dynamic environment of stream channels. These factors include flooding and variations in seasonal runoff of the rivers and streams that transport the gold-bearing sediments; changes in the stream system through time; the nature of panning in the active stream environment; and the overprint on the small-scale mining sites caused by the modern use of dredges (*dragas*) as a result of increased gold prices—therefore it is extremely difficult, if not impossible, to accurately locate ancient placer sites. Placer mining techniques, such as the use of mercury for gold amalgamation or the *aventadero* method in which gold-bearing sediments are dried and tossed into the air so that the wind removes the non-gold material, have thus far, received little attention.

Vein and hard-rock gold occurrences were also exploited and Petersen indicates that tailings from these sources may be found at Hoabamba (Petersen 2010: 24 [1970]). Also, the location of ancient rock-crushers, or *quimbaletes*, is evidence for ore treatment at or near the ancient mine sites—this method of ore treatment is still used today in Peru (Cánepa 2005; Ministerio de Energía y Minas 2000: 15). Patáz, in northern Peru, is one of the few sites that has been exploited as a hard-rock gold and small-scale gold mining site since ancient times (Schreiber et al. 1990; Wilson and Reyes 1964); however, for the most part these descriptions are general, rare, and the sites have not been excavated. Additionally, many of the mines were shallow and may have collapsed; the mines and tailings may have eroded or become overgrown. Therefore, documenting specific gold mining sites and the techniques that were used, specifically mercury amalgamation, remain challenging.

Mercury in the Ancient World

Mercury was mined and widely used as an industrial metal by ancient people in Africa, Asia, Central America, Europe, Mexico, the Middle East, and South America (Brooks 2012). For example, more than 8,000 years ago, cinnabar [HgS], the ore of mercury, was mined in Turkey and was used as a pigment and also retorted to obtain mercury (Barnes and Bailey 1972; Yildiz and Bailey 1978: 5). In ancient China, cinnabar was also used for pigments and the tomb of Emperor Ch'i-Huang-Ti, who died in 210 BC, contained a relief map of China in which the oceans and rivers were represented by mercury (Schuette 1931: 3). In Spain, mercury was known before the Christian era and the Moorish name of the mine "al-Ma'din" (Almadén), and the name of the metal *azogue*, are still known and used in Latin America. By AD 77, Rome imported 4–5 metric tons of mercury annually from the mines at Almadén that were used for gold amalgamation and small-scale mining (D'Itri and D'Itri 1977: 7).

The earliest written description of gold amalgamation was by al-Biruni, an eleventh century Persian scientist—the gold was processed from crushed ore, mercury was added to amalgamate the gold, the mercury–gold amalgam was recovered and burned in order to volatilize the mercury and recover the gold (Al-Hassan and Hill 1986: 247). Craddock (2000: 233) questioned the use of mercury for gold amalgamation in the ancient world, specifically at Sardis, and proposed that if mercury had been used, then low levels of mercury would be found in the analyses of the processed gold. Therefore, gold from ancient Sardis was analyzed (SEM-EDX), and since no mercury was detected, Ramage and Craddock (2000: 103–107) concluded that mercury had not been used.

Mercury in the New World

Cinnabar mining and mercury retorting date to 1000 BC in Mexico (Langenscheidt 1986: 139; Consejo de Recursos Minerales 1992: 27) and Barba and Herrera (1988) describe ancient cinnabar mining, processing, and details of the configuration of the ceramic vessels used to retort mercury at San José Ixtapa, in Querétaro, Mexico. In Central America, cinnabar occurrences are also known (Roberts and Irving 1957: 169) and native mercury was found in a tomb in Belize that dates to AD 900–1000 (Pendergast 1982). There are also cinnabar and mercury occurrences in Bolivia (Ahlfeld and Schneider-Scherbina 1964: 122; Barba 1640: 82 [1923]); Chile (McAllister et al. 1950); Colombia (Lozano 1987; Wilson 1941; Wokittel 1958);

Ecuador (Truhan et al. 2005: 197); and Peru (Nuñez and Petersen 2002: 136; Petersen 2010: 29 [1970]). And, one of the world's largest cinnabar occurrences is at Huancavelica, Peru (Brown 2001: 467; Yates et al. 1955: 10).

In ancient Peru, cinnabar was widely used as a decoration on gold masks (Gordus and Shimada 1995); as a pigment on murals (Bonavia 1985: 79; Brooks et al. 2008); as a cosmetic for elite Inca women (Brown 2001: 477); and for funeral preparations (Shimada and Griffin 2005). Powdered cinnabar was found in a mollusk shell that had been used to mix pigments at a precontact archaeological site near Ica, southern Peru (Petersen 2010: 80 [1970]).

Isotopic data on mercury in lake sediments, combined with ^{14}C geochronology, indicate that cinnabar mining at Huancavelica began around 1400 BC and that production peaked at approximately 500 BC and again at AD 1450, corresponding to (1) the heights of Chavin and Inca prominence, respectively, in the region (Cooke et al. 2009), and (2) a peak in Inca gold mining (Petersen 2010: 23 [1970]). Fragments of precontact retorts were described (Burger and Matos 2002; Petersen 2010: 49 [1970]; Rivero and Tschudi 1853: 25) and detrital native mercury, weathered from mercury-bearing outcrops, was found in drainages near Huancavelica (Arana 1901: 6; Petersen 2010: 29 [1970]).

The Inca (~AD 1300–1532) recognized the health hazards of mercury and that exposure to mercury and cinnabar during mining and retorting would cause the ancient miners "to shake and lose their senses" and, therefore, the use of mercury by the Inca declined (Larco Hoyle 2001: 135). At about the same time in Europe, Agricola (1556/1912: 427) described methods for retorting mercury and workers were warned to turn their backs to the sweet smelling mercury fumes that would loosen their teeth.

After the Conquest, Spain transported mercury from Almadén to be used mainly for silver amalgamation in the New World and Spanish shipwrecks that still contain mercury from Almadén are known in Colombia and the Dominican Republic (Petersen 1979: 851). Lechtman (1976: 23) credited the introduction and use of mercury for both gold and silver amalgamation in the New World to the Spanish in 1571; however, this present research shows that mercury was likely used for gold amalgamation much earlier.

In the earliest work on metallurgy in the New World, Barba 1640: 142 [1923]) describes how silver ores at Potosí, Bolivia, were processed by either amalgamation or smelting. However, upon "re-discovery" of the mercury occurrences at Huancavelica, in 1566–1567 (Arana 1901: 77; Larco Hoyle 2001: 135; Nuñez and Petersen 2002: 137), which had earlier been mined by ancient Peruvians (Cooke et al. 2009), imported Spanish mercury was replaced by mercury that had been retorted from the rich cinnabar ore at Huancavelica. In Colombia, in the 1750s, mercury was used to amalgamate ore from the gold-platinum placers in the Choco district (Juan and De Ulloa 1748/1807: 606).

The dangerous mining conditions, cold, elevation of 4,000 m, and exposure to the toxic mercury fumes caused Huancavelica to become known as the *mina de la muerte* to the ancient Andean miners (Brown 2001: 468). Later, during Colonial times, Huancavelica mercury was also used in the "patio process" for silver amalgamation in Bolivia, Mexico, and Peru (Nuñez and Petersen 2002: 142; Robins

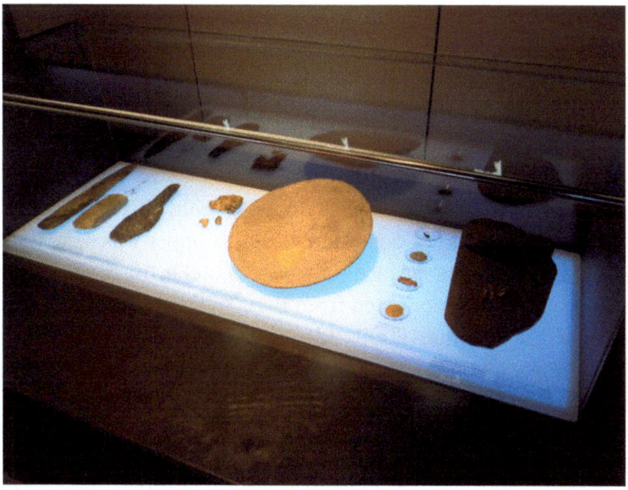

Fig. 10.1 Ancient mining tools Museo de Oro, Bogota (028)

2011). Salt, mercury, and vitriol (mixed copper and iron sulfates) were mixed with crushed silver ore in a large open area, or patio, which was basically a flat surface impervious to mercury. The silver-containing amalgam was then burned to volatilize the mercury and recover the silver. A variation, the "buytron process" was also used at Colonial silver mines at Potosí, Bolivia, where the cold climate required heat from below to speed amalgamation and therefore, silver production (Craddock 1995: 216; Crozier 1993). Mercury was essential for silver processing and adding mercury, *el azogado*, was an important step in Colonial silver production (Del Busto Duthurburu 1996: 98). Mercury that was originally mined at Almadén, or from Huancavelica, has been retorted from Colonial silver amalgamation tailings near Zacatecas, Mexico (Ogura et al. 2003).

Quimbalete and Gold Processing in Ancient Peru

Based on the presence of large, 1–2 ton, rocker-like crushing stones (*quimbaletes* or Inca mills) and tailings found at Hoabamba and other archaeological sites in Peru, both Cabrera La Rosa (1954) and Posnansky (in Petersen 2010: 24 [1970]) proposed that mercury was retorted and used for amalgamation as a part of *quimbalete* gold-processing from altered rock and vein occurrences that contained fine-grained gold. The ore was crushed by the rocking motion of the massive stone *quimbalete* and mercury was added to the watery slurry at the base of the *quimbalete*. The action and weight of the *quimbalete* brought the gold and mercury into contact to form an amalgam nugget (*perla* or *pepita*) from which the gold could be recovered by burning the amalgam (*refogado*) to volatilize the mercury and leave the gold as a *charpita* (Fig. 10.1).

Mercury is widely used for present-day *quimbalete* gold-processing in Peru's Costa Sur Media (Cánepa 2005: 49) and elsewhere in Peru (Ministerio de Energía y Minas 2000: 15; El Comercio 2009)—therefore, it is logical that mercury was also used for *quimbalete* gold-processing in ancient Peru. Examples of ancient *quimbaletes* are shown in Petersen (2010: 83 [1970]) and thus far, use of the *quimbalete* appears unique to Peru.

Gold was also recovered from porphyry copper–gold–silver occurrences, as *tumbaga*, and through the process of depletion gilding, copper was selectively removed, thereby, leaving a rich golden surface on a base-metal artifact. Gold nuggets (*pepitas*) would have easily been recovered from Peru's widespread placer deposits (e.g., Ministerio de Energía y Minas 2000: 43) by using a gold pan (*batea*) and water. Mercury or plant juices would have also been used to recover the fine-grained gold (*chispitas*) remaining in the black sand concentrate in the *batea*.

Mercury was known to the Chavin (700–500 BC; Petersen 2010: xix [1970]) and Larco Hoyle (2001: 128) indicated that mercury was used by the Moche (~100 BC–AD 800) to amalgamate placer gold. Alternatives to mercury, though not widely used, include certain plant juices (*cedro playero*), for example, in Colombia (Castillo Espitia 2007: 305) and Peru (Larco Hoyle 2001: 138). In the *aventadero* method, gold-bearing sands were dried, tossed into the air, which left a gold concentrate (Petersen 2010: 26 [1970]). Reminiscent of Jason and the Golden Fleece, rough carpet or animal skins may be placed in the sluices and streams to trap the fine-grained gold in some parts of Peru; however, the gold may then be further concentrated using mercury.

Fabrication and Mercury Content

In the central Andean metalworking area, which includes Colombia, Ecuador, and Peru, gold objects were more commonly shaped by hammering rather than by casting (Lechtman 1988; Plazas 2007). For example, anvils, gold foils, and stone hammers were found at a site in south-central Peru that dates to 1490 ± 100 BC (Grossman 1972); however, unfortunately, the gold foils were not analyzed.

Using spectrographic analysis, Petersen (Petersen 2010: 25, 57 [1970]) was the first to provide analytical data on the composition of gold from several placer gold occurrences in Peru and a gold Chimú lip ornament. Simply hammering the metal does not change its composition, therefore, if the ornament had been hammered directly from placer gold, then the mercury content of the placer gold and the fabricated ornament should be the same. However, there is a significant decrease in the mercury content of the gold Chimú ornament (<100 ppm), when compared to the mercury content of the placer gold samples (1,000–10,000 ppm), that can only be explained by heating and volatilization of the mercury.

Petersen's (Petersen 2010: 25 [1970]) spectrographic analyses would ideally provide the background mercury content of the placer gold; however, the mercury content of the placer gold may be the result of contamination from several sources that include (1) detrital native mercury released from mercury-containing outcrops over millions of years; (2) mercury released by ancient Andeans for gold amalgamation for several thousands of years ((Brooks et al. 2009); 3) an overprint of mercury released from widespread use of mercury for Spanish Colonial silver production; (4) regional volcanism; or (5) mercury volatilized from underground coal fires and then precipitated across the region. Therefore, it is difficult to determine background mercury content using placer gold composition and it is not possible to date the mercury enrichment in the placer gold.

Analytical Techniques

Because of the proximity of gold and mercury on the periodic chart and the low levels of mercury in the worked gold, neither SEM–EDX (Scanning Electron Microscopy–Energy Dispersive X-ray Spectroscopy) (e.g., Ramage and Craddock 2000: 103–107) nor XRF (X-ray fluorescence) (e.g., Aldenderfer et al. 2008) would have provided sufficient analytical discrimination to detect the potentially low levels of mercury in the gold samples. In another study, elements in concentrations of less than 100 ppm in gold were detected by SEM–EDS (Scanning Electron Microscopy–Energy Dispersive X-ray Spectroscopy); however, data for these elements were not reported (Rehren and Temme 1994). Replication of Petersen's sampling, now using the low-level detection capability of Inductively Coupled Plasma (ICP) analysis for mercury and other elements would provide more precise data on the composition of the gold for comparison and interpretation.

In Table 10.1, analyses of amalgam, *refogado* gold, a placer gold nugget, and three *charpitas* of gold are presented. In order to establish the background mercury and other elemental content of gold-mineralized rock samples directly from the outcrop, samples from several small-scale gold mines in Peru were also analyzed (Table 10.2). The samples are from Patáz, in northern Peru, which is known to have produced placer gold since ancient time (Haeberlin et al. 2004; Schreiber et al. 1990; Wilson and Reyes 1964) and the other is from Costa Sur Media, Peru (Cánepa 2005)—no visible gold was present in either sample.

These crushed samples were also panned in order to obtain a gold concentrate with which to determine background elemental composition of the gold; however, the samples contained very fine-grained gold that would not yield a pan concentrate as fine-grained gold cannot break the surface tension of the water, clings to the lower surface of the water, and is washed out of the gold pan (Oyler 1997; West 1974: 21). However, background mercury content can be inferred from microprobe analyses of placer gold which show that the cores of placer gold do not contain mercury even though there is a rim of amalgam which is contaminant mercury likely released from placer mining in the region (McCready et al. 2003).

In Table 10.3 data are compiled on the mercury and elemental content of three samples of precontact gold foils from northern Peru and four samples of precontact gold foils from Colombia. These samples were also analyzed by ICP in order to establish

Table 10.1 ICP analyses of modern gold samples from Peru and Venezuela

	PEa	PEr	PEc1	PEc2	PEc3	VZn	VZa	VZr
Ag	1,130	818	38,700	>100	>100	2,470	2,510	2,000
Al	339	120	153	1040	948	819	177	582
As	8	4	28	2.3	2.0	3	5	7
Au	368,000	368,000	300,000	>400	>400	296,000	513,000	742,000
B	228	163	<20	1555	1425	110	186	221
Ba	4	0	7	17	12	0	2	2
Bi	0	0	<3	0	0	0	0	7
Ca	43	35	<100	394	344	24	47	54
Cd	0	0	<0.5	0	0	0	0	0
Co	1	1	<1	0	0	0	0	0
Cr	1	1	<1	3	3	1	3	3
Cu	7	6	126	>1,000	>1,000	1,150	4,580	280
Fe	215	118	338	974	656	1,360	94	239
Hg	312,000	822	8.3	37	1.2	3,690	248,000	96,400
K	7	0	<100	395	243	1	1	0
La	2	0	3	2	1	0	1	2
Mg	9	34	<100	1	1	30	36	27
Mn	1	2	3	1	1	1	3	1
Mo	0	0	<1	1	1	0	0	3
Na	9	8	184	>100	>100	9	8	12
Ni	3	6	3	3	5	5	4	6
P	27	7	16	1	1	5	7	21
Pb	83	514	<3	1	1	5	58	61
S	9	5	<500	10	10	18	8	29
Sb	9	16	<3	1	14	3	7	9
Se	1	1	<5	3	7	0	0	4
Sr	5	13	44	4	2	12	7	12
Th	2	2	<3	1	1	0	1	2
Ti	38	11	<100	6	4	36	2	8
Tl	0	0	<5	1	1	0	0	0
U	0	0	<8	1	1	0	1	1
V	1	0	<1	1	1	3	1	1
W	2	2	<2	1	1	1	2	2
Zn	4	2	<1	2	2	3	3	6

PEa—unburned amalgam, Las Quebradas mining camp, Huepethue, Madre de Dios, Peru; PEr—*refogado* gold after two burns in the field at Las Quebradas; PEc1-3—*charpitas*: (1) from a gold shop in Puerto Maldonado, Madre de Dios, Peru, (2–3) from a gold dealer in Lima, also from Madre de Dios, No gold nuggets were available in the Madre de Dios area; VZn—gold nugget from the Km 88 mining district, Estado Bolivar, Venezuela (Brooks et al. 1995); VZa—unburned amalgam from Km 88; VZr—gold that has been burned (*refogado*) once in a gold shop in Km 88 Inductively Coupled Plasma analysis, in parts per million (ppm), by American Assay Laboratories, Sparks, NV

Table 10.2 ICP analyses of gold ore from mines at Nazca and Pataz, Peru

	NZ	PTZw	PTZo
Ag	59	8	51
Al	6,350	4,910	4,070
As	1,238	3,900	13,400
Au	7	23	26
B	<20	<20	<20
Ba	5	27	19
Bi	51	3	<3
Ca	1,333	17,400	7,300
Cd	3	10	54
Co	700	9	11
Cr	15	11	12
Cu	76,439	78	120
Fe	239,111	63,628	176,806
Hg	0.373	0.098	0.404
K	244	2,680	2390
La	2	6	3
Mg	3,636	1,776	2,700
Mn	724	392	1,030
Mo	3	12	11
Na	521	435	208
Ni	65	6	11
P	250	204	249
Pb	365	939	17,983
S	1,170	48,122	164,783
Sb	<3	<3	19
Se	<5	<5	<5
Sr	12	15	8
Th	<3	<3	<3
Ti	<100	<100	<100
Tl	<5	<5	<5
U	<8	<8	<8
V	23	<1	<1
W	<2	19	<2
Zn	189	1,120	5,890

NZ—outcrop sample, Nazca, Costa Sur Media, Peru and has not been in contact with mercury; PTZw—outcrop sample, Pataz, La Libertad, northern Peru and has not been in contact with mercury; PTZo—ore concentrate, also from Pataz

Inductively Coupled Plasma analysis, in parts per million (ppm), by American Assay Laboratories, Sparks, NV

the mercury content of precontact gold ornaments and to refine the analytical data initially provided by Petersen (2010: 57 [1970]). The low levels of mercury in the gold foils from Peru and Colombia (Table 10.3) are consistent with Craddock's (2000: 233) proposal that if mercury had been used to amalgamate placer gold in the ancient world, then low levels of mercury would be detected in the end-product gold.

Table 10.3 ICP analyses of pre-contact gold from Peru and Colombia

	MNA1	MNA2	MNA3	CP1	CP2a	CP2b	CP3
Ag	28,500	41,300	32,700	44,972	26,562	16,062	6,382
Al	645	135	121	118	354	308	179
As	14,800	5,900	15,500	1	1	1	1
Au	190,000	208,000	178,000	332,947	897,992	896,677	329,997
B	301	117	92	200	200	200	200
Ba	11	3	4	1	1	1	1
Bi	3	165	95	7	1	1	1
Ca	1,500	1,260	983	139	239	186	1,214
Cd	<0.5	<0.5	<0.5	0.02	0.03	0.11	0.02
Co	14	4	56	1	1	1	1
Cr	<1	<1	<1	0.5	1.9	1.0	0.5
Cu	665,000	700,000	800,000	610,337	38,500	49,400	659,461
Fe	93,817	<100	<100	111	418	263	103
Hg	13.8	12.3	13.9	1	2	9	12
K	413	202	<100	134	129	130	312
La	<1	<1	<1	1	1	1	1
Mg	323	336	<100	12	28	21	26
Mn	<2	4	1028	0.2	3.2	2.0	0.5
Mo	<1	<1	10	1	1	1	1
Na	3,180	2740	753	877	965	1145	1675
Ni	280	208	322	4	4	3	2
P	24	<10	<10	5	5	5	1,675
Pb	114	126	107	7.1	12.6	4.1	8.5
Pt	na	na	na	46	12	3	43
S	2,190	<500	<500	58	14	93	229
Sb	59	120	111	5	12	9	6
Se	40	48	46	2	2	2	2
Sr	8	17	8	0.7	1.1	0.8	5.8
Th	5	8	5	1	1	1	1
Ti	<100	<100	<100	0.7	3.1	3.4	2.1
Tl	<5	<5	<5	2	1	1	1
U	<8	<8	<8	2	2	2	2
V	47	26	24	2.1	0.4	0.6	2.9
W	<2	<2	<2	1	5	4	1
Zn	51	13	10	11	22	14	7

MNA1-3, fragments of Middle Sicán (AD 900–1200) gold foils, Huaca de la Ventana, Lambayeque, Peru (Carcedo Muro and Shimada 1985: 62); all Peru samples were obtained from Dra. Carmen Arellano Hoffman, Museo Nacional de Arqueología, Antropología e Historia del Peru, Lima

CP1—fragment of a pre-contact nose ornament, location unknown; CP2a-b—fragment of a pre-contact bead, Calima region, 200 BC–AD 1000, southwestern Colombia; CP3—fragment of a pre-contact disc, Tairona region, AD 800–1500, northeastern Colombia; all Colombia samples were identified by and obtained from Dra. Clemencia Plazas, Universidad Nacional de Colombia, Bogotá

Inductively Coupled Plasma analysis, in parts per million (ppm) by American Assay Laboratories, Sparks, NV,—indicates at or below detection, *na* not analyzed

Contamination

Analyses of 872 samples from 364 ancient Peruvian gold artifacts, with conspicuous surficial cinnabar, from a precontact burial showed that the mercury content ranged from 100,000 to 300,000 ppm (Gordus and Shimada 1995). They concluded that the elevated mercury content of the samples was contaminant-mercury from the cinnabar and was not an integral part of the gold.

The mercury content of the precontact Peruvian and Colombian gold foil samples analyzed for this study is far below the contaminant-mercury range of 100,000–300,000 ppm established by Gordus and Shimada (1995) and is consistent with, and comparable to, the amount of mercury in *charpita* gold from the amalgamation-*refogado* process in use today. Cinnabar was used in ancient burials and tombs in Peru but is not known to have been used in ancient tombs in Colombia.

The mercury content of the gold Chimú ornament, <100 ppm, analyzed by Petersen (Petersen 2010: 57 [1970]) is also far below the contaminant range established by Gordus and Shimada (1995). Microprobe analyses of placer gold indicate that the native mercury released during small-scale mining in eighteenth century Argentina is not pervasive and is limited to the rim of the placer gold grain (McCready et al. 2003). This indicates that any tomb-related mercury or cinnabar contamination would have been similarly limited to the surface of the gold artifact and would have been removed using standard cleaning and restoration techniques at the respective museums.

Interpretation and Discussion

Modern *refogado* gold, from Madre de Dios, that was amalgamated and then burned twice in the field, with a third and final burn at the gold shop, resulted in *charpitas* of gold with an average mercury content of approximately 15 ppm. The low mercury content of these samples, with an average of approximately 10 ppm mercury, is consistent with amalgamation and burning (*refogado*) the gold samples in the ancient Andes as is done today at small-scale gold mines in Peru, Colombia, and Ecuador. The low mercury content (<100 ppm) of the Chimú sample analyzed by Petersen (2010: 57 [1970]) is also consistent with this interpretation.

The decrease in mercury content from 1,000–10,000 ppm mercury in placer gold from Peru (Petersen 1970/2010: 25) to 12.3–13.9 ppm in the Sicán samples from this study, or from ~300,000 ppm in unburned amalgam to 8 ppm after three burns as indicated by the gold shop sample, can be explained by re-interpretation of the technique shown in a well-known Benzoni woodcut (Petersen 2010: 84 [1970]) (Fig. 10.2).

The Benzoni woodcut, which dates to the late 1500s, provides graphic detail on the final stages of gold processing in the New World. It is commonly assumed that the blowpipes or *sopletes* were used to blow on the coals at the base of the flames; however, this would not beautify the gold in the crucible, and perhaps only melt it.

Fig. 10.2 Benzoni woodcut showing sopletes in use with a crucible

As proposed herein, the *sopletes* were used as blowpipes to focus the flame directly onto the gold in the crucible. This technique would have produced a very hot, focused flame, ~1,500 °C, just beyond the visible portion of the flame, and would have produced gold that is very bright, reflective, and beautiful (Hurlbut 1971: 205). This would certainly have added to the aesthetic appeal of gold which, in the ancient Andes, was believed to be *lágrimas del sol*, or tears of the sun. This technique would also have reduced the mercury content of the gold.

In another Benzoni woodcut, molten gold was shown to have been poured down the throats of the Spaniards (Bray 1978: 28; van de Goot et al. 2003). This woodcut shows that *sopletes* were also used to melt the gold and this permits the inference that the melting temperature of gold, 1,063 °C, was readily achieved using the *soplete* method as shown in the woodcut.

Also, in ancient Colombia, the *soplete* was used to achieve the temperatures needed to sinter gold, which melts at 1,063 °C, and platinum, which melts at >1,700 °C (Petersen 2010: 55, 63 [1970]). Ancient Colombian and Ecuadorian metallurgists routinely worked gold and platinum using only a *soplete*, yet platinum was not smelted in Europe until the late 1800s (Bergsoe 1937; Bray 1988; McDonald 1960; Scott and Bray 1994). And, in an ancient storehouse near Cusco, Peru, copper, and ceramic *sopletes* of many different sizes were found (Petersen 2010: 37 [1970]).

The low mercury content of the Chimú gold sample (<100 ppm) (Petersen 2010: 57 [1970]) is consistent with volatilization of the mercury by burning the gold. Today, this *refogado* process is routinely used by many small-scale miners in Peru, where a gas torch, also called a *soplete*, is used to volatilize the mercury and improve the appearance of the gold (Brooks et al. 2007: 15). The *refogado* process is widely used by small-scale miners in many other regions of South America.

Conclusion

The abundance of precontact gold artifacts that were produced in the Andes prior to the arrival of the Europeans is consistent with a global chronology of cinnabar mining, mercury retorting, and small-scale gold production by mercury amalgamation. Given that cyanide methods for gold recovery were not developed in the USA until the late 1880s the amalgamation technology offers the best explanation for how gold foils were produced in the ancient Andes. The availability of cinnabar and native mercury in the region, specifically at Huancavelica where precontact mining took place and retorting vessels have been found, also supports this conclusion. Gold mining is known to have peaked during Inca times; and now, isotopic and geochronologic evidence from lake sediments also indicates that cinnabar mining at Huancavelica also peaked at the same time. The presence of ancient *quimbaletes* and tailings in Peru permit the inference that mercury was used then, as it is used now, to amalgamate the fine-grained gold from the *quimbalete*-crushed ore. Re-consideration of the ancient *soplete* method to direct a hot flame onto the amalgam to volatilize the mercury and beautify the gold has a present-day analog in the use of the gas torch, also called a *soplete* (Fig. 10.3).

Results of this study show that the multistep *refogado* process used by small-scale miners in present-day Peru is very efficient in reducing the mercury content of unburned amalgam from approximately 300,000 ppm in the field to approximately 15 ppm after the final burn in the gold shop. The quantity of mercury in the present-day *charpitas* is comparable to the low mercury content of ancient gold foils from Peru, Colombia, and Ecuador and is consistent with the interpretation that a similar mercury amalgamation-*refogado* process was used in the ancient Andes for small-scale gold mining and production before European contact.

Fig. 10.3 Modern batea by small scale miners. Note silvery gold–mercury amalgam in the bottom of the batea and small white plastic bottle of mercury along the cement wall of the muddy tank

Acknowledgments The authors wish to thank Prof. Georg Petersen, posthumously, for his insights into the relationship between ancient humans and the use of geological resources such as gold, copper, and silver found in the Andes. His research and use of analytical chemistry in his 1970 publication, Mineria y Metalurgia en el Antiguo Peru serves as a benchmark contribution to understanding ancient mining and metallurgy in the New World. Sincere thanks are also expressed to Dra. Carmen Arellano Hoffman, Museo Nacional de Arqueología y Antropología, Lima, Peru, and Dra. Clemencia Plazas, Universidad Nacional de Colombia, Bogotá, Colombia, for their enthusiasm for this project, and most importantly, providing the gold samples used for this research.

References

Agricola, G. (1556/1912). *De Re Metalica [H. C. Hoover and L. H., Trans.]*. New York: Dover Publications.

Ahlfeld, F., & Schneider-Scherbina, A. (1964). *Los Yacimientos Minerales y de Hidrocarburos de Bolivia* (Ministerio de Minas y Petroleo, Departamento Nacional de Geologia, Boletin 5). Bolivia: La Paz.

Aldenderfer, M., Craig, N. M., Speakman, R. J., & Popelka-Filcoff, R. (2008). Four thousand year old gold artifacts from the Lake Titicaca Basin, Southern Peru. *Proceedings of the National Academy of Sciences of the United States of America, 105*(13), 5002–5005.

Al-Hassan, A. Y., & Hill, D. R. (1986). *Islamic technology, an illustrated history*. Cambridge: Cambridge University Press.

Arana, P. P. (1901). *Las Minas de Azogue del Peru*. Lima: Imprenta El Lucero.

Barba, A. A. (1640). *El arte de los metales [R. E. Douglass and E. P. Mathewson, Trans., (1923), The art of metals]*. New York: Wiley.

Barba, L., & Herrera, A. (1988). San José Ixtapa, Un Sitio Arqueológico Dedicado a La Producción de Mercurio. *Anales de Antropología, Instituto de Investigaciones Antropologicas, 23*, 87–104.

Barnes, J. W., & Bailey, E. H. (1972). Turkey's major mercury mine today and how it was mined 8000 years ago. *World Mining, 25*(4), 49–55.

Bergsoe, P. (1937). The metallurgy and technology of gold and platinum among the pre-Columbian Indians. *Nature, 137*(29), 50–72.

Bonavia, D. (1985). *Mural painting in ancient Peru*. Bloomington: Indiana University Press.

Boyle, R. W. (1979). *The geochemistry of gold and its deposits* (Canadian Geological Survey Bulletin 280). Ontario: Ottawa Press.

Bray, W. (1988). Resolving the platinum paradox. *Americas, 40*(6), 45–49.

Brooks, W. E. (2012). Industrial use of mercury in the ancient World. In M. A. Bank (Ed.), *Mercury in the environment: Pattern and process* (pp. 19–24). Los Angeles: University of California Press.

Brooks, W. E., Guerra, A., & Nunez, F. (1995). Gold resources of the Rio Chivao Study Area, Estado Bolivar, Venezuela. *U. S. Geological Survey Bulletin, 2124*, H1–H5.

Brooks, W. E., Kent, J. D., & Willett, J. C. (2004). Mineral, energy, and fertilizer resources of the North Coast of Peru. U.S. Geological Survey Open-File Report 2004–1024. Online at http://pubs.usgs.gov/of/2004/1024.

Brooks, W. E., Piminchumo, V., Suarez, H., Jackson, J. C., & McGeehin, J. P. (2008). Mineral pigments from Huaca Tacaynamo, Chan Chan, Northern Peru. *Instituto Francés de Estudios Andinos (Lima), 37*(3), 1–10.

Brooks, W. E., Sandoval, E., Miguel, Y., & Howard, H. (2007). *Peru mercury inventory 2006*. U.S. Geological Survey Open-File Report 2007–1252. http://pubs.usgs.gov/of/2007/1252/. Accessed 27 Aug 2007.

Brooks, W. E., Schwörbel, G., Luis, E., & Castillo, L. E. (2009). Mercury and small-scale gold mining in the Ancient Andes. *Geological Society of America Abstracts with Programs, 41*(7), 435.

Brown, K. W. (2001). Worker's health and colonial mercury mining at Huancavelica, Peru: the Americas. *The Academy of American Franciscan History, 57*(4), 467–496.
Burger, R. L., & Matos, R. (2002). Atalla, a center on the periphery of the Chavin Horizon. *Latin American Antiquity, 13*(2), 10–25.
Cabrera La Rosa, A. (1954). *Situación Actual de la Minería del Mercurio en el Peru* (pp. 3–12). Madrid: Minería y Metalurgia.
Cánepa, C. (2005). *Minería a Pequeña Escala en la Costa Sur Media del Peru* (Dirección de Geología Económica y Prospección Minera, Bol 3, Ser E). Lima: Mineria,Instituto Geológico Minero y Metalurgíco.
Carcedo Muro, P., & Shimada, I. (1985). Behind the gold mask, Sicán gold objects from Batan Grande, Peru. In J. Jones (Ed.), *The art of Precolumbian gold* (The JanMitchell collection). New York: Graphic Society.
Castillo Espitia, N. (2007). Minería Aurífera en el Noroeste Andíno de Colombia, Etnografía de la Técnica. In R. Lleras Perez (Ed.), *Metalurgía en la Améríca Antigua*. Bogotá, DC: Fundación de Investigaciones Arqueológicas Nacionales, Banco de la República, Instituto Francés de Estudios Andinos.
Consejo de Recursos Minerales. (1992). *Geological mining monograph of the state of Queretaro, Mexico*. Mexico, DF: Consejo de Recursos Minerales, Secretaria de Energía, Minas E Industria Paraestatal, M-4e.
Cooke, C. A., Balcom, P. H., Biestar, H., Alexander, P., & Wolfe, A. P. (2009). Over three millennia of mercury pollution in the Peruvian Andes. *Proceedings of the National Academy of Sciences of the United States of America, 106*(22), 8830–8834.
Craddock, P. T. (1995). *Early metal mining and production*. Washington, DC: Smithsonian Institution Press.
Craddock, P. T. (2000). Early history of the amalgamation process, appendix 3. In A. Ramage & P. T. Craddock (Eds.), *King Croesus' gold*. London: British Museum Press.
Crozier, R. D. (1993). Silver processing in Spanish America, the Patio and Buytron processes. *Canadian Mining and Metallurgical Bulletin, 86*(972), 86–91.
D'Itri, P. A., & D'Itri, F. M. (1977). *Mercury contamination, a human tragedy*. New York: Wiley.
Del Busto Duthurburu, J. A. (1996). *La Plateria en el Peru. Colección Enrico Poli, Banco Sur del Peru*. Lima: Cuzzi y Cia, S.A.
El Comercio. (2009). Lucha Contra Minería es Desigual. *El Comercio*, Lima, 31 de mayo. http://elcomercio.pe/impresa/notas/lucha-contra-mineria-desigual/20090531/293934. Accessed 27 Aug 2009.
Elton, C. (2001). Vaivenes de una industria milenaria. *Americas, 4*, 6–15.
Gordus, A. A., & Shimada, I. (1995). Neutron activation analysis of microgram samples from 364 gold objects from a Sicán burial site in Peru. In P. B. Vandiver et al. (Eds.), *Materials issues in art and archaeology IV* (Materials Research Society Proceedings, 352, pp. 127–142). Pittsburgh: Materials Research Society.
Grossman, J. W. (1972). An ancient gold worker's tool kit. *Archaeology, 25*(4), 270–275.
Haeberlin, Y., Fontebote, L., & Moritz, R. M. (2004). Carboniferous orogenic gold deposits at Pataz, eastern Andean Cordillera, Peru. *Economic Geology, 99*, 73–112.
Hemming, J. (1970). *The conquest of the Incas*. London: Macmillan.
Hurlbut, C. S. (1971). *Dana's manual of mineralogy* (18th ed.). New York: Wiley.
Juan, J., & De Ulloa, A. (1748). *A voyage to South America [J. Adams J, Trans., (1807) of Relación Historica del Viaje a la America Meridional, [Madrid]*. London: Antonio Marín.
Langenscheidt, A. (1986). El Cinabrio y el Azogue en el Mexico Antiguo. *Revista Minero-Americana, Mexico City, 17*(1), 24–29.
Larco Hoyle, R. (2001). *Los Mochicas, t. II*. Lima: Museo Arqueologíco Rafael Larco Herrera.
Lechtman, H. (1976). A metallurgical site survey in the Peruvian Andes. *Journal of Field Archaeology, 3*(1), 1–42.
Lechtman, H. (1988). Traditions and styles in Central Andean metalworking. In R. Maddin (Ed.), *The beginning of the use of metals and alloys*. Boston: Massachusetts Institute of Technology Press.

Lozano, H. (1987). Minerales de Mercurio. In A. Villegas Betancourt (Ed.), *Recursos Minerales de Colombia, t. I, Metales* (pp. 275–294). Bogotá, DC: Publicaciones Geológicas Especiales del INGEOMINAS.

McAllister, J. F., Flores, H., & Ruiz, C. (1950). Quicksilver deposits of Chile. *U. S. Geological Survey Bulletin, 964-E*, 361–400.

McCready, A. J., Parnell, J., & Castro, L. (2003). Crystalline placer gold from the Rio Neuquen, Argentina, implications for the gold budget in placer gold formation. *Economic Geology, 98*(3), 623–633.

McDonald, D. (1960). *A history of platinum*. London: Johnson Matthey and Co.

Ministerio de Energía y Minas. (2000). *Atlas de la Pequeña Minería en el Peru 2000*. Lima: Ministerio de Energía y Minas, Dirección General de Minería.

Noble, D. C., & Vidal, C. (1994). Gold in Peru. *Society of Economic Geologists Newsletter, 17*, 6–13.

Nuñez, E., & Petersen, G. (2002). *Alexander von Humboldt en el Peru, Diario de Viaje y Otros Escritos*. Lima: Banco Central de Reserva del Peru, Goethe Institut Inter Nationes.

Ogura, T., Ramirez-Ortiz, J., Arroyo-Villasenor, Z. M., Martinez, S. H., Palafox-Hernandez, J. P., Garcia de Alba, L. H., & Fernando, Q. (2003). Zacatecas (Mexico) companies extract Hg from surface soil contaminated by ancient mining industries. *Water, Air, and Soil Pollution, 148*, 167–177.

Oyler, K. (1997). Screen mesh sizes. *The Gold Nugget*, April. http://www.lornet.com/prospector/articles/screensmesh.htm. Accessed 15 June 2010.

Pendergast, D. M. (1982). Ancient Maya mercury. *Science, 217*(6), 533–535.

Petersen, M. (1979). Graveyard of the quicksilver galleons. *National Geographic, 156*(6), 850–876.

Petersen, G. (2010). *Mining and metallurgy in Ancient Peru [W. E. Brooks, Trans., 1970]* (Minería y Metalurgia en el Antiguo Peru, Arqueologicas 12). Boulder, CO: Publicaciones del Instituto de Investigaciones Antropologícas, Museo Nacional de Antropología y Arqueología, Pueblo Libre, Lima, Peru.

Petersen, U., Vidal, C. E., & Noble, D. C. (1990). A special issue devoted to the mineral deposits of Peru. *Economic Geology, 85*(7), 1–289.

Plazas, C. (2007). La Metalurgía del Area Intermedia Sur Dentro del Panorama Americano. *International Journal of South American Archaeology, 1*, 33–38.

Ramage, A., & Craddock, P. T. (2000). *King Croesus' gold*. London: British Museum Press.

Rehren, T., & Temme, M. (1994). Pre-Columbian gold processing at Putushio, Southern Ecuador. In D. A. Scott & P. Meyers (Eds.), *Archaeometry of Pre-Columbian sites and artifacts* (pp. 267–284). Los Angeles, CA: UCLA Institute of Archaeology and the Getty Conservation Institute.

Rivero, M. E., & Tschudi, J. J. (1853). *Peruvian antiquities*. New York: G.P. Putnam.

Roberts, R. J., & Irving, E. M. (1957). *Mineral deposits of Central America* (U.S. Geological Survey Bulletin 1034). Washington, DC: U.S. Government Printing Office.

Robins, N. A. (2011). *Mercury, mining, and empire*. Bloomington: Indiana University Press.

Schreiber, D. W., Fontbote, L., & Lochman, D. (1990). Geologic setting, paragenesis, and physicochemistry of gold-quartz veins hosted by plutonic rocks in the Pataz region. *Economic Geology, 95*(3), 1328–1347.

Schuette, C. N. (1931). *Quicksilver* (U.S. Bureau of Mines Bulletin 335). Washington, DC: US Department of Commerce.

Scott, D. A., & Bray, W. (1994). Pre-Hispanic platinum alloys: Their composition and use in Ecuador and Colombia. In D. A. Scott & P. Meyers (Eds.), *Archaeometry of Pre-Columbian sites and artifacts* (pp. 285–322). Los Angeles, CA: UCLA Institute of Archaeology and the Getty Conservation Institute.

Shimada, I., & Griffin, J. A. (2005). Precious metal objects of the Middle Sicán. *Scientific American, 15*(1), 80–89.

Shimada, I., & Merkel, J. F. (1991). Copper-alloy metallurgy in ancient Peru. *Scientific American, 265*(1), 80–86.

Truhan, D. L., Burton, J. L., & Bruhns, K. O. (2005). *El cinabrio en el mundo Andino* (Revista de Antropología 18, pp. 193–205). Cuenca: La Casa de La Cultura Ecuatoriana Benjamin Carrion.

Van de Goot, F. R. W., ten Berge, R. L., & Vos, R. (2003). Molten gold was poured down his throat until his bowels burst. *Journal of Clinical Pathology, 56*(2), 157–170.

Vaughn, K. J., Linares, M., & Eerkens, J. W. (2007). Hematite mining in the ancient Americas, Mina Primavera, a 2000 year old Peruvian mine. *Journal of Metals, 115*(4), 16–20.

West, R. C. (1952). *Colonial placer mining in Colombia*. Baton Rouge: Louisiana State University. 157 p.

West, J. M. (1974). *How to mine and prospect for placer gold* (*U.S. Bureau of Mines Information Circular* 8517). Denver, CO: U.S. Bureau of Mines Information.

Wilson, F. (1941). Colombia. *Enineering and Mining Journal, 142*(8), 154–155.

Wilson, J. J., & Reyes, L. (1964). *Geologia del cuadrangulo de Patáz* (*Dirección General de Minería, Bol.* 9). Lima: Ministerio de Energía y Minas.

Wokittel, R. (1958). *La mina de mercurio La Esperanza, Municipio Aranzazu, Departamento Caldas: Servicio Geologico Nacional*. Informe: Bogotá. 1282.

Yates, R. G., Kent, D. F., & Concha, J. F. (1955). *Geology of the Huancavelica quicksilver district*. Peru: U.S. Geological Survey Bulletin. 975A.

Yildiz, M., & Bailey, E. H. (1978). *Mercury deposits in Turkey. U.S. Geological Survey Bulletin 1456*. Washington, DC: U.S. Geological Survey Bulletin.

Chapter 11
Silver Mines of the Northern Lake Titicaca Basin

Carol A. Schultze

Introduction

Archaeological silver and copper mining and production sites are prominent features of the cultural landscape of the Andes (Emmerich 1965; Jones 2005; Lechtman 1984; Chaps. 12 and 13). Prehispanic and colonial silver mines and processing centers have been identified in the Department of Puno in the Northern Lake Titicaca Basin (Fig. 11.1), including some of the largest and earliest centers of ore acquisition and metalwork (Table 11.1). Significant tribute demands in silver and gold were made of the Altiplano communities as documented by both the *visita* of Garci Diez de San Miguel (1567 [1964]: 146) and Toledo's *tasa* of 1570 [1975]. Puno Bay along the northern shore of the Lake Titicaca Basin is reported as the location of one of the largest bonanza silver mines in the Colonial period. The Laykakota find by the brothers José and Gaspar Salcedo in 1657 is said to have, for a time, surpassed the production of Potosí in Bolivia (Cusi et al. 1992; CTAR 1999; Parodi Isolabella 1995). The mines and old townsite are located southeast of the current city of Puno in the shadow of Cerro Cancharani. The original town was called San Luis de Alba and is estimated to have housed some 10,000 inhabitants (Nuñez Mendiguri 2001). Industrial-scale mining pits have been recorded in the Laykakota/Salcedo mine in the hills directly above Puno Bay (Fig. 11.2).

The earliest recovered precious metal artifacts in the Andes are from excavated contexts at Jiskairumoko, in Lake Titicaca's Ilave drainage (Aldenderfer et al. 2008). Hammered beads of native gold were associated with contexts radiocarbon dated to 3733±43 years BP, or calibrated from 2155 to 1936 BC. These Terminal Archaic finds predate by approximately 600 years the previously earliest find of thin gold sheets on the Peruvian north coast in the Lurín Valley, which were dated to a calibrated 1410 to 1090 calendar BC. (Burger and Gordon 1998).

C.A. Schultze (✉)
Historical Research Associates, Inc. 1904 3rd Avenue, Seattle, Washington 98101
e-mail: cschultze@hrassoc.com

N. Tripcevich and K.J. Vaughn (eds.), *Mining and Quarrying in the Ancient Andes*, Interdisciplinary Contributions to Archaeology, DOI 10.1007/978-1-4614-5200-3_11,

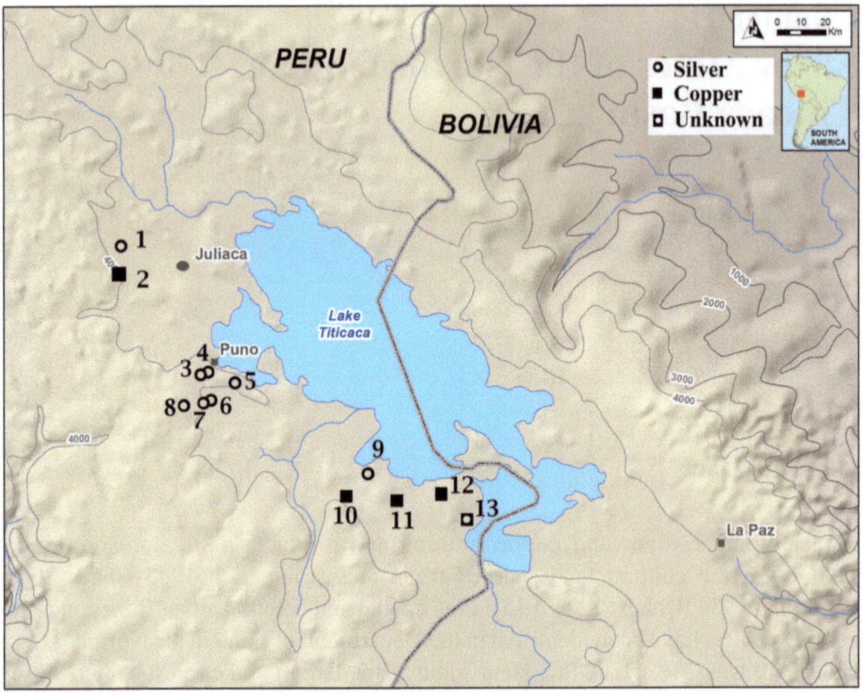

Fig. 11.1 Department of Puno with UTM locations of archaeological mining sites mapped in ArcView GIS 10. 1=Paratía [Ag-Pb]; 2=Mawka Paratía ; 3=Itapalluni [Ag]; 4=Laicacota [Ag]; 5=Platería [Cu]; 6=Chiluyo [Ag]; 7=Andamarca [Ag]; 8=San Antonio de Esquilache (ag); 9=Nairanaqqe [Ag]; 10=Jaruni [Cu]; 11=Llaquepa [Cu]; 12=Batalla-Rinconada [Cu]; 13=Trapiche-El Molino (Co. Capia) (Lechtman 1976)

Table 11.1 Lake Titicaca cultural phases (after Levine 2012; Janusek 2004; Moseley 2001) compared to the Ica valley sequence (Menzel 1977, Moseley 2001)

South Central Andean phases	Approximate dates	North and coastal phases	Approximate dates
Archaic	10,000–1500 BC	Lithic	10,000–3000 BC
		Preceramic	3000–2000 BC
Lower Formative	2000–1300 BC	Initial Period	2000–700 BC
Middle Formative	1300–500 BC	Early Horizon	700 BC–0 BC
Upper Formative	500 BC–AD 500	Early Intermediate Period (EIP)	AD 0–600
Middle Horizon	AD 500–1100	Middle Horizon	AD 600–1100
Late Intermediate Period (LIP)	AD 1100–1450	Late Intermediate Period (LIP)	AD 1100–1400
Late Horizon	AD 1450–1532	Late Horizon	AD 1400–1532

Fig. 11.2 Overview of P117 locus of Laykakota/Salcedo mine (*left*), single mine pit in P117 (*right*), both facing northwest

The earliest silver workshop debris in the Andes is the Puno Bay finds from the site of Huajje (Schultze et al. 2009). At this U-shaped temple mound, workshop debris from silver ore cupellation was found associated with radiocarbon dates of 1960±40 BP and 1870±40 BP, or a calibrated range overlapping at AD 60 to 120 (Schultze et al. 2009: 17281). The Huajje metallurgical debris demonstrates that ore reduction (employing temperatures greater than 1,000 °C) was known and in use by Andean people in the first millennium AD. These same technologies continue with only minor changes into the Spanish Colonial period. Three independent lines of evidence established the chronological integrity of the deposit: (1) a ceramic sequence in uninterrupted stratigraphic layers, (2) absolute radiocarbon dates, and (3) absolute ceramic thermoluminescence (TL) dates. While the absolute dates confirm an Upper Formative workshop, the ceramic sequence shows that the site was occupied in some form as early as the Middle Formative (Schultze 2008: 324).

Lake sediment studies have confirmed these early dates for silver production in the Lake Titicaca Basin (Abbott and Wolfe 2003; Cooke et al. 2007). The lead content found in lake sediment profiles has been measured, where increases in deposition of volatized, initially airborne, lead is interpreted as a by-product of silver smelting. Sediment profiles in Lake *Taypi Chaka* on the Bolivian side of Lake Titicaca registered an increase in sedimentary lead levels circa AD 400. This is significantly earlier than similar increases in settled lead deposits from the more northern Lake Pirhuacocha in Junín and southern Lake Lobato near Potosí. These register significant lead deposition only after circa AD 1000 (Abbott and Wolfe 2003; Cooke et al. 2008).

Additionally, reconnaissance survey by the author in 2010 has confirmed San Antonio de Esquilache to be a large scale mining center of the Inca and Colonial periods. Significant labor investment in the Late Horizon is apparent in the carefully constructed mine shaft galleries and workers' residences (Fig. 11.3).

Fig. 11.3 A stone mine shaft gallery in the caldera surrounding the San Antonio de Esquilache Inca mining settlement with a characteristically Prehispanic lintel design. Both Colonial and earlier diagnostic ceramic forms were found across the site

Taken together, these finds indicate that some of the earliest and most enduring exploitation of native and ore-based precious metals in the Andes took place in the Lake Titicaca Basin. This paper discusses recent systematic archaeological work in Puno Bay regarding the development of mining technologies in the Northern Lake Titicaca Basin of Southern Peru.

While the primary focus of this report is on the mining center at Laykakota/Salcedo, it is discussed as part of a larger economic system for the utilization of a spatially discrete set of "Puno Group" intrusive volcanic deposits. This paper explores the linkages between silver, copper, and andesite production within Puno Bay, through an examination of archaeologically identified mines, quarries, and workshop sites.

The Study Area

Physiographically, the Lake Titicaca Basin falls within the central-volcanic zone (~3° to 33°S), and is a volcanically active steep slab formation. It is built on a variable-age Precambrian and Palaeozoic basement, which uplifted in the Late Miocene epoch, some 10.3 to 6.7 million years ago (Ghosh et al. 2006). The Altiplano-*puna* plateau is approximately 700 km long and 200 km wide, with an average elevation of 3,700 m. Precious metals in the highest regions (>6,300 m) are associated with complexes of Late Miocene to Pliocene andesitic to dacitic stratovolcanic and giant dacitic ignimbrite rock formations. Important tin and silver deposits are additionally

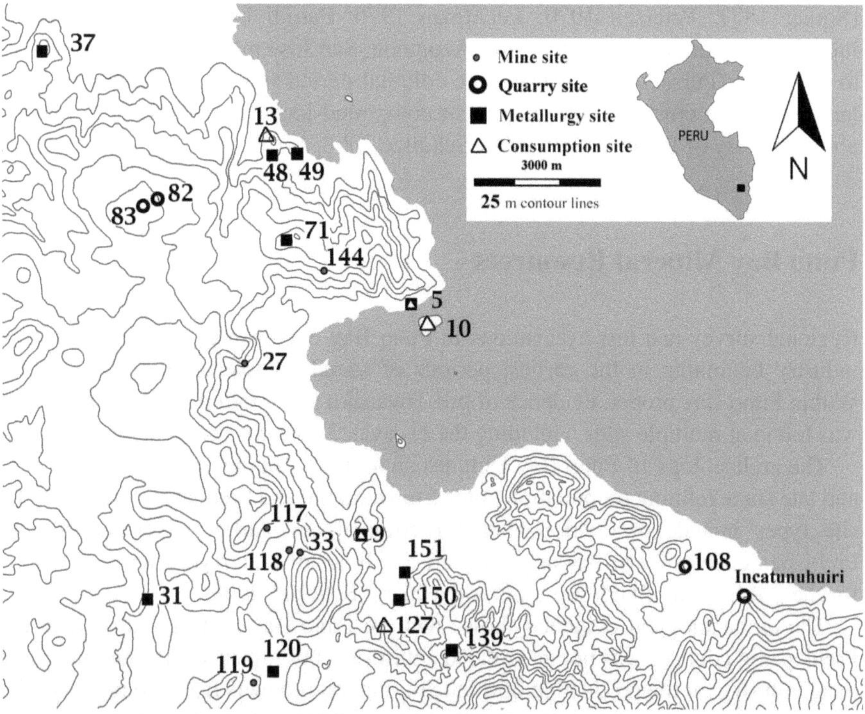

Fig. 11.4 Location of the Puno Bay metal and mineral production sites. All numbers are Puno Bay site numbers referred to in the text as P# (e.g., 31 on map is P31 in text)

found in Tertiary volcanic concentrations along both branches of the Cordilleras (Kay et al. 2005).

The Puno Bay project area includes the city of Puno as well as drainages to the north and south. The city of Puno is located along the northwest shoreline of Lake Titicaca and is the capital of the department of the same name. The project elevation ranges from the lakeshore at 3,800 m to the highest peak of Cerro Cancharani at 4,330 m. General topographic features depicted in Fig. 11.4 (east to west) include: (1) lower altitude flatlands along the lake and river drainages; (2) steep sideslopes; (3) mountain peaks; (4) high altitude pampas; and (5) foothills of the western Cordillera. The lower elevation river drainages and lakeshore flats are highly productive agricultural zones in the 3,800–3,850 m range.

The geology of the Puno Bay region includes areas of Quaternary alluvium along the shorelines and riverbeds, with the Quaternary Sillapaca Volcanic formation comprising the plateau overlooking the city. Agriculture is most optimal along the alluvial pampas, although terracing of soils along the Huancané and Puno Group hillsides is also commonly practiced.

Minerals of economic importance (silver, copper, andesite, and others) occur with Tertiary period intrusives. Historic and modern period copper, mercury, and silver mines have been reported in the intrusive zones in the hills surrounding Puno

(Squier 1877; Petersen 1970; Lechtman 1976; Parodi Isolabella 1995). These include Cancharani/Laicacota, Cerro Azoguine, San José mines, and smaller mines to the south (Purser 1971). During the colonial period Chorrillos Itapalluni was a large-scale ore crushing and refinement compound located on the eastern plateau above the city of Puno (CTAR 1999; Schultze 2008: 136).

Puno Bay Mineral Resources

Regional survey and test excavations in Puno Bay documented a silver-producing industry beginning in the earliest periods of social complexity (Schultze 2008). Within Puno Bay proper, evidence of pre-Tiwanaku silver mining and ore reduction was found at multiple sites, including the excavated contexts in the site Huajje (P5).

The archaeology of Puno Bay includes an array of mining, quarrying, smelting, and late stage refining (cupellation) sites from colonial and Prehispanic periods. The site types include production sites for colonial and earlier silver-ore mining, Prehispanic silver ore processing, andesite and limestone quarrying, and workshops for the production of andesite tools and ashlars (Schultze 2008).

The archaeological evidence shows that the Puno Bay mining industry, in the Prehispanic context, was not an isolated enterprise, but rather formed an integrated local industry for the production of a suite of mineral products from the raw material present in the surrounding hills. The Tertiary-era "Puno Group" intrusive volcanic deposits present in the bay (Acosta et al. 2010; INGEMMET 1970) correlate with silver metal ores, but also include abundant and economically important mineral resources including copper, andesite, mercury, granite, rhyolite, quartz-monzonite, monzonite, granodiorite, dacite, and diorite (Petersen 1965: 420).

Elemental Analysis of Archaeologically Recovered Silver Ore

Speculation about the potential ores exploited in the Andean world has tended to center on lead-base galenas (Pb-S) and gossans containing impurities of silver sulfide (Petersen 1970), or alternatively, it has been speculated that silver was produced using less-technologically advanced reduction-smelting of native silver with lead contaminants (Patterson 1971).

The colonial history of Puno makes it clear that whatever original ore bodies were present in Puno Bay, these have long since been removed. As a testament to the immeasurable richness of the geology of the Andes, individual miners continue to opportunistically exploit copper veins found in Puno's surrounding hillsides. For example, two open pit mine sites in the northern portion of the project area, P32 and P144, are locations of modern copper extraction. A vein of copper ore was visible in a hard rock gallery on Cerro Azoguine (P32). However, it is uncertain how similar these ores are to those available in prehistory.

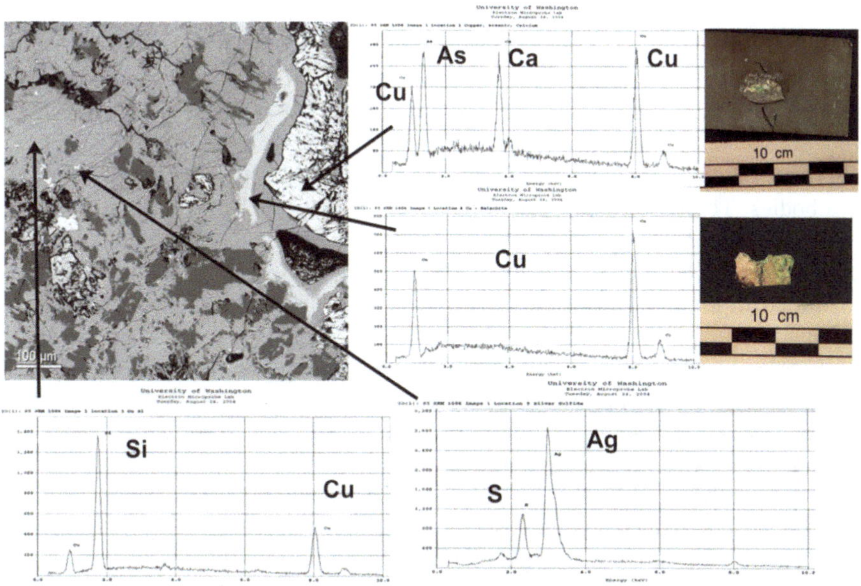

Fig. 11.5 Scanning election microscope spectra for ore rock sample SEM 1007, Huajje P5, Puno Bay, Peru

To discover more about the native materials, *Proyecto Wayruro* collected and assayed two ore samples from archaeological contexts. One is from a secured stratified context at the silver workshop site at Huajje. Sample SEM 1007 was recovered from 3.6 to 3.7 m below ground surface in the monumental terraced mound site and sunken court Huajje P5. These levels correspond to the earlier portion of the Middle Horizon or Tiwanaku deposits. SEM work conducted by the author and Scott Kuehner of the Department of Earth and Space Sciences at the University of Washington in 2004 showed this rock to contain zones of arsenical copper, copper silicate, pure copper metal, and small inclusions of silver sulfide (Fig. 11.5).

The second sample was collected from the surface of a densely occupied workshop site of Punanave (P9). We collected one (7.91 g) green-colored rock from a central flat terrace that was littered with a high-density scatter of ceramic fragments, lithic debitage, and crucibles with slag-covered surfaces.[1] Punanave was occupied in all ceramic periods, so no chronological information is available regarding this sample.

[1] Surface density estimates are defined as "low" (0–2 artifacts/1 m^2), "medium" (2–10 artifacts/1 m^2), and "high" (10+ artifacts/1 m^2).

Analysis at the XRF laboratory at the University of California at Berkeley was conducted by the author and Nicholas Tripcevich in 2011. This identified the presence of copper at 10,171.66 ppm and silver at 155,372.28 ppm. Barium and antimony are the only other elements present at similarly high orders of magnitude (357,544.60 ppm and 26,158.38 ppm, respectively).

These two samples demonstrate that copper and silver were present in the same ore bodies. These rich mineral and metal deposits would have been amenable to multi-crafting for the production of a variety of outputs, including both silver and copper metal. Thus far, no archaeological evidence of copper production has been identified in the bay. However, there is evidence for the consumption of copper products in all periods.

The Puno Group intrusive deposits presented the people of Puno Bay with abundant opportunities for multi-craft production (Shimada 2007: 5). Silver production was one facet of a larger mineral industry, which also centered around andesites and basalts. As early as the middle to upper Formative periods (Table 11.1), the andesitic volcanics were fashioned into construction blocks (ashlars), agricultural tools (*hachas),* and monoliths or figural sculptures (Hastorf 2005). Limestone, sandstone, and clay deposits are also available in Puno Bay and were utilized throughout the cultural sequence for construction materials, sculpture, and ceramics (Kidder 1943; Schultze 2008).

In Puno Bay, there are many different types of sites related to mining, quarrying, and mineral processing. In order to understand how these spatially and functionally distinct sites might have been integrated into a single productive sequence, a brief description of the silver reduction sequence is provided.

The Silver Ore Reduction Process

Silver purification is a complicated, multi-stage process. Each step requires specific technical requirements be met, such as the creation of oxidizing versus reducing conditions, the availability of water, the presence of ores themselves, and in the Andean case, predictably high winds (van Buren and Mills 2005). Tasks performed at production sites (and their products) can be grouped into (1) mining/direct extraction (ores); (2) ore crushing (rocks); (3) smelting by *huayrachina* (Pb slag, Cu matte); (4) further reprocessing of slags (Pb slag, Cu matte, Ag–Pb metal), (5) cupellation to remove silver from lead bullion (PbO, Ag metal), and (6) fabrication of finished silver objects (hammered metal sheets of *tumbaga,* Ag alloys). Each of these steps needs to be carried out in specific types of locations.

Obviously, mining would have to be conducted at the source. Smelting requires temperatures up to 400 °C (Cohen 2008: 59). Traditionally, these temperatures would have been attained in traditional *huayrachina* furnaces, in locations with reliably heavy winds. *Huayrachina* are large ceramic vessels, with vent holes along the side (Lechtman 1976: 9; Peele 1893). They were the technology of "the Aymara (who) smelted gold in pottery ovens exposed to the wind on hilltops" (Tschopik

1946: 537). In each successful instance, smelting produces a slag-covered crucible, chunk slag of lead glass (PbO with impurities), a matte cake of base metals (CuS with impurities), and a bullion nugget containing pure lead metal surrounding a globule of pure silver metal.

The lead–silver bullion piece would then have been taken to a cupellation furnace, called *tocochimbo* in the Andes, and heated to temperatures greater than the melting point of silver, around 1,000 °C in oxidizing conditions (Cohen 2008: 62; van Buren and Mills 2005). This process oxidizes and separates the lead, leaving lead oxide (called litharge) and purified silver metal. Reverberatory furnaces (*tocochimbos*) have been documented in ethnographic conditions in contemporary Potosí, Bolivia (Cohen et al. 2006, van Buren and Mills 2005).

The silver would then have been taken to yet another location to be alloyed and fashioned into final products for consumption (ornaments, *keros*, or other high status - low weight items). Thus, transformation of silver-bearing ore to purified silver requires at least three different physical settings: the mine, the windy ridge, and the protected cupellation furnace. Alloying and fabrication facilities could represent a forth location.

In Puno Bay, we have documented all but the final two steps in this productive chain (Rehren et al. 2010; Schultze et al. 2009). Also sites which are likely candidates for cupellation facilities and hammered metal workshops have also been identified. The sites that have been recorded, and some discussion of their potential role in the larger quarrying and mining industry of the Bay, are discussed in the following sections.

Archaeological Mining and Metallurgical Sites in Puno Bay

The Huajje silver metallurgical finds are best viewed within a broader regional context of mining and extraction. Regional survey identified a total of seventeen mines and metal production sites which will be discussed in greater detail below (see Fig. 11.4). The known archaeological landscape of Puno includes mines for silver (P32, P33, P117, P118), mines for copper and other minerals (P27, P144, Purser 1971, Squier 1877; Petersen 1970; Lechtman 1976; Parodi Isolabella 1995), smelting locations (P9, P37, P48, P49, P71, P120, P139, P150, P151), cupellation facilities (P5), and evidence for consumption in the form of finished hammered metal sheets (P10, P13, P127, Julien 1983). Final-stage fabrication workshops have not been found, but may have been present in the palatial compound on the island site Isla Esteves P10, at least during the Middle Horizon (de la Vega 1998). In addition to these metal-related sites a number of andesite and basalt quarries (P71, P82, Kidder 1943) and workshops for finishing andesite and basalts (P13, P36, P83, P108) were recorded by the 2002 survey.

Three principal silver-related sites were occupied in all ceramic periods: the silver mine of Laykakota at Cerro Cancharani P118, and smelters and reprocessing at Punanave P9 and Huajje P5. The following sites were found on survey using a stratified sample that took in more than 50 % of all major elevation zones. The results are therefore representative, rather than comprehensive. All site numbers refer to locations shown in Fig. 11.4.

The survey identified sites related to mining, smelting, cupellation, and consumption of precious and utilitarian metals. The remainder of this paper describes the salient features of each of these sites to give the reader an overview of the archaeological record of Puno Bay as it relates to the mineral industries in which silver ore extraction and processing were embedded.

Consumption Sites

The evidence for local consumption in the archaeological evidence consists of fragments of hammered copper sheets found in ritual contexts. Finished silver has not been found archaeologically, likely the result of looting. Copper metals had utilitarian and ornamental uses. Silver was typically alloyed with gold and/or copper into *tumbaga* hammered metal sheets, from which would have been formed ornaments, high status serving vessels, and items of display. Export is clearly a goal of production in the later periods, and may be able to be demonstrated for the Formative period. But the presence of hammered copper metal on sites of all periods demonstrates that a range of processed metal products did make their way into the local economy.

Copper metal was found in the lower sunken court excavated at P13 (Schultze 2008: 342–385). This site is located on the top and terraced sideslopes of a hill at elevation of 3907 m amsl. This 20 hectares covers the entirety of Cerro Chincheros. At the summit is a large (25 m N-S × 26 m E-W × 4 + m deep) sunken court. A basalt-monolith was recorded in the 2002 season but has subsequently been looted away. The interior of the lower sunken court was composed of a cultural midden including Qaluyu serving bowls and bits of disintegrating copper. Across the site was a high density of artifacts from which were recovered fragments of a neckless olla forms, Qaluyu and Pucara decorated pieces, as well as Tiwanaku fineware. These diagnostic artifacts indicate the consumption of metal products in ritual contexts beginning in the Middle Formative period.

The site was also surrounded by workshop terraces which included evidence for the production of andesite agricultural tools. There were andesite flakes in all stages of reduction consistent with primary production of Formative plough blades and the fine finishing of andesite ashlar blocks for monumental construction. The blocks of the lower sunken court at P13 were made of smoothly ground or "finished" andesite and basalt ashlars. This finishing technology may have developed early in the Middle Formative with the grinding of exteriors and bits of the plough blades, or "-'*tqlla*" (Seddon 1994) in order to flatten high spots and make for a more attractive or durable surface (Hintzman 2004).

Copper consumption is also seen at the U-shaped monumental structure Huajje (P5). Copper metal was found adhered to the tooth of a Tiwanaku IV incensario puma recovered during salvage operations associated with road construction by

Fig. 11.6 Tiwanaku IV puma incensario recovered from Huajje (P5) that had copper metal adhered to tooth of puma head

Félix Palacios in the 1970s (Fig. 11.6). This site is also a smelting location and is described more fully below.

During the Middle Horizon, finished metal was consumed at the near-shore island site of Isla Esteves P10. Worked copper artifacts were found in excavations at this elite residential enclave and ceremonial location of the Tiwanaku state. The central feature of this site is an Akapana-shaped artificial mound (pyramid) on the north end of the island. Excavations in the 1970s and 1990s argued for occupation exclusively during expansive Tiwanaku phases IV and V (de la Vega 1998). These excavations revealed elite domestic compounds. Beautiful fineware ceramics were recovered, along with worked copper artifacts, projectile points and a variety of stone tools (Schultze 1999). Very fine Tiwanaku *kero*s were recovered during this construction, which strongly suggest the presence of a sunken court and elite ceremony.

Evidence for Late Intermediate period consumption of finished metal comes from an unnamed site P127 found at an elevation of 3,990 m amsl. These were two looted large "beehive" burial chambers, 3 m in diameter that were built into the space below a rock overhand at the edge of the cliff-face contour. Two remarkably well-preserved textiles were recovered (Schultze 2008: 539). In addition, there were rounded plate rims, a thumb-sized basalt utilized flake, one (8.2 g) piece of shaped wood, and a disk of hammered copper (2.7 cm diameter, <0.1 cm thickness).

For the Late Horizon, a sheet of hammered copper was found in a burial in the Azoguine neighborhood of the city of Puno (Julien 1981). Ceramic types were similar to Hatuncolla series 3 (Julien 1983), reflecting a period well into the Inca Horizon and a northern (*Colla*) affiliation for this portion of Puno during the Late Horizon. The burial goods included two bowls, one small bottle, three basalt flakes, and one T-shaped copper knife (~9 cm blade).

The preponderance of the archaeological evidence recovered in Puno Bay records the consumption of copper and andesite/basalt stone. Local consumption of the silver metal in these same mineral deposits is inferred with support from a series of mining and smelting sites in the surrounding landscape.

Silver Mines

The principal mining region in the Puno Bay study area is the Cancharani/Laykakota complex, corresponding to the Salcedo bonanza mine discussed above (von Tschudi 1865; Squier 1877). Three loci of mining pits were recorded (P33, P117, P118), surrounding a small modern community and cemetery in the level plain between Cerros Negro Peque and Cancharani.

The northern locus of this mining complex, P117 is a 15.75 hectare site consisting of more than 50 quarry pits with associated tailing piles. Some of the pits are as large as 10 m across with depths up to 7 m. Some have chambers that head off in different directions. Large mill stones for grinding ore are located near the center of the site. There is a medium density artifact scatter from all periods, including neckless olla fragments indicative of the Formative period, kero fragments from the Middle Horizon, and local Sillustani Inca plate fragments with a characteristic fern-leaf black on orange slip motif.

Site P118 is located along the sideslope of Cerro Cancharani. It is a 0.5 hectare mining site consisting of 6+ mining pits and spoils piles. There is also a large (2–3 m diameter) looted slab cist tomb onsite. Downslope from this slab cist tomb is a mining pit (4 m diameter) that branches into two perpendicular chambers at the bottom. Water is seeping in at the ends of the chambers, which measure 7 m long and 15 m deep. No surface artifacts other than quarrying debris were seen. A smaller (0.25 hectare) locus of mining pits was recorded as P33 at the eastern slope of Cerro Cancharani, amid a low density scatter of Late Intermediate, and Formative ceramic sherds.

To the southeast, along the high altitude flats overlooking the city, another mining locus was recorded as Minas Pumperia P119, an 8 hectare site that consists of more than 20 large piles of mine tailings that range in size from 10 to 30 m^2. No obvious pits can be seen. These tailing piles are of an order of magnitude larger than that seen at P117. There was a low-density ceramic scatter of only Colonial forms. Nearby, at site P120, was found a crude unslipped ceramic crucible with black and white residue on the interior.

These sites were likely part of the large-scale ore processing complex associated with the San Luis de Alba colonial mining district that includes the Rio Itapalluni P31 ruins. The San Luis de Alba colonial mining district was the subject of a survey and mapping project to evaluate its potential as a tourist attraction (CTAR 1999). The Laykakota complex at Cerro Cancharani (P33, P117, P118) and Chorrillos Itapalluni P31 were located and described. Archival research on documents from the 17th through 19th centuries identified other mines and ore-processing *haciendas* in the region. These are listed as El Manto, Cerro Ponperia, San Antonio de

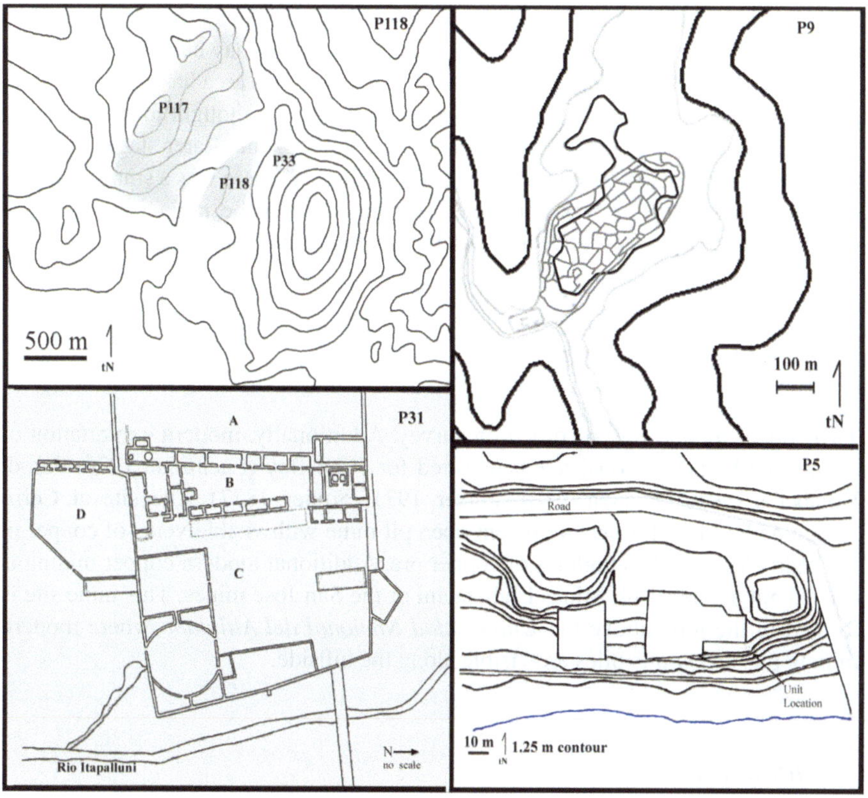

Fig. 11.7 Principle mining and metal ore processing locations in Puno Bay, Peru [clockwise from upper left P118—Laykakota, P9—Punanave (showing workshop/domestic terraces), P5—Huajje, P31—Chorillos Itapalluni] (after Schultze 2008; Nuñez Mendiguri 2001)

Esquilache, San Miguel de Totorani, and Rio San Miguel. Mario Nuñez M. published excavation of the Colonial silver mining complex at Chorrillos Itapalluni (CTAR 1999: Nuñez Mendiguri 2001).

The town of San Luis de Alba was razed in the Colonial era, but is thought to have been located in the plain between Cerro Cancharani and Cerro Negro Peque. Little remains of the settlement. The main ore processing facility at Chorrillos Itupalluni still exists in a relic condition (CTAR 1999). In addition to the archival work, Peruvian crews cleared, mapped, and surface collected major structures associated with the Salcedo mining operations (Fig. 11.7).

Proximity to Rio Itupalluni was a principal criterion for the location of this ore-processing site. The *hacienda* was completely enclosed by adobe walls as high as 4 m. Several activity areas were identified, and are labeled on the map as zones A–D (Fig. 11.7). Zone A housed laborers. Zone C was the primary production area. There were large grinding stones (1.2 × 1 m) that would have been both water and manually powered. Scoria was abundant in this area, suggesting firing or processing of materials fired in Zone A. There was a water reservoir, canals for water, and for drainage. The open patio space measures 22 m×93 m and may have had an interior water canal. Smelting furnaces were also located here. This area has a series of two-person dormitory rooms as large as 4.7×2.8 m. Zone B is thought to be an administrative area. The primary furnace was located in this zone. Later stages of silver purification were also carried out in the center patio, and there was a small chapel at the south end. Zone D was completely separated from the main structure. The smaller rooms at the west end suggest additional dormitory facilities.

Copper Mines

Two copper mines were recorded on survey. Additionally, modern exploitation of copper ore bodies is consistently reported for Puno Bay (Lechtman 1976; Parodi Isolabella 1995; Petersen 1970; Purser 1971; Squier 1877). The site at Cerro Azoguine P27 is a 1.5 hectare recent open pit mine with visible veins of copper in the bedrock, as well as nodules of copper ore. Additional modern copper mining is carried out at site P144, Minas Llallahuani or the San José mines. This mine site is located in the hills above the *Universidad Nacional del Altiplano* where modern mining pits and spoils piles are visible along the hillside.

Smelting Sites

The primary silver smelting sites of the Prehispanic era were Huajje P5 and Punanave P9 (Fig. 11.7). Both sites had abundant surface ceramics from the Formative through Colonial periods. In addition to slag-encrusted crucibles and bits of copper ore, the surface collections included ceramic spindle whorls, a variety of lithics including red ochre, andesite basalt, rhyolite, obsidian, limestone, sandstone, an unusually wide variety of cherts and chalcedonies, and a biconically drilled disc shell bead. Given the ridge-line location of Punanave, it is hypothesized that this was the location of *huayrachina* processing. The ridge is completely open and subjected to winds coming off the lake at regular and predictable intervals.

By contrast, Huajje (P5) is in a sheltered location along the shoreline of Lake Titicaca, at an elevation of 3,825 m amsl. It is protected from the winds to the north. This location is more conducive to the third stage of silver ore purification, an oxidization step that could have been carried out in *tocochimbo*-style reverberatory furnace.

Excavation data provide evidence that this stage of smelting did take place at Huajje. In addition to a variety of smelting debris types found throughout the unit, at 370–480 cm below surface a circular stone structure was uncovered. This structure had an elevated stone fire-hearth constructed into the wall, making it a plausible candidate for a cupellation furnace. These levels of the site correspond to the Middle and Upper Formative occupations in the ceramic chronology. In addition to the circa AD 60–120 dates obtained from the lowest levels, carbon from the level of this firebox (4 m below surface) returned a calibrated radiocarbon date in the range of AD 220–530 (1,690 ± 70 BP), corresponding largely to the post-Pucara Upper Formative period of the Northern Basin (Schultze 2008: 227). This structure was filled in with workshop midden and ritually closed with a camelid offering at 370 cm, a level corresponding to the appearance of Tiwanaku (Middle Horizon) ceramic forms.

Excavation revealed a 5+ m deep cultural deposit with abundant artifacts. Silver smelting debris was found in every level from 30 cm to 480 cm. The 3,447 (7544.9 g) pieces of smelting debris collected included palm-sized plainware crucibles, crucibles encrusted with glassy slag, chunks of glassy slag, heavy circular matte cakes, vesicular glassy debris, limestone chunks, and small bits of green ore rocks. These were analyzed through Scanning Electron Microscopy at the University of Washington by Scott Kuehner and the author. They were found to contain lead, silver, copper compounds, and mineral impurities consistent with the refining of silver (Schultze 2008: 398–440). This also appeared to be a multi-purpose site. Bone weaving tools were also found in addition to evidence for *incensario* ritual and food consumption (Schultze 2008: 177–341).

The end product of the refining activity seen at Huajje would have been a nugget of lead and silver bullion, which would require an additional process (cupellation) for the final separation of the silver metal from the litharge. This could have been done at Huajje or at another site. In some periods, these products would have been immediately transported to a secure location.

Laykakota, Punanave, and Huajje were the key silver producing sites of the region during the entire occupational sequence of the bay. But even these sites incorporated smelting activities among a broader suite of industrial, ritual, and domestic activities. Other loci of smelting during the Formative period appeared even less intensively focused on metal ores.

To the north of Huajje, evidence for Formative period smelting was found at a cluster of sunken court and residential sites on Cerros Chincheros and Jallupata (sites P13, P48, and P49). This entire group of low hilltop domestic terraces comprises a regional center for the Formative period of some 20 hectares. At P13, there was evidence for a stone working industry in andesite and basalt. Smelting evidence was found on Cerro Jallupata P49 and Huerta Huaraya P48. At site P49 A, a large number of ceremonial and fine serving vessel forms were found spanning the Middle and Upper Formative through Middle Horizon period diagnostic types. Abundant Qaluyu and Pukara finewares including bowls and jars were identified on this site, as well as metavolcanic agricultural artifacts (Schultze 2008: 558–559).

Smelting debris is also found among a cluster of ritual sites at Cerro Tancane (P137, P138, and P139). These are part of a cultural district locally known as Llusqha Llusqhani Sector, south of Puno in the western Jallihuaya pampa. This area has Formative period pictographs (P148) and petroglyphs (P93) (Schultze 2008: 520, 550), as well as sandstone geoglyphs across the sector. The mural rock art depicts human hunters and camelids (Schultze 2008: 551). The cluster of sites atop Cerro Tancane includes an intensively used modern *pago* ritual site. Survey recovered a slag-encrusted crucible among a moderate density scatter of Formative artifacts. These included Pucara polychrome-incised trumpet and bowl fragments (Schultze 2008: 589). The trumpet body fragment was zone-incised fugitive black and red slip with straw or cane impressions visible on the interior of the vessel. These ritual finewares, geoglyph formations, and modern *pago* structures point to a long tradition of ceremonial observance in the Cerro Tancane area.

A major administrative complex and ore refining center of the Inca and Colonial periods is located directly east and downslope of Cerro Cancharani. On the confluence of the drainages from Cerros Pitiquillia and Cerro Tancane, this zone (recorded as sites P150 and P151) includes the building foundation of a substantial Inca structure, as well as anvils, a milling stone and a 1.5 m high furnace. The last of these is surrounded by prepared fields in which Incan fine masonry ashlars of finished andesite have been re-used for fieldstone markers and foundation stones for modern adobes. This site could have been an Inca *tambo* or regional administrative building which then served as a Colonial period silver refining and production center. At the south edge of the site is an approximately 500 year old Colonial church. While no artifacts remain on the surface, there is a moderate density of artifacts seen in the adobe walls above these older foundations. An additional circular mill stone made of locally available marine conglomerate was located out of primary context further to the north, as site P151. It is roughly 2 m in diameter with a square center hole (Schultze 2008: 553).

Smelting was also in evidence at the 0.5 hectare hilltop site at Cerro Llallahuani P71. There was a quarry on the western edge of this site that was used in building the *chulpas,* with ceramics from Formative and Late Intermediate periods. Industrial uses are indicated by a basalt tool and a crucible fragment with adhered slag. There was also a bowl fragment with a white, chalky material resembling calcium carbonate on the interior. This caliche-like material might have been used to slow the reaction between the melting ore and the material of the ceramic crucible.

Additionally, crucibles were in evidence at the 7.0 hectare Formative and Inca period site Cerro Santa Vela P37 (Arkush 2005: 496). This *pukara*/walled fortress site is located on a flat mesa-top with a substantial east-facing natural defensive buffer outcrop. This natural defense has been augmented along the western slopes by segments of concentric defensive walls. Arkush reports Formative ceramics, including Pucara Incised, Sillustani-Inca, and Inca styles. Lithic andesite *hachas* are present along with groundstone, flakes, and slag-adhered metal smelting crucibles. Evidence of modern mining was noted along the eastern slope. The mesa-flat would have caught strong winds and been a practical location for wind-driven smelters. A river runs along the NW base of the hill 500 m away.

Metavolcanic Quarries and Adjacent Workshops

Silver ore bodies in Puno Bay occur within the same volcanic deposits as are found andesites and basalts. Evidence for very early elaboration of technologies for andesite tool production shows that these resources were part of a larger economic system utilizing the available geologic resources. Andesite workshops are most commonly located near quarries, although exceptions to this pattern have been noted (Arkush 2005). The largest workshop noted in the study area was Cerro Ichur P108, a 6 hectare site near the previously recorded andesite source, Incatunuhuiri (Kidder 1943; Stanish et al. 2005: 109). Incatunuhuiri is a quarry and ceremonial center located just south of the current project boundaries. Incatunuhuiri is primarily a Formative site, with a lighter occupation in subsequent periods. It was a substantial residence or workshop, with evidence for primary reduction of stone into "blanks" for transport.

Directly across the Ichur *pampa* from Incatunuhuiri, *Proyecto Wayruro* recorded a 6-hectare workshop and residential site Cerro Ichur P108 on a flat natural ridgeline terrace. The site has a very dense scatter of lithic artifacts and ceramics on a series of terraces which run along the top and side slopes of the ridge. There is a carpet of andesite flakes and *hacha* fragments in all stages of production. From the lithic materials collected on the surface, it appears that this site is a locus for processing raw andesite, either into finished products or into blanks for transportation. Debris from this production occurs in abundance along the lake-facing terraces and at the base of the hill. Artifacts from all ceramic periods are present, although the Formative presence is most strongly represented (illustrations in Schultze 2008: 581–582).

Another andesite workshop complex is found at Cerro Calechejo (sites P82 and P83). The 0.2 hectare quarry is located on the eastern edge of Cerro Calechejo P82 is an outcrop of bedrock andesite that has been extensively quarried. The vicinity is covered with a high density of andesite flakes, many of which are patinated, indicating some antiquity. No artifacts were found in association with the quarry.

Above this quarry, on the flat summit of Cerro Calechejo is a 2.5 hectare lithic workshop site P83 with a small central clearing which may have served as a public gathering area (or miniature "plaza"). The cleared area measures 40 m × 30 m and is delineated by a semi-circular rock wall along the south and southwest edge. This circle is continued by a semi-circular length of intentionally augmented outcrops and multiple mound burials which encircle the plaza.

In the center of the "plaza" area is a large upright andesite monolith embedded into the ground. The exterior of the semi-circular mound/outcrop is surrounded by a series of cleared terraces and fieldstone walls several courses high. One of these terraces, located 40 m at 340 ° (NW) from the andesite monolith, showed evidence of being a lithic workshop. There was a dense concentration of andesite lithic debris, including *hacha* fragments, tabular fragments, and flakes in all stages of production. Andesite groundstone was also recovered which may have been used to fine finish other blocks of andesite. Ceramics in the central area are Formative period types, while those on the burial mounds are of Late Intermediate types. The location is extremely windy, making it a potential location for the operation of wind smelters.

At Mesa Carihua P56 there is an Inca site situated on a mesa-top above an andesite quarry. No evidence to connect the site and quarrying activity was found. The quarry was in use in the modern period and was the source of building material for corrals as well as field- and path-demarcating walls. Obsidian and metavolcanic flakes are present. One Sillustani-Inca plate fragment (black and white on red) was collected from the surface for thermoluminescence dating. These tests were conducted by James Feathers and the author in the University of Washington thermoluminescence lab in 2003 (TL#7-UW 926). This sample returned a minimum age of 1488+55 AD, consistent with a Late Horizon occupation (Schultze 2008: 498).

Conclusions: Social Organization of Production in Puno Bay

This brief description of the mineral extraction and processing sites in Puno Bay serves to identify the area as a primary region for the quarrying, mining, and secondary processing of intrusive volcanic minerals. Silver ore was extracted from the principal mining site at Laykakota during all periods. Secondary processing through smelting was carried out immediately downslope at Punanave, with additional reprocessing and/or cupellation being done on the bay shore site of Huajje. In the Formative period, smelting debris was found at four other sites or site clusters. Concurrently, industries for the production of finely finished andesite and basalt agricultural tools and construction blocks/ashlars are found in association with outcrops across the bay.

The Puno Group intrusive minerals formed the basis of a Formative period craft production industry which continued through all ceramic periods and into the colonial period. At this stage of research, it is only possible to make broad inferences regarding the political economy of silver production in Puno Bay. A transition from holistic to more prescriptive forms of labor organization (Franklin 1983; Liu 2007) in the Middle Horizon may be inferred by an increase in scale and focus on metal producing sites. However, this evidence is preliminary. Similarly, data are insufficient to indicate if the work was carried out across the landscape in a modular or sequential manner (e.g., Shimada 2005), although it is clear that multiple sites were concurrently involved in mining and processing ores during all periods. Specialization by *ayllu* (aka lineage-based) corporate groups with some autonomy is anticipated, based on surveys and case studies of craft specialization traditions from across the Andes (Bernier 2010; Ignacio Angiorama and Taboada 2007; Janusek 1999; Ramírez 1994). More investigation of the mining and workshop sites discussed above will be needed to clarify how these traditions found expression in the local environment.

Acknowledgments This work was supported by the UCLA Department of Anthropology and Cotsen Institute of Archaeology, University Research Expeditions Program (UREP), and the National Science Foundation (NSF #BCS 9905138). Thanks also to the great help provided by Charles Stanish, Mark Aldenderfer, Edmundo de la Vega Machicao, Fernando John Sosa Alcón, Rolando Paredes, Karen Doehner, Cecilia Chávez Justo, Anita Piña, Mourik Bueno de Mesquito, Elizabeth Arkush, Aimee Plourde, Elizabeth Klarich, Javier Challcha Saroza, Amadeo Mamani Mamani, David Oshige Adams, Barbara Carbajal, Sarah Abraham, Ulli Green, Keith Carlson, Ilana Johnson, Javier Pilco Quispe, Yan de Julia, Norfelinda Cornéjo Gallegos, Michael Timm, Ian Stonerock, Howard Tsai, Robin Beck, Karin Klemic, Ana González, and many others. Post-field analyses were supported by Thilo Rehren, Scott Kuehner, James Feathers, David Scott, and Nicholas Tripcevich.

References

Abbott, M., & Wolfe, A. (2003). Intensive pre-Incan metallurgy recorded by lake sediments from the Bolivian Andes. *Science, 26*, 1893–1895.

Acosta, J., Rivera, R., Valencia, M., Chirif, H., Huanacuni, D., Villarreal, E., Paico, D., Santisteban, A., & Neyra, A. (2010). *Mapa metalogenético del Perú: operaciones y proyectos mineros*. Peru: Instituto Geológico Minero y Metalúrgico (INGEMMET).

Aldenderfer, M., Craig, N. M., Speakman, R. J., & Popelka-Filcoff, R. (2008). Four-thousand-year-old gold artifacts from the Lake Titicaca basin, southern Peru. *Proceedings of the National Academy of Sciences of the United States of America, 105*(13), 5002–5005.

Arkush, E. (2005). *Colla Forts* (Unpublished Ph.D. dissertation). University of California, Los Angeles.

Bernier, H. (2010). Craft specialists at Moche: organization, affiliations, and identities. *Latin American Antiquity, 21*(1), 22–43.

Burger, R., & Gordon, R. B. (1998). Early Central Andean metalworking from Mina Perdida, Peru. *Science, 282*, 1108–1111.

Cohen, C. R. (2008). *The winds of change: An archaeometallurgical study of silver production in the porco-potosi region, southern Bolivia AD 1500–2000* (Unpublished Ph.D. dissertation). Institute of Archaeology, University College, London.

Cohen, C., Rehren, Th., & Van Buren, M. (2006). *Indigenous Bolivian silver-smelting using Huayrachina furnaces: New evidence revealed.* Paper presented at the 71st meeting of the Society for American Archaeology, San Juan, Puerto Rico.

Cooke, C., Abbott, M., & Wolfe, A. (2008). Late Holocene atmospheric lead deposition in the Peruvian and Bolivian Andes. *Holocene, 18*, 353–359.

Cooke, C., Abbott, M., Wolfe, A., & Kittleson, J. (2007). A millennium of metallurgy recorded by lake sediments from Morococha, Peruvian Andes. *Environmental Science and Technology, 41*, 3469–3474.

CTAR (Concejo Transitorio de Administración Regional). (1999). *Puesta en Valor del Centro. Histórico Colonial San Luis del Alba.* Estudio Realizado Por El Consejo De Administración Regional Puno. Publication of Proyecto Especial Binacional Lago Titicaca (PELT), Puno, Peru.

Cusi Ponce, M., Checalla Suárez, B., & Jamachi Vilca, M. (1992). *Geográfica 1, Base Puno.* Puno: Colegio Profesional de Profesores del Peru.

de la Vega Machicao, E. (1998). *Informe de Investigaciones: Proyecto de Excavación Arqueológica en el sitio de Isla Esteves, Puno.* Report presented to the Ministerio de la Cultura. Oficina del Ministerio de Cultura, Puno, Peru.

Emmerich, A. (1965). *Sweat of the sun and tears of the moon: Gold and silver in Pre-Columbian art.* Seattle: University of Washington Press.

Franklin, U. (1983). The beginnings of metallurgy in China: A comparative approach. In G. Kuwayama (Ed.), *The great bronze age of China: A symposium.* Los Angeles: Los Angeles County Museum of Art.

Garci Diez de San Migue, D. d. S. M. (1567 [1964]). *Vista Hecha a la Provincia de Chucuito. .. en el Año.* Lima: Documentos Regionales para la Etnología y Etnohistoria Andinas I, Casa de la Cultura del Perú.

Ghosh, P., Garzinoa, C., & Eiler, J. (2006). Rapid uplift of the Altiplano revealed through 13C-18O bonds in Paleosol carbonates. *Science, 311*, 511–514.

Hastorf, C. (2005). The upper (middle and late) formative in the lake Titicaca Basin. In C. Stanish et al. (Eds.), *Advances in Titicaca Basin archaeology* (Vol. 1). Los Angeles: Cotsen Institute of Archaeology at UCLA.

Hintzman, M. (2004). *A technological consideration of ground-bit tools from the neolithic settlement of 'Ain Ghazal'.* Paper Presented to the Meetings of the American Schools of Oriental Research, Neolithic Archaeology Session, November, San Antonio, TX.

Ignacio Angiorama, C., & Taboada, C. (2007). Producción metalúrgica en un contexto doméstico de Los Amarillos (Jujuy, Argentina). In R. Lleras Pérez (Ed.), *Metalurgía en la América*

Antigua: Teoría, Arqueología, Simbología y Tecnología de los Metales Prehispánicos. Perú: Instituto Francés de Estudios Andinos, Lima.
INGEMMET (Instituto Geológico Minero y Metalúrgico). (1970). *Ministerio de Energía y Minas Instituto de Geología y Minera Mapa Geológico del Departamento de Puno 1:250,000, Lima.* Publication of the Instituto Geológico Minero y Metalúrgico, Lima, Peru.
Janusek, J. (1999). Craft and local power: embedded specialization in Tiwanaku cities. *Latin American Antiquity, 10*(2), 107–131.
Janusek, J. (2004). Tiwanaku and its precursors: recent research and emerging perspectives. *Journal of Archaeological Research, 12*(2), 121–183.
Jones, J. (2005). Innovation and resplendence: Metalwork for Moche Lords. In *Moche art and archaeology in ancient Peru* (Studies in the History of Art, first issued 2001, Vol. 63). Washington DC: National Gallery of Art.
Julien, C. (1981). A late burial from Cerro Azoguini, Puno. *N awpa Pacha, 19*, 129–163.
Julien, C. (1983). *Hatunqolla: A view of Inca rule from the Lake Titicaca region* (Vol. 15). Berkeley: University of California Publications in Anthropology.
Kay, S., Mpodozis, C., & Ramos, V. (2005). Andes. In R. C. Selley et al. (Eds.), *Encyclopedia of geology*. Oxford: Elsevier Academic Press.
Kidder, A. (1943). *Some early sites in the Northern Lake Titicaca Basin* (Papers of the Peabody Museum of American Archaeology and Ethnology). Cambridge, MA: Harvard University.
Lechtman, H. (1976). A metallurgical site survey in the Peruvian Andes. *Journal of Field Archaeology, 3*(1), 1–42.
Lechtman, H. (1984). Andean value systems and the development of prehistoric metallurgy. *Technology and Culture, 25*(1), 1–36.
Levine, A. R. (2012). *Competition, cooperation, and the emergence of regional centers in the northern Lake Titicaca Basin, Peru* (Unpublished Ph.D. dissertation). University of California, Los Angeles.
Liu, Y.-t. (2007). Specialization, context of production, and alienation in the production process: comments and afterthoughts. *Archaeological Papers of the American Anthropological Association, 17*(1), 169–180.
Menzel, D. (1977). *The Archaeology of Peru and the work of Max Uhle*. Berkeley: RH Lowie Museum.
Moseley, M. (2001). *The Inca and their ancestors* (2nd ed.). London: Thames and Hudson.
Nuñez Mendiguri, M. (2001). *Minas y Mineros del Siglo XVII in San Luis de Alva y Sus Efectos Socio-Economicos*. Puno, Peru: Universidad Nacional del Altiplano, Oficina Universitaria de Investigación.
Parodi Isolabella, A. (1995). *El Lago Titicaca: Sus Características Físicas y Sus Riquezas Naturales Arqueológicas y Arquitectónicas*. Banco del Sur: Editorial Industria Gráfica Regentus, Arequipa.
Patterson, C. C. (1971). Native copper, silver, and gold accessible to early metallurgists. *American Antiquity, 36*(3), 286–321.
Peele, R. Jr. (1893). A Primitive Smelting-Furnace. *The School of Mines Quarterly 15*, 8–10.
Petersen, U. (1965). Regional geology and major ore deposits of central Peru. *Economic Geology, 60*(3), 407–476.
Petersen, G. (1970). Minería y Metalurgia en el Antiguo Perú. *Arqueológas, 12*, 1–140.
Purser, W. F. C. (1971). *Metal-mining in Peru, past and present* (Praeger special studies in international economics and development). New York: Praeger Publishers.
Ramírez, S. (1994). Ethnohistorical dimensions of mining and metallurgy in sixteenth-century Northern Peru. In A. K. Craig & R. C. West (Eds.), *In quest of mineral wealth: aboriginal and colonial mining and metallurgy in Spanish America* (Geoscience and man 33). Baton Rouge: Louisiana State University.
Rehren, Th., Schultze, C., & Stanish, C. (2010). *The mining-metallurgical landscape near Puno, Lake Titicaca: a major pre-Spanish silver producing centre in southern Peru.* Paper presented at the First International Meeting on Prehispanic Mining in America, San Pedro de Atacama, Chile.

Schultze, C. A. (1999). *Lithic analysis of collections from the sites of Isla Esteves and Sillumocco-Huaquina, Juli, Peru*. Paper presented to the annual meetings of the Society for American Archaeology, New Orleans.

Schultze, C. A. (2008). *The role of silver ore reduction in the Tiwanaku state expansion into Puno Bay, Peru* (Unpublished Ph.D. dissertation). University of California, Los Angeles.

Schultze, C. A., Stanish, C., Scott, D., Rehren, T., Kuehner, S., and Feathers, J. (2009). Direct evidence of 1,900 years of indigenous silver production in the Lake Titicaca Basin of southern Peru. *Proceedings of the National Academy of Sciences of the United States of America, 106*(41), 17280–17283.

Seddon, M. (1994). Lithic artifacts. In C. Stanish & L. Steadman (Eds.), *Archaeological research at Tumatumani, Juli, Peru* (Fieldiana series in anthropology, no 23). Chicago: Field Museum of Natural History.

Shimada, I. (2005). Late Moche urban craft production: A first approximation. In J. Pillsbury (Ed.), *Moche art and archaeology in ancient Peru*. Washington DC: National Gallery of Art.

Shimada, I. (2007). *Craft production in complex societies: multicraft and producer perspectives*. Salt Lake City: University of Utah Press.

Squier, E. (1877). *Peru. Incidents of travel and exploration in the land of the Incas*. London: MacMillan and Co.

Stanish, C., Frye, K., de la Vega, E., and Seddon, M. (2005). Tiwanaku expansion in the Western Titicaca Basin. In C. Stanish et al. (Eds.), *Advances in Titicaca basin archaeology* (Vol. 1). Los Angeles: Cotsen Institute of Archaeology at UCLA.

Toledo, F. d. (1570–1573 [1975]). *Tasa de la Visita General de Francisco de Toledo* (Introducción y versión paleográfica de Noble David Cook). Lima: Universidad Nacional Mayor de San Marcos.

Tschopik, M. (1946). *Some notes on the archaeology of the department of Puno, Peru* (Papers of the Peabody Museum of American Archaeology and Ethnology, Vol. 27(3)). Cambridge, MA: Harvard University.

van Buren, M., & Mills, B. (2005). Huayrachinas and Tocochimbos: traditional smelting technology of the Southern Andes. *Latin American Antiquity, 16*(1), 3–25.

von Tschudi, J. J. (1865). *Travels in Peru, on the coast, in the Sierra, across the Cordilleras and the Andes, into the Primeval forests (Thomasina Ross, Trans.)*. New York: A.S. Barnes and Co.

Chapter 12
Mining, Commensal Politics, and Ritual under Inca Rule in Atacama, Northern Chile

Diego Salazar, César Borie, and Camila Oñate

Introduction

The Atacamenian territory, in what is today northern Chile, is part of the most arid desert in the world (Fig. 12.1). The scarcity of water resources and vegetation in this extensive area contrasts with its rich abundance of metallic and nonmetallic ores. Within its mineral diversity, copper plays an outstanding role given the wealth of deposits and the importance that its exploitation has had for the economic development of individuals, groups, companies, and nations.

Current archaeological evidence indicates that exploitation of copper ores as semiprecious stones began in the Atacamenian region as early as the Late Archaic Period (ca. 3500 BC), when it was mainly used for the production of copper mineral beads (Núñez 2006; Soto 2010). From the Formative Period onward (ca. 1000 BC) the production of these beads increased significantly (Núñez et al. 2006; Rees 1999; Soto 2010), but this increase was also contemporaneous with the first appearance of copper metallurgy as suggested by finished objects as well as scarce evidence of slag and metallurgical waste both in the Salar de Atacama and the Loa River Basin (Agüero 2005; Figueroa et al. 2010; Muñoz 1989; Núñez 2006; Salazar et al. 2011a). Since this time, crushed copper minerals have also been used as ritual offerings (Berenguer 2004; Sinclaire 1994), as decoration in wooden or textile ornaments, or as pigment in rock art (Sepúlveda and Laval 2010) and pottery decoration, among other uses.

Notwithstanding, copper was not only an important resource in Atacamenian craft production, it had an important role in many ritual activities and ceremonies, including burials, and by the Formative Period also played a key role in local political economy. It has been stated that copper ores were among the main commodities

D. Salazar (✉) • C. Borie • C. Oñate
Departamento de Antropología, Universidad de Chile,
Ignacio Carrera Pinto 1045, Piso 2, Ñuñoa, Chile
e-mail: dsalazar@uchile.cl; cesarborie@gmail.com; camila.aos@gmail.com

N. Tripcevich and K.J. Vaughn (eds.), *Mining and Quarrying in the Ancient Andes*, Interdisciplinary Contributions to Archaeology, DOI 10.1007/978-1-4614-5200-3_12,

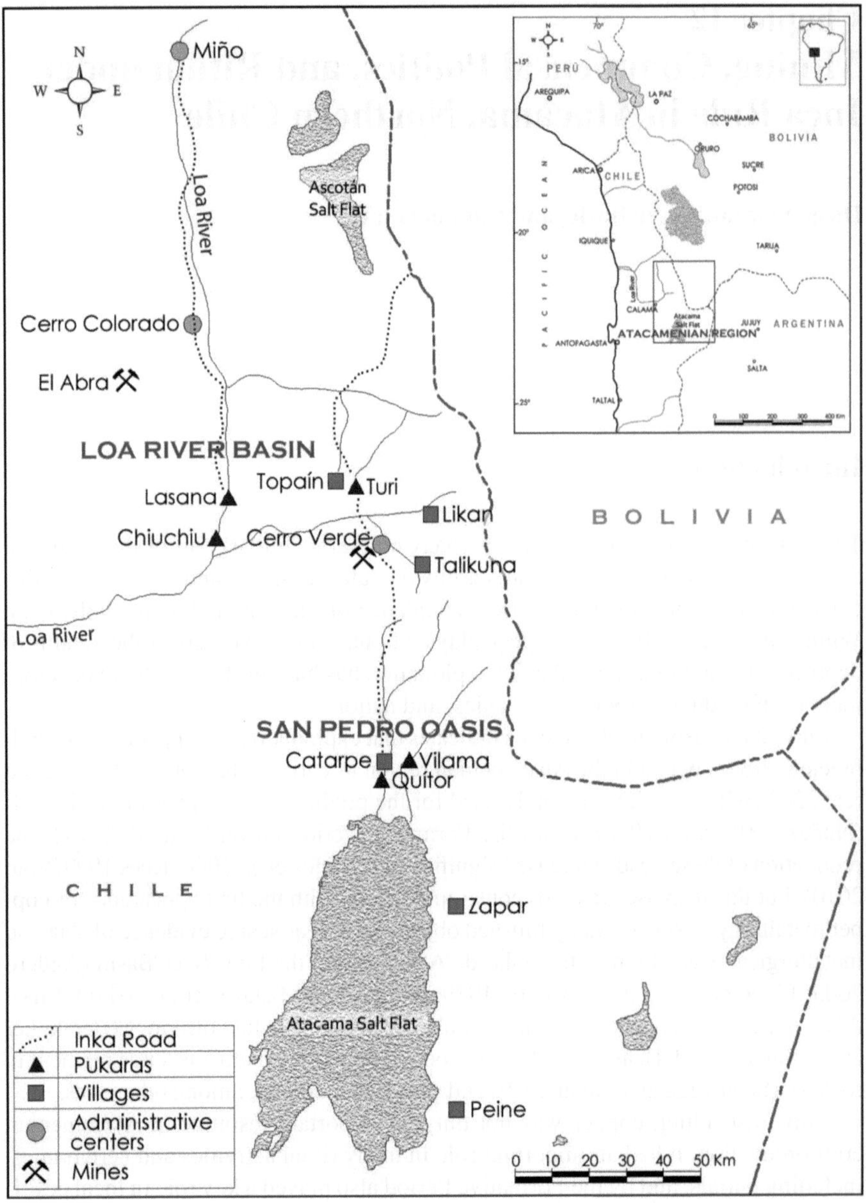

Fig. 12.1 Atacamenian territory in the context of the South-Central Andes. The San Pedro de Atacama Oasis and the Upper Loa River drainage and main archaeological sites during the Late Period

exchanged by Atacamenian polities in their long-distance economic trade with societies of the Bolivian altiplano, northwestern Argentina, and the valleys of Tarapacá farther north (Núñez 2006), and also that copper objects were important symbols of political authority, hierarchy, and ethnic affiliation (Llagostera 2006; Núñez 2006).

When the Incas conquered what is today northern Chile, they found in Atacama a millenary tradition of copper mining technology as well as a complex social and economic system organized around the production and distribution of this precious resource.

Most scholars agree that one of the main interests of the Inca state in northern Chile was precisely the control of these copper production and distribution systems (Adán and Uribe 2005; Berenguer 2004, 2007; Llagostera 1976; Raffino 1981; Niemeyer and Schiappacasse 1998; Núñez 1999, 2006; Núñez et al. 2005; Salazar 2002–2005; Uribe 1999–2000; among others). Diverse kinds of evidence support this claim (see Salazar et al. 2011b). Here, we focus on a copper mining complex located in San José del Abra, in the Upper Loa River Basin, where we have conducted research for almost a decade (Núñez 1999; Salazar 2002–2005, 2008; Salazar and Salinas 2008).

Our previous studies have centered on the technological and economic dimensions of mining in these copper extraction complexes, but hitherto we had not searched for nor found clear evidence of ritual activities or of sociopolitical control exerted by the Incas over the local Atacamenian miners. Given the well-known symbolism of indigenous mining in the South-Central Andes during the Colonial Period (Bouysse-Cassagne 2008; Platt and Quispe 2008; Cruz 2009; see also Chap. 13) and until the present day (Absi 2005; Godoy 1985), as well as the control exerted by Inca administration over gold and silver mines (Berthelot 1986), it seemed reasonable to suspect that the lack of positive evidence for ritual activities and a hierarchical sociopolitical organization in Atacamenian mining complexes controlled by the Inca was due to the absence of systematically oriented research rather than absence of archaeological data. Our most recent research in the San José del Abra mining complex, begun in 2010, was aimed precisely to explore these important dimensions of Inca mining in Atacama. In this paper we present our first results, which confirm our presuppositions of ritual and hierarchical organization within the mining community placed by the Incas at El Abra. We present this evidence and further discuss the interrelatedness of technology, ritual, and sociopolitical control over Inca mining in the Atacama Desert.

The Incas in Atacama

During the Late Intermediate Period (ca. 950–1400 AD), and just before Inca expansion into Atacama, the area was inhabited by a local culture known as Atacamenian. As was the case with many other contemporary Andean societies, the Atacamenian economy was based on large-scale agriculture, camelid herding, and the collecting and processing of local algarrobo and chañar fruits, and husks (Adán and Uribe

1995; Schiappacasse et al. 1989; Uribe 2002). Since water was a critical resource in the arid Atacama Desert, settlement systems were organized around the San Pedro de Atacama oasis and the Loa River Basin (Fig. 12.1). Through extensive caravan trade, goods from distant provinces were available to local populations (Núñez and Dillehay 1978). In exchange, copper ores was one of the main commodities exported (Núñez 2006). The settlement system was organized mainly around villages of ca. 100–200 structures and *pukaras* (fortified hilltop sites) of up to 600 structures, strategically located next to principal water sources and grasslands. In the Upper Loa River basin, three contemporary *pukaras* are known, as well as three smaller villages (Urbina 2007; Uribe et al. 2004). Besides these, there was a wide scatter of smaller sites destined to more specific productive activities, given the well-known variability of Andean ecosystems due to differences in altitude and microclimates (Aldunate and Castro 1981; Adán and Uribe 1995; Adán and Uribe 2005). One of these smaller sets of sites has been studied in the Santa Barbara area of the Upper Loa River Basin, recovering information of at least four Late Intermediate hamlets of ca. 40 structures located in the bottom of the Loa canyon (Berenguer 1994, 2004). Evidence of small-scale agriculture, herding, and hunting activities demonstrate a stable occupation at these sites, though it is likely that they were not used on a year-round basis but were "*estancias*" used during a few months of the year (Berenguer 1994). Nearly 25 km to the west of these hamlets, Late Intermediate period copper mines have been identified in the El Abra locality, some of which were exploited at least since Late Formative times (Salazar et al. 2010) and continued in use until the Incas arrived in Atacama.

Current archaeological evidence suggests that Inca conquest of Atacama occurred sometime during the first half of the fifteenth century (Berenguer 2010; Uribe 2002). Tawantinsuyu made its dominion over Atacama through diverse means. We lack space here to discuss details, but it is important to note that the state's material expression included at least two segments of the Inca road or *Qapaqñan* (Berenguer et al. 2005; Castro et al. 2004; Varela 1999), Inca-style architecture intruding local villages and Pukaras (Aldunate 1993; Castro et al. 1993; Cornejo 1999; Gallardo et al. 1995; Lynch 1977), new sites built entirely with Inca-style architecture and layout (Adán 1999; Berenguer 2007; Salazar et al. 2011b; Uribe and Urbina 2009), Inca Cuzco and Provincial Inca pottery, some metal objects and textile styles (Uribe et al. 2002; Uribe et al. 2004). Thus, unlike what was assumed some decades ago (Llagostera 1976), today it is clear that Inca dominion over Atacama was based on a territorial strategy that implied a highly controlled extraction model (Alconini 2008; D'Altroy 2002). According to most scholars, and given the lower "agro-herding" productivity of the Atacamenian valleys as compared to neighboring areas, copper minerals were among the most important commodities extracted by the Incas in Atacama.

To accomplish this task, the Incas had to control and reorganize previous systems of copper production and distribution. Current evidence supports this claim (Salazar 2002–2005, 2008; Salazar et al. 2011b), and in this paper we want to analyze this further, exploring how the Incas managed to control copper production at El Abra using a complex strategy that combined ritual, sociopolitical, economic, and technological dimensions of mining.

Inca Mining at el Abra

El Abra is a locality well known for its mineral riches until today. Physiographically it is dominated by a rocky massif known as *Sierra del Medio* (mean 4,000 masl), running in a North–South direction to the west of the Andean Range. The dominant climate is that of a High-Altitude Marginal Desert, though higher zones may be classified as High-Altitude Steppes. Thus, the mean year-round temperatures are low, even though there are great oscillations in daily temperature, and year-to-year precipitation is variable, though always moderate. Given its climatic and orographic conditions, the study area presently exhibits little vegetation other than that found at the bottom of quebradas. From a mineralogical point of view, El Abra's main mineral deposit is a porphyry copper composed of a significant layer of copper oxides on top of an even larger deposit of sulfides. The oxide layer, focus of Prehispanic mining operations, presents a wide variety of minerals, predominantly chrysocolla, brochantite, pseudo-malaquite, and turquoise.

During the first phase of the Late Intermediate period (hereafter LIP), there is evidence for a relatively stable occupation at El Abra, centered around the main water sources, and with evidence of at least one copper mine and two smelting areas (Salazar et al. 2010). Smelting sites are small and do not show evidence for the use of furnaces. On the other hand, during the second phase of the LIP (ca. 1200–1400 AD), human occupation seems less permanent, but there are at least five small-scale mines scattered throughout the locality, which are usually associated with one or two habitation structures with scarce archaeological deposit (Salazar et al. 2011b). Smelting sites seem to have been abandoned. Previously, we have interpreted these data as the outcome of task-groups of miners visiting El Abra for short periods of time in order to extract copper ores to be smelted elsewhere (Salazar 2008). It seems likely that these miners came from hamlets in the Santa Barbara region or even from the most distant villages and Pukaras of the Loa Basin. In any case, Inca dominion over Atacama brought significant changes to this productive system (Salazar 2002–2005; Salazar 2008; Salazar et al. 2011b).

Once Inca rule was established in Atacama during the Late Period (ca. 1400–1540 AD; hereafter LP), the number of sites at El Abra, as well as their sizes and the overall volume of associated archaeological deposits and materials, increased considerably. Over 20 sites have evidence of LP occupation. But, unlike the scattered settlement pattern of the LIP, during the LP most of these cluster around two mining complexes, one for the extraction of turquoise and chrysocolla (San José del Abra mining complex) and the second for chrysocolla and pseudomalachite (San Pedro de Conchi mining complex). Each of these two mining complexes includes copper mines, mineral-crushing areas, storage structures, corrals, habitational sites, probable administrative structures, and ritual platforms (Fig. 12.2).

In the vicinities of the San José del Abra complex, where surveys have been more intensive, other secondary sites have been identified, predominantly next to water springs and in the routes connecting this mining complex to the Loa River Basin. These locations were used for activities that complemented the functioning of the

Fig. 12.2 Main campsite of the San José del Abra mining complex. At the *center* of the photograph the main prehistoric mines of the complex

Fig. 12.3 Main copper mines of the Late Period at the El Abra locality. (**a**) Site AB-22/39 at the San José del Abra mining complex. (**b**) Site AB-99 at the San Pedro de Conchi mining complex

mining complexes, such as food preparation, hunting, and herding, but also for the control of llama caravans.

The mines themselves include different kinds of open-cast pits and trenches, tailings, retention walls, and stock areas (Fig. 12.3). Site AB-22/39 is the largest open-cast mine yet known. Covering an area of approximately 3,000 m^2, we have so far recorded 4 large pits, 5 shafts, and nearly 20 small extraction and quarrying areas, with their respective tailings. The largest of these pits is more than 20 m × 25 m.

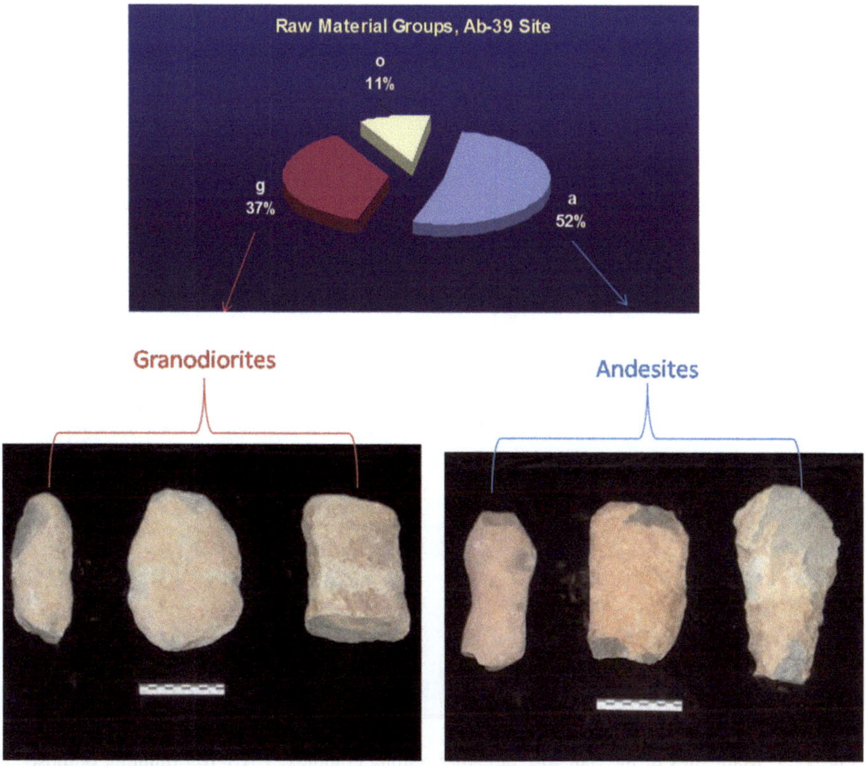

Fig. 12.4 Hammerstones used for copper mining during the Late Period at El Abra

Its depth is not known because of postdepositional fill but it was no less than 10 m. Given that the mines are located on a steep slope, at least eight retention walls were built in order to prevent rainwater, colluvium, and/or waste rock from up-slope operations from filling in the mines. The main mineral extracted at the site was turquoise and, to a lesser extent, chrysocolla. Both in the mines and tailings, as well as in the crushing areas and the habitational structures, nearly 1,000 lithic hammerstones have been found both on surface and in stratigraphic deposits (Fig. 12.4). They were made both of local (granodiorite) and non-local rocks (mostly andesites coming from ca. 10 km away), and show clear evidence of hafting, even though the haft has not been preserved in these specimens. Hafting can be inferred from the natural and/or pecked groove present in the medial segment of most hammerstones.

We focus here on the San José del Abra mining complex (Fig. 12.5), which is made up of six main sites including a habitational camp with more than 40 structures (Inkawasi-Abra); the mining area of ca. 3,000 m^2 described above (AB-22/39); two platforms (AB-39/5 and AB-40) aligned toward the highest peaks of the area; secondary crushing areas (AB-37); probable storage structures (AB-48); and other structures as yet functionally undetermined (AB-38). As stated before, a few more sites located in nearby quebradas played complementary roles to the functioning of

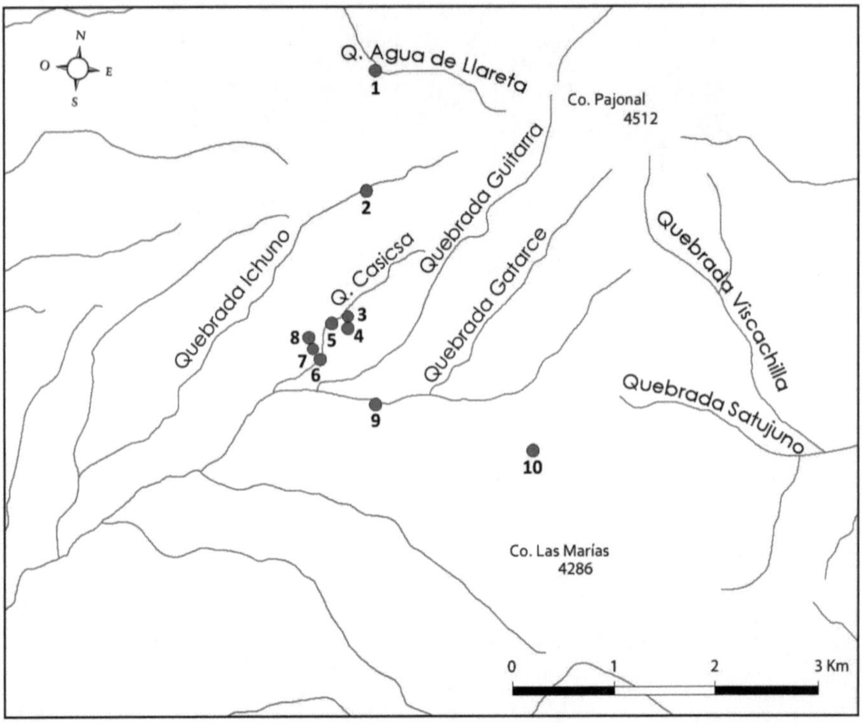

Fig. 12.5 Main sites of the San José del Abra mining complex. *1* AB-105 (hunting outpost); *2* Ichunito (control post); *3* AB-38; *4* AB-37; *5* AB-22/39; *6* Inkawasi-Abra or AB-36; *7* AB-48; *8* AB-40; *9* AB-44 (control post); *10* VMA-3 (function not determined); *11* AB-73 (hunting camp)

the mining complex (Salazar 2008). All of these sites were in use during the Late Period, but a few of them also show evidence of LIP occupations, thus suggesting the transformation of a pre-Inca mining system. Our data further demonstrate that Inca administration radically changed this previous system while employing local Atacamenian workforce with an efficient knowledge of local geological resources and mining technology.

Transformations during Inca rule are indicated by the new sites appearing during this period, the bigger sizes in all sites and the significant increase in production volumes between the LIP and LP (Salazar 2002–2005). The overall size of the operations is clear evidence of the latter. For example, the Cerro Turquesa mine was exploited between 200 and 1200 AD using a similar technology and on a similar geological setting as compared to the LP mines of AB-22/39. But whereas in Cerro Turquesa, we estimated an extraction volume of ca. 180 m^3 of rock, extraction activities during LP at Pits 1 to 4 (site AB-22/39) removed up to 5,000 m^3 or more. Besides the transformation in production volumes, there is evidence of change from a dispersed pattern of production seen during the LIP to a nucleated pattern (*sensu* Costin 2001, see also Zori 2011) during the LP. During the second half of the LIP (ca. 1200–1450 AD) there were at least five mines in exploitation in San José del Abra, all of them of similar sizes and associated with low investment and small-scale habitational structures. On the contrary, during the LP, mining at El Abra was nucleated on site AB-22/39 and a mining camp of more than 40 structures was built less than 500 m away from the mines.[1] These changes seem to be associated to an increase in specialization, as indicated by a spatial segregation of initial extraction and crushing from secondary crushing activities, as well as the correlation between these discrete activity areas and hammerstones with specific and regular attributes such as size and extreme-edge shape (Salazar and Salinas 2008). We interpret these data as transformations introduced by Inca administration in order to achieve an increase in production volume, while at the same time decreasing the costs in control and administration that would require a highly scattered productive system. The spatial segmentation of the mining production stages, and the use of adequate instruments for the specific technical requirements of each of them, must have allowed the Incas to gain the benefit of an economy of scale.

Current evidence suggests that the productive process ended at El Abra after the secondary crushing and sorting of the ores, at least during the LP. Ores were then transported elsewhere for the next phases of the productive process, namely extractive metallurgy and/or the elaboration of lapidary goods.

Ritual and Commensal Politics in a Mining Context

Despite the transformation introduced by the Inca in the organization of production at El Abra, there is evident continuity in mining know-how and technology between LIP and LP (Salazar and Salinas 2008). On the other hand, almost all pottery styles in the sites of the San José del Abra mining complex are local Atacamenian and local-Inca ware (Fig. 12.6). Architecture is also consistent with local "styles," with the exception of some local-Inca architecture to be discussed below. Archaeological data thus shows that Atacameños were in charge of mining copper ores for the Incas at El Abra during the LP. But as seen above, the scale of these operations changed dramatically when compared to LIP mining. How did the Incas organize and control these intensive copper-extraction operations?

Our evidence suggests that a diversity of means were at play at the same time, integrating economic, political, and ideological dimensions. On the one hand, we have already suggested that the state was in charge of financing this operation (Salazar 2008). In order to achieve such volumes of extraction, full-time dedicated workers were required. This means that the Inca had to manage the constant supply of hammerstones, food, and water for the miners. The workers dwelt in a campsite specially built by the Inca administration, though using mostly local-style architecture (for exceptions to this rule, see below). The site is known as Inkawasi-Abra,

[1] LP mining was also concentrated at the San Pedro de Conchi complex, even though in this case we have knowledge of only one pre-Inca mine.

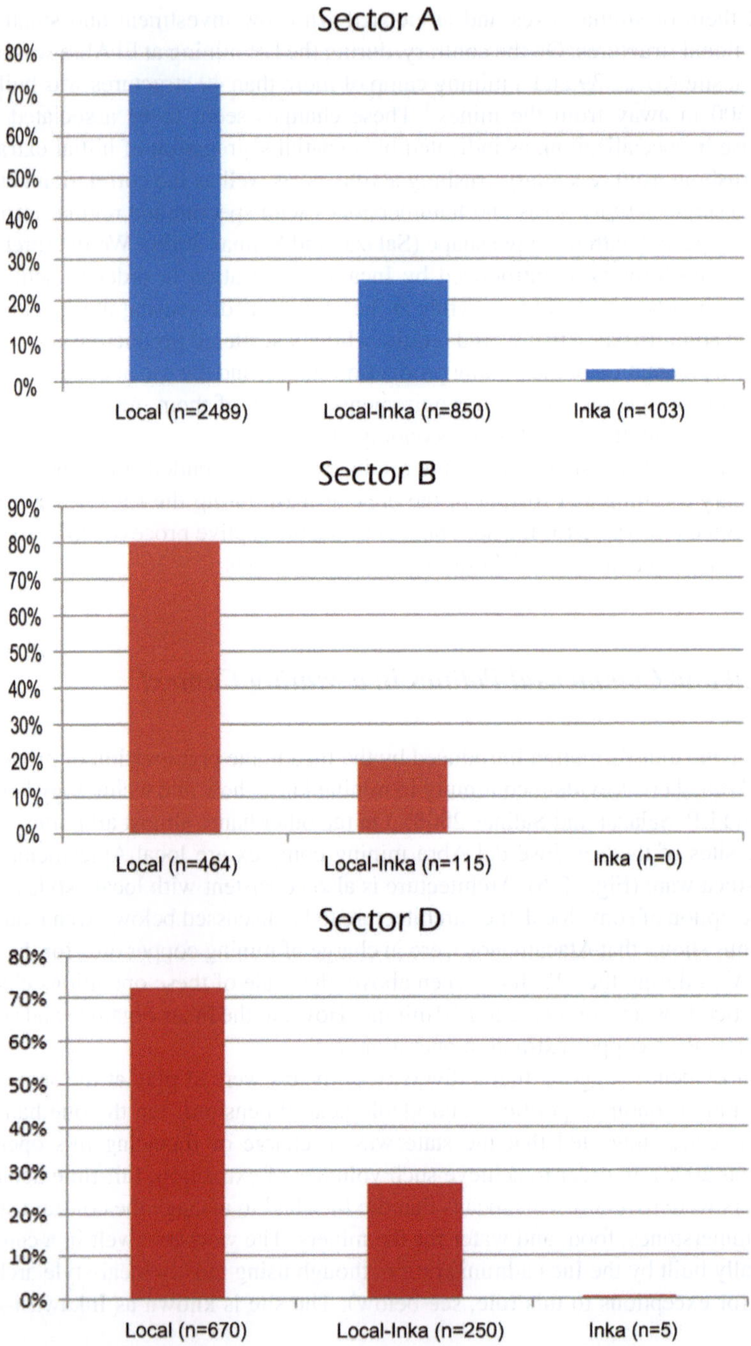

Fig. 12.6 Main pottery styles found at structures A2, B8 and D11 (the latter two habitational structures) in the Inkawasi-Abra site

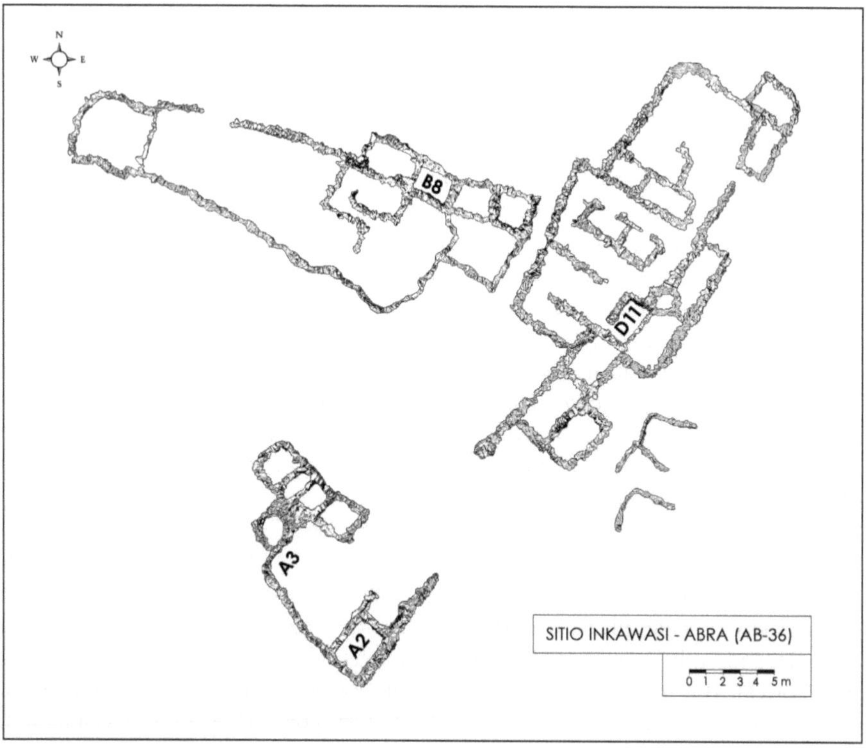

Fig. 12.7 Plan view of the mining campsite (Inkawasi-Abra) of the San José del Abra mining complex during the Late Period

and it is made of more than 40 structures organized in three main sectors, two of which have perimetral walls. In the middle of these areas there is an open place, most probably a small plaza (Fig. 12.7).

Our excavations at the site show a clear standardization in the distribution of local pottery throughout all structures. The same local and local-Inca cooking pots, large containers and jars were used by everyone at the mining camp, and there were very few plates available. Food resources were also evenly distributed, and include large quantities of camelid meat, chañar, and algarrobo fruits and, to a lesser degree, maize. This standardization in food and pottery distribution strongly supports Murra's (1989) early claim that according to Andean reciprocity obligations, the state was responsible for providing food, tools, and clothing for the local workers serving their tribute in Inca-sponsored work. Considering the presence of wild camelids in the archaeozoological sample (both guanaco and vicuña), including high- and low-utility units (Labarca and Calás 2010), together with a few projectile points found during excavations at the site, it seems that the Inca arranged for local hunting to provide for at least part of the meat supply of the mining community. It is thus probable that not only miners occupied the campsite. Surely water had to be fetched daily as well, and there is some evidence of local plants used especially for fuel that had to be gathered on a daily basis.

At the same time, the Incas organized space and activities in such a way as to make evident hierarchical relationships between those who inhabited the main campsite. First of all we have the architecture. Even though most of the sites that make up the mining complex were built with low-investment local-style architecture, a few structures made in Inca-style were strategically placed. Site AB-48 is remarkable in this regard. Located on an elevated position just middle way between the mining camp and the mining operations, on a spot that can be constantly seen and from which all activities in the complex can be observed, this structure seems to have served storage purposes, even though we cannot rule out the possibility of ritual offerings as well. It has Inca architectural features completely absent from most structures in the complex and in contemporary local Atacamenian sites (Salazar 2008). Within the campsite, on the other hand, two other structures also stand out for their Inca-like architecture. One of them, structure D1, is placed in the middle of local-style structures, in such a way that in order to go in and out of this part of the campsite, one has to pass inevitably through an Inca-style structure, thus reinforcing the omnipresence of the state.

The third Inca-style structure is located on sector A, separated from the rest of the site by a small quebrada, now completely dry (Fig. 12.6). There was a clear intention to separate this area from the rest of the site, locating it just in front of the public space or small plaza. Unfortunately, sector A has been severely modified by Colonial and Early Republican herders and miners, so all the architecture seen today is historical. But in our excavations we found evidence of a 4 m × 3 m-Inca-style semi-subterranean structure (A2) under the historic corral (Fig. 12.8). Even though the historical occupation may have removed some of the LP deposits in this structure, under a thick layer of historical animal dung we found evidence of hearths and nearly 20 cm deep of ashy layers throughout the structures' floor. Immediately to the east of A2 there is a big secondary refuse area, formed as a result of continuous cleaning of the hearths and cooking areas inside structure A2. Compared to the two habitational structures we have excavated on the other side of the campsite (structures B8 and D11), A2 is bigger in size and especially denser in materials and stratigraphic deposit. High amounts of local cooking pots and containers have been found here, together with camelid bones (NISP = 11,117), maize, chañar (*Geoffreoa decorticans*), and algarrobo (*Prosopis* sp.) fruits and few remains of *Lagenaria* sp. (Uribe and Oñate 2010; McRostie 2010; Labarca and Calás 2010). It is evident that massive preparation of food took place inside A2, as we said on a bigger scale when compared to what is happening within the other habitational structures of the site.

It is interesting to consider once again that this area for preparing food on a greater scale than the domestic level was placed within an Inca-style structure separated from the rest of the camp. Furthermore, in the hearths and refuse of A2 we found the higher amounts of Inca pottery styles (Local and Imperial styles) of the whole mining complex (Fig. 12.6), including *aríbalos* usually used for serving maize chicha (Bray 2008) and *ollas de pedestal*, considered personal cooking pots for state representatives in the provinces (Bray 2004). Our most recent excavations in sector A have identified more LP structures under the historical corral. At least two of these structures show evidence of domestic use, so they may have housed the

Fig. 12.8 An Inca structure for the production of food for local banquets found during excavations under an historical corral

State's representative(s). Here we have found the highest frequencies of maize in all Inkawasi-Abra, as well as higher frequencies of Provincial-Inca and local-Inca plates (we have to consider that local miners were not using pottery plates for food consumption).

Thus, our interpretations indicate that sector A may have been an area destined to the residence of state representative(s) as well as for the preparation of food and serving of both food and drink in collective, feasting activities.[2] It is well known that feasting was an integral part of the mit'a (corvée labor) system organized by the Inca (Murra 1989; Morris 1995), an example of commensal political strategies employed by the state in its relation with local populations within an ideology of redistribution and asymmetric reciprocity (Bray 2008, Dillehay 2003). Thus, we have archaeological evidence of one of the most common political strategies used by the Inca to ensure loyalty to the state and state-sponsored tasks inside the provinces.

[2] It is unlikely the preparation of food in sector A served day-to-day purposes, since each domestic structure at the site (in sectors B and D) has its own hearth with kitchen refuse (cooking pots, jars, camelid bones, chañar, and algarrobo fruits).

Fig. 12.9 Platform AB-39/5 overlooking the Inca turquoise mines and the retention walls in the Complejo Minero San José del Abra

Besides these communal feasts held inside the campsite, most probably in the central plaza, there is evidence of more restricted state ritualism associated directly with the mines. We have mentioned before the presence of two platforms in the San José del Abra complex. They are located on the higher slopes of quebrada Casicsa, overlooking the LP mines, but given their size and location it is evident that not everyone had access to these loci. However, platform AB-39/5 can be seen from every point in the quebrada, including each and every LP site found therein (Fig. 12.9). So, even though it was an exclusive place, it was meant to be constantly seen by local workers.

Platform AB-39/5 is oriented toward Cerro Las Marías (4286 masl), while platform AB-40 is oriented toward Cerro Pajonal (4512 masl). These are the highest altitudes in the El Abra locality and between both of them there is an *abra* (natural pass between high mountains) that served as one of the only two natural connections between El Abra and the Loa River. Furthermore, recent excavations at platform AB-39/5 showed evidence of Spondylus shell fragments (likely *Spondylus princeps* and *Spondylus calcifer*), which clearly support the idea of a ritual context, given the well-known symbolic importance of this shell and its relation with fertility in Inca ideology (Murra 1989). A radiocarbon date of 350 ± 30 BP (Cal AD 1440–1530 and Cal AD 1560–1630) from a small hearth near the spondylus shells demonstrates these ritual offerings date to the LP. We must keep in mind that these shells were probably brought from coastal Ecuador and that in the LP northern Chile they appear almost exclusively associated with rituals performed at mountain-peak sanctuaries, usually including child sacrifice or Capacocha (Checura 1977). The presence of specimens of Spondylus shells strongly suggests state-sponsored rituals performed in honor of the local mining *huaca* at El Abra.

Thus, our findings suggest that the Inca state not only performed ceremonial banquets and feasts in the mining campsite but also controlled some of the most important rituals associated with the fertility of the mines through the supply of non-local products such as the Spondylus shells that were offered in this secluded yet all-encompassing *loci*. Most likely, the rituals performed at these platforms were permeated by state ideology as well and/or conducted by state representatives acting as mediators between local miners and the supernatural being that control the fertility of the mines and the safety of the workers (see Chaps. 8 and 13).

Inca Regional Administration and the Mining Landscape

The mining complexes studied at El Abra were controlled by the Inca through diverse means that included economic organization, social institutions, commensal politics, and rituals. The sites were connected to other Inca settlements in the region, and specifically to the Inca road and its associated infrastructure running parallel to the Upper Loa River. More than 150 km of Inca road and different categories of sites that run along it have been previously documented there (Berenguer et al. 2005; Berenguer 2007). We are currently investigating the road and caravan trail network that connected the mining complexes to the Loa River *Qhapaqñan*.

Our preliminary data are interesting in relation with the control strategies exerted by the Incas: There are two natural connections between the San José del Abra mining complex and the Loa River Basin, considering the interference of topographic barriers. And in both cases, the Incas built small structures with the purpose of controlling what came in and out of the mining complex (Figs. 12.5 and 12.10). Once the caravans left these control posts, they had one or two days travel before reaching the Inca *tambos* and *chaskiwasis* running along the Qhapaqñan. Exactly 11 sites serving this function have been located along the Inca road running parallel to the Upper Loa River Basin (Berenguer et al. 2005, Berenguer 2007). But among them, foremost in size, architectural complexity and density of Inca Cuzco and provincial Inca ceramics, is the site of Cerro Colorado (Fig. 12.11). This is not just an ordinary *tambo*, but an administrative center as well (Berenguer 2007). It is located 25 kilometers east of the San José del Abra mining complex, but more than 50 kilometers away from the nearest Atacamenian pukara or town. It is also located next to a natural *huaca* or sacred site (Salazar et al. 2011b), the Cerro Sirawe, a very particular dune formation in the hillside, and where other factors important to Inca sacred geography in the South Andes (Cruz 2006) are also present.

It has been stated that one of the main functions of the Cerro Colorado site was the administration of Inca mining at El Abra and the sponsoring of ritual celebrations according to *mit'a* calendar (Berenguer et al. 2005, Berenguer 2007). It is well known that throughout the empire, the Incas built ceremonial and administrative centers where the main rituals were held. During these celebrations state ideology was reproduced, the sacredness of the Inca was strengthened, and reciprocity bonds between the state and local populations were reinforced (Morris 1995, De Marrais et al. 1996).

Fig. 12.10 Site AB-44, a control post for llama caravans

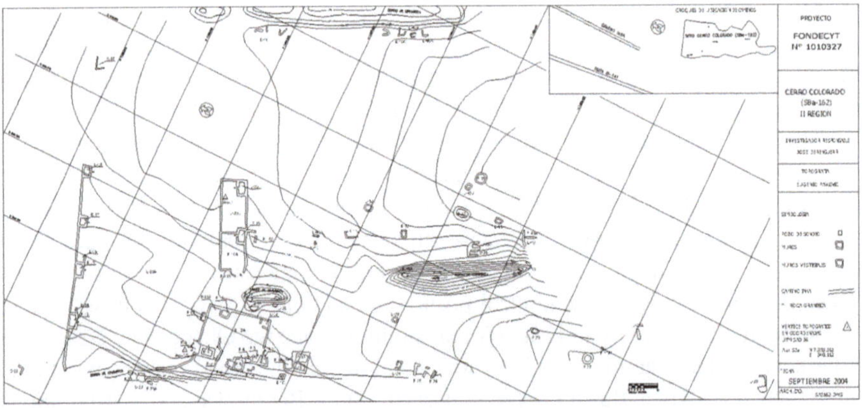

Fig. 12.11 Plan view of the Cerro Colorado site (after Berenguer 2010)

Cerro Colorado is one of the three administrative and ceremonial sites especially built by the Incas in Atacama. The two others were Cerro Verde (Adán 1999; Adán and Uribe 2005) and Miño (Berenguer 2010). These three sites share important characteristics (Salazar et al. 2011b). First, the three sites are located in areas highly significant according to Inca sacred geography. Second, none of these sites had significant previous LIP occupations. Third, the most visible traits of Inca settlement planning and architecture of the whole region are present at the sites, including the only *ushnu* of Atacama found at Cerro Verde (Fig. 12.12). Fourth, in these sites we have the highest frequencies of Imperial and provincial Inca pottery styles of the region. And most interestingly, all these sites are associated with the administration of Inca mining centers. Large public spaces and high quantities of aríbalos found in excavations attest to the importance of feasting at these sites, most likely related to

Fig. 12.12 The Ushnu at the Cerro Verde site

mit'a working obligations at the nearby mines. There is evidence of small-scale smelting near the Miño sites (not securely dated to the LP), but not around Cerro Colorado nor Cerro Verde.

Local and regional data reviewed in this article indicate that copper mining was not only of economic importance for the Incas. The mines themselves were probably considered *huacas* (Cruz 2009; Platt and Quispe 2008). And in Atacama it was close to these mining *huacas* that the Incas built their main administrative and ceremonial centers, thus creating a new sacred landscape in the region, structured around copper extraction and the sacredness of the mines.

Conclusions

It has been recently stated that Berthelot's (1986) doctoral research remains perhaps the most all-encompassing analysis of Inca mining yet published (Van Buren and Presta 2010; Chap. 9). According to Berthelot, there were mines owned directly by the Inca and the state, and others owned by local communities and their leaders. Inca mines were distinct insofar as they were concentrated in certain areas and usually separated from the smaller scale and scattered pattern of local mines. Moreover, Inca mines were directly supervised by the state's authorities and the control over the mines included the control over ritualism associated with their fertility. Our data from San José del Abra fits quite comfortably with Berthelot's criteria for identifying Inca-owned mines. The only criterion that is not present is the purported technological difference between Inca and local mines, with the former being more sophisticated. We believe in our study area the Incas did not need to introduce technological changes in copper extraction as Atacamenian were probably among the most experienced miners in the South Andes. Thus, since both technologies and technical strategies do not show significant changes after Inca rule at El Abra, we may conclude the state took advantage of several millennia of mining expertise by the local Atacamenian populations and reorganized it to fit imperial demands.

The control exerted directly by the state over the El Abra mines that we have suggested in this paper, reformulates our previous interpretations on this topic (i.e., Salazar 2008) but at the same time gives further credit to prior suggestions regarding the importance of copper mining in the Inca conquest and administration of Atacama.

In the Upper Loa River Basin, we have studied two mining complexes aimed at the exploitation of turquoise, chrysocolla, and pseudomalachite ores. There were other mines as well in Atacama, notably in Cerro Verde and in the Miño-Collahuasi area to the north of El Abra (Salazar et al. 2011b). But in San José del Abra, which is the best studied yet, the impact of Inca expansionism is most clearly seen. When compared to Late Intermediate Period mining, in El Abra there is clear evidence for the increase in overall volumes of extracted ores during Inca administration. This was made possible by a nucleated system of exploitation that included higher amounts of workers in the mines and increased dedication to productive activities, while being financed by the state according to traditional Andean rules of reciprocity.

Nevertheless, the latter were mostly changes in the economic organization of mining. Under Inca rule, transformations included a new sociopolitical organization of extractive operations as well. This is most evident at the Inkawasi–Abra campsite that housed the mining community. It is well known how the Incas used architecture in order to promote and reproduce social relations. El Abra was no exception. Here, the Inca placed state symbols in the center of the socio-spatial organization of Inkawasi–Abra and of the mining complex as a whole, thus reproducing ideas of hierarchy and control through the use of architectural devices. Furthermore, the Incas controlled the production of food and the distribution of chicha at the main communal gatherings and feasts of this mining community. They supplied some of the most important ritual offerings destined to favor the fertility of the mines (e.g., Spondylus shells brought from Ecuador) thus making their intervention essential for the success of the operation. They also built small posts in strategic locations in order to control goods and people coming in and out the mining complexes. And last but certainly not the least, in order to administer these mining operations at a regional level, they rebuilt local sacred landscapes situating mining in a privileged position and charged this new mining landscape with state symbols and rituals.

Thus, during the LP the Inca state completely restructured mining production in Atacama in order to control surplus. To achieve this they organized the work force through state institutions such as the *mit'a*, financed the mining operations, feasted the workers, and controlled the symbolic technology and productive rituals (*sensu* Van Kessel 1989) indispensable for the functioning of Andean productive systems.

Inca mining at Atacama was indeed a complex system that deeply related technology to economy, sociopolitical organization, and religion. Most likely this system was context-specific, so it may have varied between regions according to imperial demands and local realities.

Acknowledgments Our research at El Abra has been funded since 1999 by Sociedad Contractual Minera El Abra. We are grateful for this support. We would also like to thank all the field and laboratory team members who have contributed to our better understanding of these sites with their encouraging discussions. Special thanks to Kevin Vaughn and Nico Tripcevich for their invitation to publish our work in this book, and to an anonymous reviewer for helpful comments that improved the manuscript.

References

Absi, P. (2005). *Los ministros del diablo*. IRD/IFEA/Fundación PIEB, La Paz, Bolivia: El trabajo y sus representaciones en las minas de Potosí.

Adán, L. (1999). Aquellos antiguos edificios. *Acercamiento arqueológico. Estudios Atacameños, 18*, 13–33.

Adán, L., & Uribe, M. (1995). Cambios en el uso del espacio en los períodos agroalfareros: un ejemplo en ecozona de quebradas altas, la localidad de Caspana. Actas del II Congreso Chileno de Antropología, Valdivia.

Adán, L., & Uribe, M. (2005). El dominio inca en la localidad de Caspana: un acercamiento al pensamiento político andino (río Loa, norte de Chile. *Estudios Atacameños, 29*, 41–66.

Agüero, C. (2005). Aproximación al asentamiento humano temprano en los oasis de San Pedro de Atacama. *Estudios Atacameños, 30*, 29–60.

Alconini, S. (2008). Dis-embedded centers and architecture of power in the fringes of the Inka empire: New perspectives on territorial and hegemonic strategies of domination. *Journal of Anthropological Archaeology, 27*, 63–81.

Aldunate, C. (1993). *Arqueología del Pukara de Turi*. Temuco: Actas del XII Congreso Nacional de Arqueología Chilena.

Aldunate, C., & Castro, V. (1981). *Las Chullpas de Toconce y su relación con el poblamiento altiplánico en el Loa Superior*. Santiago: Ediciones Kultrún.

Berenguer, J. (1994). Asentamientos caravaneros y tráfico de larga distancia: el caso de Santa Bárbara. In M. E. Albeck (Ed.), *Taller de costa a selva*. Jujuy: Instituto Interdisciplinario de Tilcara.

Berenguer, J. (2004). *Caravanas, interacción y cambio en el desierto de Atacama*. Santiago: Sirawi Ediciones.

Berenguer, J. (2007). El camino inka del Alto Loa y la creación del espacio provincial en Atacama. In A. E. Nielsen, M. C. Rivolta, V. Seldes, M. M. Vásquez, & P. H. Mercolli (Eds.), *Producción y circulación prehispánicas de bienes en el sur andino* (pp. 413–443). Córdoba, Argentina: Editorial Brujas.

Berenguer, J. (2010). Estrategias ocupacionales incaicas en dos espacios internodales del norte de Chile. *Actas del XVII Congreso Nacional de Arqueología Chilena, 2*, 1389–1401.

Berenguer, J., Cáceres, I., Sanhueza, C., & Hernández, P. (2005). El Qhapaqñan en el Alto Loa, norte de Chile: un estudio micro y macromorfológico. *EstudiosAtacameños, 29*, 7–39. San Pedro de Atacama.

Berthelot, J. (1986). The extraction of precious metals at the time of the Inka. In J. V. Murra, N. Wachtel, & J. Revel (Eds.), *Anthropological history ofAndean polities* (pp. 69–88). Cambridge: Cambridge University Press.

Bouysse-Cassagne, T. (2008). Minas del sol, del Inka y de la gente. Potosí en el contexto de la minería prehispana. In P. Cruz & J. Vacher (Eds.), *Minas y Metalurgias en los Andes del Sur, entre la época prehispánica y el siglo XVII* (pp. 278–301). Sucre: Instituto Francés de Estudios Andinos e Institute de Recherche pour le Développement (IRD).

Bray, T. L. (2004). La alfarería imperial inka: una comparación entre la cerámica estatal del área de Cuzco y la cerámica de las provincias. *Chungara, 36*(2), 365–374.

Bray, T. (2008). The role of Chicha in Inca state expansion: A distributional study of Inca Arībalos. In J. Jennings & B. Bowser (Eds.), *Drink, power, and society in the Andes* (pp. 108–132). Gainesville: University of Florida Press.

Castro, V., Maldonado, F., & Vasquez, M. (1993). Arqueología del "pukara" de Turi. *Boletín Museo Regional de la Araucanía, 4*(1), 79–106.
Castro, V., Varinia, V., Aldunate, C., & Araneda, E. (2004). Principios Orientadores y Metodología para el Estudio del Qhapaqñan en Atacama: Desde el Portezuelo del Inka hasta Río Grande. *Chungara, 36*(2), 463–481.
Checura, J. (1977). Funebria Incaica en el cerro Esmeralda (Iquique, I Región). *Estudios Atacameños, 5*, 12–141.
Cornejo, L. (1999). El Inka en la Región del río Loa: lo local y lo foráneo. In *Actas del XIII Congreso Nacional de Arqueología Chilena* (pp. 203–212). Antofagasta: Sociedad Chilena de Arqueología/Universidad de Antofagasta.
Costin, C. (2001). Craft production systems. In G. M. Feinman & T. Douglas Price (Eds.), *Archaeology at the millennium* (pp. 273–327). New York: Kluwer.
Cruz, P. (2006). Mundos permeables y espacios peligrosos. Consideraciones acerca de punkus y qaqas en el paisaje altoandino de Potosí, Bolivia. *Boletín del Museo Chileno de Arte Precolombino, 11*(2), 35–50.
Cruz, P. (2009). Huacas olvidadas y cerros santos. *Apuntes metodológicos sobre la cartografía sagrada en los Andes del sur de Bolivia. Estudios Atacameños, 38*, 55–74.
D'Altroy, T. N. (2002). *The Incas*. Massachusetts y Oxford: Blackwell Publishers.
De Marrais, E., Castillo, L. J., & Earle, T. (1996). Ideology, materialization and power strategies. *Current Anthropology, 37*(1).
Dillehay, T. (2003). El colonialismo Inka, el consumo de chicha y los festines desde una perspectiva de banquetes políticos. *Boletín de Arqueología PUCP, 7*, 355–363.
Figueroa, V., Montero, I., & Rovira, S. (2010). Estudio tecnológico de objetos de cobre procedentes de Cerro Turquesa (San José del Abra, II Región). In *Actas del XVII Congreso Nacional de Arqueología Chilena, Tomo II* (pp. 1135–1147). Valdivia.
Gallardo, F., Uribe, M., Ayala, P. (1995). Arquitectura inka y poder en el pukara de Turi. Gaceta Arqueológica Andina 24, Lima.
Godoy, R. (1985). Mining: anthropological perspectives. *Annual Review of Anthropology, 14*, 199–217.
Labarca, R., & Calás, E. (2010). Informe arqueofaunístico de los yacimientos Inkawasi Abra (Ab-36). Inkawasi Loa y Cerro Colorado 1, Región de Antofagasta, Chile. Research report.
Llagostera, A. (1976). Hipótesis sobre la expansión incaica en la vertiente occidental de los Andes meridionales. En Homenaje al Dr. R. P. Gustavo Le Paige, L. Núñez, Compilador (pp. 203–218). Antofagasta: Universidad del Norte.
Llagostera, A. (2006). San Pedro de Atacama y el sistema reticular de interacción puneña. In H. Lechtman (Ed.), *Esferas de interacción prehistóricas y fronteras nacionales modernas: los Andes sur centrales* (pp. 303–323). Lima: Instituto de Estudios Peruanos.
Lynch, T. (1977). Tambo incaico Catarpe-Este: informe de avance. *Estudios Atacameños, 5*, 142–147. San Pedro de Atacama.
McRostie, V. (2010). Informe arqueobotanico. Sitios Incahuasi Abra, Incahuasi Loa, Cerro Colorado. Research report.
Morris, C. (1995). Symbols to power. Styles and media in the Inka state. In C. Carr & J. Neitzel (Eds.), *Style, society and person: Archaeological and ethnological perspectives*. New York: Plenum Press.
Muñoz, I. (1989). El Período Formativo en el Norte Grande. In J. Hidalgo, V. Schiappacasse, H. Niemeyer, C. Aldunate, & I. Solimano (Eds.), *Prehistoria. Desde sus orígenes hasta los albores de la conquista* (pp. 107–128). Santiago: Editorial Andrés Bello.
Murra, J. (1989). *La organización económica del estado Inca*. Madrid: Siglo XXI Editores.
Niemeyer, H., & Schiappacasse, V. (1998 [1987]). Patrones de asentamiento incaicos en el Norte Grande de Chile. In T. D. Dillehay & P. J. Netherly (Eds.), *La frontera del Estado Inca* (pp. 114–152). Quito: Fundación Alexander Von Humboldt / Editorial Abya-Yala.
Núñez, L. (1999). Valoración minero-metalúrgica circumpuneña: menas y mineros para el Inka Rey. *Estudios Atacameños, 18*, 177–221.

Núñez, L. (2006). La orientación minero-metalúrgica de la producción atacameña y sus relaciones fronterizas. In H. Lechtman (Ed.), *Esferas de interacción prehistóricas y fronteras nacionales modernas: los Andes surcentrales* (pp. 205–260). Lima: IEP-IAR.

Núñez, L., Cartagena, I., Carrasco, C., de Souza, P., & Grosjean, M. (2006). Emergencia de comunidades pastoralistas formativas en el sureste de la Puna de Atacama. *Estudios Atacameños, 32*, 93–117.

Núñez, L., & Dillehay, T. (1978). *Movilidad giratoria, armonía social y desarrollo en los Andes Meridionales: patrones de tráfico e interacción económica.* Antofagasta: Universidad del Norte.

Núñez, L., Grosjean, M., & Cartajena, I. (2005). The expansion of the Inka empire into the Atacama Desert. In M. Smith & P. Hesse (Eds.), *23S. Archaeology and Environmental History of the Southern Deserts* (pp. 324–332). Canberra: National Museum of Australia Press.

Platt, T., & Quispe, P. (2008). Sobre las huellas del silencio: Potosí, los incas y el virrey Francisco de Toledo (siglo XVI). In P. Cruz & J. Vacher (Eds.), *Minas y metalurgias en los Andes del Sur, entre la época prehispánica y el siglo XVII* (pp. 231–277). Sucre: Instituto Francés de Estudios Andinos e Institut de Recherchepour le Développement (IRD).

Raffino, R. (1981). *Los Inkas del Kollasuyu.* La Plata: Ramos Americana Editora.

Rees, C. (1999). Elaboración, distribución y consumo de cuentas de malaquita y crisocola durante el período Formativo en la vega de Turi y sus inmediaciones, subregión del río Salado, norte de Chile. In C. Aschero, A. Korstanje, & P. Vuoto (Eds.), *Los Tres Reinos: Prácticas de recolección en el Cono Sur de América* (pp. 85–98). Tucumán: Instituto de Arqueología y Museo, Universidad Nacional de Tucumán.

Salazar, D. (2002–2005). Investigaciones arqueológicas sobre la minería incaica en San José del Abra (II región, norte de Chile). *Xama, 15–18*, 101–118.

Salazar, D. (2008). La organización de la producción minera en San José del Abra durante el Período Tardío Atacameño. *Estudios Atacameños, 36*, 43–72.

Salazar, D., Figueroa, V., Mille, B., Morata, D., Manríquez, G., & Cifuentes, A. (2011a). Metalurgia en San Pedro de Atacama Durante el Período Medio: Nuevos Datos, Nuevas Preguntas. *Revista Chilena de Antropología, 23*(1), 123–148.

Salazar, D., Berenguer, J., & Vega, G. (2011b). Paisajes minero-metalúrgicos inkaicos en Atacama (norte de Chile). Chungara (submitted for publication).

Salazar, D., & Salinas, H. (2008). Tradición y transformaciones en la organización de los sistemas de producción mineros en el norte de Chile prehispánico: San José del Abra, siglos I al XVI d.C. In P. Cruz & J.-J. Vacher (Eds.), *Mina y metalurgia en los Andes del Sur desde la época prehispánica hasta el siglo XVII* (pp. 163–200). Sucre: Instituto Francés de Estudios Andinos.

Salazar, D., Salinas, H., McRostie, V., Labarca, R., & Vega, G. (2010). Cerro Turquesa: Diez siglos de producción minera en el extremo norte de Chile. In *Actas del XVII Congreso Nacional de Arqueología Chilena, Tomo II* (pp. 1085–1097). Valdivia.

Schiappacasse, V., Castro, V., & Niemeyer, H. (1989). Los desarrollos regionales en el norte grande. In J. Hidalgo, V. Schiappacasse, H. Niemeyer, C. Aldunate, & I. Solimano (Eds.), *Prehistoria. Desde sus orígenes hasta los albores de la conquista* (pp. 107–128). Santiago: Editorial Andrés Bello.

Sepúlveda, M., & Laval, E. (2010). Uso de minerales de cobre en la pintura rupestre de la localidad de río Salado (II Región, norte de Chile). In *Actas del XVII Congreso Nacional de Arqueología Chilena, Tomo II* (pp. 1111–1122). Valdivia.

Sinclaire, C. (1994). Los sitios de "muros y cajas" del río Loa y su relación con el tráfico de caravanas. In M. E. Albeck (Ed.), *Taller de Costa a selva.* Jujuy: Instituto Interdisciplinario de Tilcara.

Soto, C. (2010). Tipología de cuentas de collar en la Quebrada de Tulán (Salar de Atacama): Nueva línea de evidencia para la transición Arcaico-Formativo. In *Actas del XVII Congreso Nacional de Arqueología Chilena, Tomo II*, pp. 1123–1134.

Urbina, S. (2007). Estudio arquitectónico del Pucara de Topaín (Región de Antofagasta. Norte de Chile). *Boletín de la Sociedad Chilena de Arqueología, 40*, 29–46.

Uribe, M. (1999–2000) La arqueología del Inka en Chile. Revista Chilena de Antropología 15, Santiago.

Uribe, M. (2002). Sobre alfarería, cementerios, fases y procesos durante la prehistoria tardía del Desierto de Atacama (800–1600 DC). *Estudios Atacameños, 22*, 7–32.
Uribe, M., Adán, L., & Agüero, C. (2002). El dominio del Inka, identidad local y complejidad social en las tierras altas del Desierto de Atacama, Norte Grande de Chile (1450–1541 DC). *Boletín de Arqueología PUCP, 6*, 301–336.
Uribe, M., Adán, L., & Agüero, C. (2004). Los Períodos Intermedio Tardío y Tardío de San Pedro de Atacama y su relación con la cuenca del Loa. *Actas del XV Congreso Nacional de Arqueología Chilena*, Chungara Volumen Especial.
Uribe, M., & Oñate, C. (2010). Cerámica arqueológica de mina San José del Abra (sitio Ab-36) y anexo sobre materiales de los sitios Cerro Colorado e Incahuasi-Loa (II Región de Antofagasta). Research report.
Uribe, M., & Urbina, S. (2009). Cerámica y arquitectura pública en el Camino del Inka del Desierto de Atacama (río Loa, Norte de Chile). *Revista Chilena de Antropología, 20*, 227–260.
Van Buren, M., & Presta, A. M. (2010). The organization of Inka Silver Production in Porco, Bolivia. In M. A. Malpass & S. Alconini (Eds.), *Distant Provinces in the Inka Empire: Toward a deeper understanding of Inka imperialism* (pp. 171–192). Iowa City: University of Iowa Press, IA.
Van Kessel, J. (1989). Ritual de producción y discurso tecnológico. *Chungara, 23*, 73–91.
Varela, V. (1999). El camino del Inca en la cuenca superior del río Loa, desierto de Atacama, norte de Chile. *Estudios Atacameños, 18*, 89–106.
Zori, C. (2011). *Metals for the Inka: Craft production and empire in the Quebrada de Tarapacá, northern Chile* (Unpublished Ph.D. dissertation). University of California, Los Angeles.

Chapter 13
Economic, Social, and Ritual Aspects of Copper Mining in Ancient Peru: An Upper Ica Valley Case Study

Hendrik Van Gijseghem, Kevin J. Vaughn, Verity H. Whalen, Moises Linares Grados, and Jorge Olano Canales

Introduction

The history of social differentiation in the Andes, as elsewhere, can be approached as a history of the increasingly complex and exclusive appropriation of goods, practices, and privileges by a limited few. Throughout the development of sociopolitical complexity in non-market societies, we witness more aggressive and stringent restrictions on ownership of certain goods, on consumption of certain foods, on a variety of specific behaviors, and on access to certain places (Costin and Earle 1989; Costin 1996; Schoenberger 2011: 14–15). These restrictions are sometimes grouped under the label of sumptuary laws, constituting a repertoire of outward material and behavioral markers of status, and they are accompanied by ritual and symbolic means of mediation of the social distance that they create. The unique status of the privileged individuals is not only explicitly heightened by these social principles but reproduced through time in culturally coherent ways, and mediated by ritual and religious ceremonies. We learn, for instance, that among the Inca, the possession

---

Manuscript prepared for the volume *Mining and Quarrying in the Ancient Andes: Sociopolitical, Economic and Symbolic Dimensions,* Nicholas Tripcevich and Kevin J. Vaughn, editors. 2011.

H. Van Gijseghem (✉)
Universitê de Montreal, Montreal, Quebec, Canada
e-mail: h.van.gijseghem@gmail.com

K.J. Vaughn • V.H. Whalen
Department of Anthropology, Purdue University, 700 West State Street, West Lafayette, IN 47907, USA
e-mail: kjvaughn@purdue.edu; vwhalen@purdue.edu

M.L. Grados
Arqueocare, Lima, Peru

J.O. Canales
Nasca Mining Project, MzE1 Lte12 Sector V, Derecho Urb. Antonia Moreno de Caceres, Ventanilla, Callao, 07, Peru

N. Tripcevich and K.J. Vaughn (eds.), *Mining and Quarrying in the Ancient Andes*, Interdisciplinary Contributions to Archaeology, DOI 10.1007/978-1-4614-5200-3_13,

and use of metal objects was among these myriad-exclusive domains (Moore 1958; Rostworowski and Morris 1999: 842). It is therefore understandable that the state tightly controlled the production and distribution of metal objects (Murra 1982), although in some cases mining was left in the hands of the local authorities on behalf of the empire (Berthelot 1986). The very locations on the landscape that yielded those resources were endowed with prominent economic, political, and ritual importance (Cobo [1653] 1890–1895; Rowe 1946; see also Chaps. 1 and 12).

In this paper we suggest that exploitation and control of mines not only conferred material wealth—broadly conceived as the appropriation and control of a commonly valued resource—but that it also positioned individuals and groups in a unique and powerful situation vis-à-vis supernatural beings. In various times in Andean prehistory this position may have been advantageous to various people depending on the immediate sociopolitical context. Control of mining therefore offered multiple scales of advantages. We outline general principles on mining and landscape in the Andes using literature from different fields and epochs, and evaluate the importance of these principles to different periods of the Prehispanic era. This study highlights that material, moral, and economic principles are not easily distinguishable as different realms in the ancient Americas, but form part of a single cognitive entity. In light of these principles we present recent survey data from the upper Ica valley of south coastal Peru, a region that is today an important mining region rich in copper and gold, in order to examine the relative importance of mining practices and infrastructure during different periods of south coastal prehistory.

Metals in the Andes

Metals have been produced and used in the Andes since at least the third millennium BC (Aldenderfer et al. 2008; Schultze et al. 2009; Chap. 11). Although at certain times and places tools occasionally were made of copper, gold, bronze, or silver, these constitute exceptions as metals were primarily used as a support for iconography and for the production of elite goods used in exchange and as gifts and rewards supporting a complex elite system (Lechtman 1993, 1996a; Owen 2001). Metal objects were highly prized for the skill and expertise involved in their manufacture, and their portability made them easily traded merchandise over long distances (Lechtman 1996b). Andean peoples, contrary to the Spaniards' lust for the market value of these materials, sought their visual properties: sound, color, symbolic meaning in the form of iconography, and the inherent political significance conferred by their ownership. These objects of cult, vessels, and adornments marked ethnic affiliation, rank, class, and relational status with authority and the sacred: "what mattered was not only *what* you wore, but what you were *allowed* to wear" (Lechtman 1996a: 30, original emphasis). The production and distribution of metals were not regulated by a market economy, but by an economy in which privilege was the main currency.

The technical procedures involved in the production of metal objects in the Andes are well known (Lechtman 1996b, for instance), but suffice it to say here that copper as an alloying agent was critical to Andean metallurgy to give silver and gold the desired properties combining malleability, rigidity, and proper color (Lechtman 1996b, Shimada and Merkel 1991). Until recently, copper has taken a backseat to silver and gold in the Andean literature, a relative neglect at least partly due to Spanish obsession with the latter two metals following the conquest. Still, evidence indicates that copper was nonetheless immensely important in the Andean attitudes toward metals (Lechtman 1993, 1996b). Andean ethnocategories linked gold with the sun and masculinity and silver with the moon and femininity, and both metals were central in Inca origin myths. In addition to its transformative quality in alloying and color management, some authors argue that copper may have been symbolically complementary to gold and silver in representing commoners, and embody hierarchical and gender associations (Rostworowski 1983; Lechtman 2007), in a profound symbolic parallel to its central importance in gold and silver metallurgy. In the following section we examine general symbolism and behaviors associated with mineral extraction in the Andes.

The Moral, Symbolic, and Material Economy of Mining in the Andes

Mining is by definition an activity deeply embedded and intertwined with landscape perceptions. It casts upon the landscape its own kinds of productive and symbolic dimensions that reflect particular notions of value and worth. Because of their geological properties, mines are often located away from areas of habitual activities like dwelling and farming. The increased importance of certain minerals affects landscape perceptions, and the hinterlands around zones of daily activities are in a sense cognitively "re-mapped" according to the location of these resources. The presence or discovery of mines may create new mythical and symbolic narratives or alter existing ones.

Andean peoples are notable for the complex and powerful relationships they have with their environment and landscape features. In a strictly environmental sense, the Andes are a complex landscape composed of a juxtaposition of ecozones that are differentially productive, cradled by sharp seasonal cycles, volcanism, and seismic activity. These parameters have had great influence on Andean sociopolitical organization (e.g., Farrington 1992). Most scholars believe that a consideration of this complexity is essential to an understanding of regional dynamics in the Andes. These principles are at the core of Andean people's understanding of and attitudes toward their landscape, which together compose a rich tapestry of beliefs and behaviors related to geographic features. Sacred mountains—or *apus*—were ascribed powerful forces, personalities, and idiosyncratic characteristics that together make up regional sacred geography, mythology, and imagined genealogies that unite humans and their natural milieu. The worship of mountains as significant places in the sacred landscape is evidenced as a generalized Andean attitude (Reinhard 1985; Farrington 1992; Ceruti 2004).

Literature pertaining to the Inca is informative in this regard, as many mountains were attributed particularly powerful *huaca* status (Gil García and Fernández Juárez 2008: 106; Farrington 1992; Reinhard 1985; Sallnow 1987). Huacas are generally conceived as anything sacred, places or objects that link people with their mythical past and act as poles of mediation between humans and the supernatural forces that bind and regulate the universe. Ethnohistoric literature indicates that landscape features that were visually striking or somewhat anomalous by an unusual shape, color, or size were often designated *huaca* (Van de Guchte 1999; Stone-Miller and McEwan 1990/1991; Cobo 1990 [1653]: 44).

By way of their special visual properties, their locations on mountains, and the minerals they hold—quite literally the substance of mountains—mines were thus ascribed a powerful status in the Andes, and they still are today. Archaeologists who work in the Andes are well aware of the perceived necessity of making amends to injuries incurred to the earth by making various offerings before excavations. This well-entrenched tradition is generalized in agriculture, canal digging, and any form of excavation or extraction of material from the ground. Ethnography shows that these ritual practices and mythical attitudes apply to post-colonial mining as well.

These notions of sacrifice and reciprocal obligations between mines and miners are central to ethnohistoric and ethnographic observations (Gose 1986; Gil García and Fernández Juárez 2008: 107). Traditionally, mining is not conceptually far removed from agriculture (Harris 2000: 24; Bouysse-Cassagne 2005: 447), as they both hinge on the reciprocal relationship between human agents and supernatural agents or, broadly conceived, the earth and mountains. Mines are also said to be more productive if periodically left to lie fallow (Harris 2000: 26). Proper respect, offerings, and sacrifices made to the temperamental beings are believed to be repaid with their sustenance: plant growth, absence of pests and disease, and sufficient irrigation water in one case; precious minerals and absence of accidents in the other. Cobo may have been first to highlight the ritual importance of mines and specify their *huaca* status. He mentions that people petitioned and "prayed" to the mines so that they would yield their metal: "festivals were held in (the mines') honor at which the miners danced and drank chicha all night" (Cobo [1653] 1890–1895, bk. 13, Chap. 11 cited in Rowe (1946: 246); Ramirez 1994: 95).

Andean attitudes thus link the extraction of a mountain's substance with a profound transgression, and ethnographic observations highlight elements that echo these deeply held beliefs. Gose describes mining as the "culminating violation of the most central manifestation of the *apu*: the mountain itself. It represents a quantum leap beyond any other productive activity in the intensity of relations between people and *apus*, and constitutes a definite strain on both" (1986: 303). For Gose, mining is fundamentally dangerous; physically, of course, but supernaturally as well. He recorded a very manifest relationship, embedded in a modernist narrative, between metallurgy and human sacrifice in popular folktales about the clandestine trade in human fat and flesh (1986: 297). Sacrifices and ceremonies are perceived as necessary because "the mountain spirits are sometimes held to have machines inside their subterranean abodes, by means of which they convert offerings into gold and silver" (Gose 1986: 301; citing Earls 1970: 70).

Some obvious Christian elements are inserted within traditional beliefs and underline how mining and the appropriation of wealth is a threat to morality. Gil García and Fernández Juárez (2008) describe the nature of historical mines' inherent dangerousness as being linked with temptation and sin. Although all mountains are potentially dangerous, the most dangerous are the ones that hide mineral wealth, because they are said to house "devils" that tempt people (see MacCormack 1984). If this wealth was to be inappropriately paid for, grave harm or death would be incurred on the miner. Central to this notion is the reciprocal relationship between humans and these "devils," often named *Supay*, or alternatively, *Tío* among a group of ambiguous spirits. According to Harris (2000: 65) these spirits are "unpredictable and very powerful." In some accounts, *Supay* is believed to be the consort of Pachamama, the benevolent earth-mother, associated in the Catholic pantheon with the "Virgin of the Mineshaft," who tempers the devil's murderous impulses (Harris 2000: 209). More generally, *Supay* represents the hidden, dark, unseen aspect of people, places, and objects (Taylor 1980), akin to one's shadow, and is also responsible for carnal desires, secret wishes, and various sins such as envy and greed. Its historical association with mines (the dark, hidden aspects of mountains, and a potential source of wealth, envy, and greed) is manifest. These modern accounts exhibit fundamental Andean cognitive categories and symbols that are consistent with what we know of the Prehispanic world (Bouysse-Cassagne 2005).[1]

Mining being an activity embedded in hierarchical sociopolitics, a question remains as to the ultimate responsibility for the avoidance of unfortunate events (or alternatively, the fruitfulness of the mining operations). Evidence indicates that the ritual responsibilities are not clearly assigned to specific individuals and groups, but that they can be borne and claimed contextually by different agents. Brewer (2003: 32–33) describes an event that highlights this ambiguity. During a certain ceremony, the miners present the mine owner with minerals of especially high quality, to which the owner reciprocates with large amounts of alcohol and festive paraphernalia. The event is in essence a feast that constitutes a symbolic reproduction of the economic and social relationships between miners and owner: miners ceremonially exchange raw minerals for food and drink. The exchange is accompanied by offerings and rituals during which participants fight and ridicule a personification of the devil and

[1] Taussig (1980), however, considers the evil personae associated with mines to strictly be a post-conquest phenomenon that incarnated profound changes in mining practices brought about by the Spanish. We may agree that the identity of an evil figure corresponds to notions of European Good/Evil duality (see Brewer 2003; Harris 2000: 23) that may not have existed in the pre-conquest world, where morals, justice, sanctity, and benevolence were contextually fluid and ambiguous notions. In the Prehispanic world, any being could be both good and evil, benevolent or malevolent, at once (Gil and Fernandez 2008: 106). Above all, Andean supernatural beings embodied the *potentiality* of good and evil that eschewed the sharp Roman Catholic dualism.

is meant to insure safety and prosperity in mining operations for the year to come. The entire performance highlights and reflects the shared responsibility between the lowly miners and the mine owners, who reap most of the economic rewards out of the operation, but also makes manifest the role of the employees. The owner or engineer, however, also bears a large part of the ritual responsibility of appeasing the mountain spirit for their admitted greed (Brewer 2003: 33). Indeed, it is not uncommon in the Andes for the highest social classes to carry the burden of responsibility for the maintenance of cosmic balance and the avoidance of catastrophes. As pointed out by Duviols (1977), the Prehispanic child sacrifices of *capacocha*—extreme events in a spectrum of devotional acts—are performed in response to political or environmental crises and, ultimately, the Inca seeks atonement for his own responsibility in these crises, highlighting the especially intimate relationships between the important mountain *huacas* and the highest elites who position themselves as mediators of cosmic forces (see also Ceruti 2004). The Inca therefore acts on behalf of all humans, and the ritual sacrifices consequently feature a powerful political component in the co-option of *huacas*. On the other hand, in the context of mining, the ritual prerogatives can also fall in the hands of the miners themselves, allowing them important agency in the symbolic conduct of daily affairs. Participants in the *ch'alla*, or weekly offering ceremony to *Tio* described by June Nash (1972) in Oruro near Potosi include individual or grouped miners. The initiative is taken by the head of the labor crew, and the ceremony sometimes, but not necessarily, involves an individual of special status who voices demands to *Tio*. This person is reputed for his talents as communicator with *Tio*, as Nash describes him as either the oldest worker or a diviner. The performance of these rituals is imbued with a certain prestige. As some authors highlight, there is a generalized correlation between the perils involved in certain types of rituals and how much prestige is involved in their performance (Swenson 2004: 148; Barth 1975; Whitehouse 1992: 150). Ethnographic evidence therefore indicates that ritual responsibilities in the mediation of earthly and supernatural affairs vary contextually. They seem to be generally shared between miners and sponsors, but evidence also suggests that some contexts require or allow responsibility to be appropriated by institutional elites, as well as in other, more intimate contexts, by non-elites. We argue that this distinction generally follows the scale of the mining operation: an independent, small-scale exploitation will require the miners to fulfill ritual mediation; while large-scale, elite-sponsored operations will create a context in which responsibilities are shared between miners and sponsors (see Vaughn et al., 2010; Chap. 8).

This long introduction serves to highlight three aspects inherent in mines, mining, and metallurgy that hinge on (1) the social and political capital that comes with metal manufacture, ownership, and exchange; (2) the physical appearance of mines and their status as *huaca*; and (3) the symbolic relational burden involved in mining and the dangers involved in mediating those relationships. Mines are unique and powerful locations on the landscape that neatly lie at the intersection of economy, religion, and politics. The visual aspects of copper-bearing deposits as bright green or blue stone make them prime candidates for *huaca* status. The economic and political dividends associated with the obtainment of raw materials used to make politically charged metal objects is yet another powerful component associated with mining. Finally, the reciprocal relationships developed between miners (or the sponsors of the mining activities) and supernatural forces provide ritual practitioners important opportunities to situate themselves as mitigating agents in critical transgressions against important, powerful, potentially terrifying supernatural entities inscribed in the cultural landscape. As Vaughn et al. (2010; Chap. 8) argue, diverse ritual agents, regardless of their status, can contextually take advantage of these opportunities in order to contest sociopolitical conditions or reproduce and legitimate these same conditions.

In the next section we examine survey results from a specific Andean region, the Peruvian south coast, as a case study for the articulation of power, economics, and symbolic behavior associated with mining activities on a regional scale at different periods.

Prehispanic Mining in the Upper Ica Valley

Many industrial minerals would have been available at relatively little cost to the ancient peoples of Ica, in addition to gold, silver, and copper (Fig. 13.1). The millennia of erosion cutting through volcanic intrusions in cretaceous deposits have in many locations exposed seams of oxidized ores on the Andean foothills. It is likely that these minerals, readily visible in weathering zones by their green and blue coloring, were harvested early in prehistory before metallurgy was developed, and eventually led to the discovery of copper ores and their metallurgical properties. Among the documented minerals that we know were used in the past in addition to metals they accompany are various oxides, sulfites, and carbonates found in cupriferous deposits, such as malachite, azurite, atacamite, sodalite, and chrysocolla (as well as, possibly, turquoise). These were crushed to produce green and blue pigments (Brooks et al. 2008; Donnan 1992), as well as for lapidary work (Chap. 9) and also used in powdered form in various rituals (Peregrine and Ember 2002: 219; Pozorski and Pozorski 2002:219; Sawyer 1966; Shimada 1994; Petersen (2010) [1970]). Some Paracas mummy bundles contained small pouches of these pulverized minerals (Muelle and Wells 1939).

As metallurgy developed, items made of copper became widely traded and used in gift-exchange, and the Ica Regional museum houses a large collection of copper vessels, ceremonial tools, and adornments, although most of these lack provenience. According to some authors (Lechtman 1996b; Rostworowski 1970), during the Late Intermediate Period (LIP; ca. AD 1000–1478) copper may have been among the principal goods that south coastal Chincha merchants carried northward on their ships to be traded for spondylus, one of the most significant ritual staples and symbols of status in the Andes (Paulsen 1974; Glowacki 2005). The provenience of the coastal minerals, however, is still unknown, as is the structure and scale of the mining operations, and the sociopolitical and religious contexts in which they were carried out.

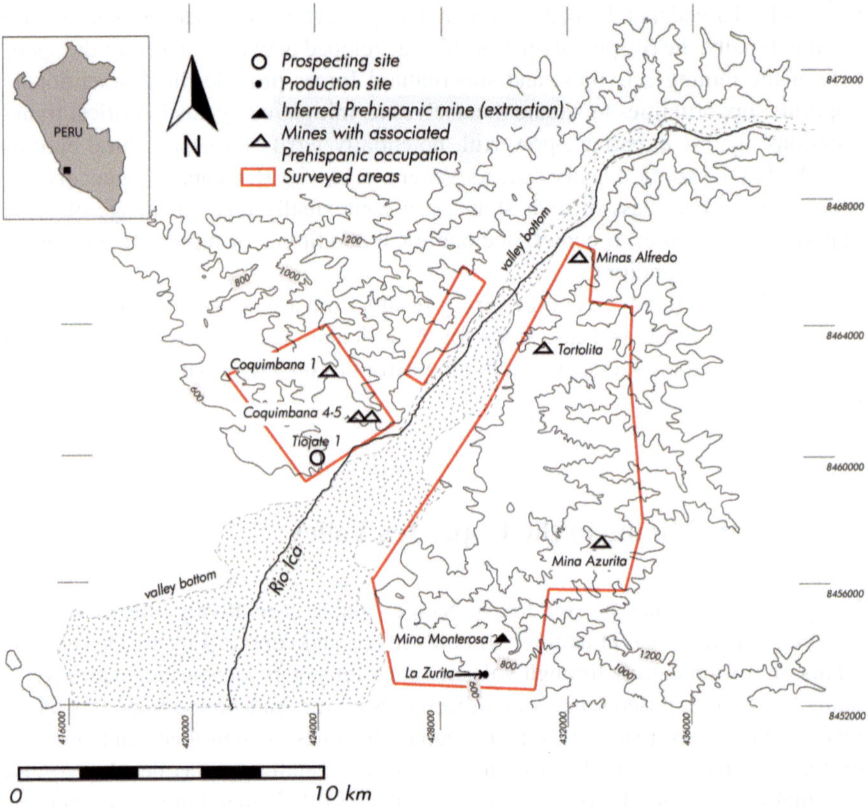

Fig. 13.1 The upper Ica valley with the location of sites discussed in the text

Our survey of the upper Ica valley first sought the discovery of remains of Prehispanic mining activities to verify if modern mineral resources were exploited in the past, indicated by the presence of tools or architectural remains in association with mines. Secondary questions that we aimed to address relate to the chronological profile of mining activities, the scale and intensity of mineral exploitation, and finally to the nature of the ritual component that accompanies mining in the Andes. Because of the inherent limitations of the archaeological record, we expect small-scale mining and ritual to have left relatively faint traces that may not be visible from survey alone. However, larger and more permanent structures and the recovery of fineware and other items of prestige would suggest a stricter political investment in mineral exploitation and closer involvement in associated ritual practices.

Our methodology consisted in the identification of locations where recent mining has taken place and to seek evidence for Prehispanic mining and metallurgy. Today many of these locations are under part-time, autonomous exploitation by independent miners and have not suffered drastic alteration, although some larger-scale enterprises exist that have presumably destroyed evidence for an ancient presence. The survey area covered about 200 km^2. We walked the entire extent of 10 *quebradas* and the foothill ranges between them surveying ancient and modern mining locations (see Fig. 13.1).

Many of the mining locations we targeted are relatively hard to reach and are located where Prehispanic settlements are not expected to be found (i.e., far away from water and agricultural fields). Since mining is currently practiced in the upper Ica valley in formal and informal manners, we were not permitted access to some locations that were jealously guarded by loosely formed coalitions of miners who were doubtful of the sincerity of our intentions. For reasons that should be obvious given current metal prices, we did not attempt to visit locations where gold is currently being mined. Many of these locations that we could not survey have been intensely mined and the archaeological record has presumably been critically altered. The survey results therefore represent an opportunistic sample of the upper Ica valley, and because most modern mining operations occur on the western side of the valley, the survey area is heavily biased toward the east, and the reduced actual pedestrian survey represents a much smaller area than the initially targeted region. Most of the valley-bottom sites had already been recorded and described by Williams and Pazos (1974), and re-visited by Massey (1986) who also augmented the area's site inventory. We also visited many of these sites for signs of activities related to mining, but in the end we only recorded and described the sites that appear on Fig. 13.1, only one of which (La Zurita) had previously been described by the aforementioned authors.

All of the sites that we have found associated with mining activities share one thing: the small quantity of artifactual remains on the surface, even when architecture is present and elaborate.[2] This can be due, depending on the context, to the rare but evident instances of flooding—when sites are on *quebrada* bottoms, or to posterior mining activities contributing to the alteration of surface remains. The scarcity of ceramic remains could also be attributed to the relatively low intensity of mining activity and occupation of these locations relative to habitation settlements, which comparatively yielded a larger quantity of trash, being occupied for longer periods of time, occupied year-round, and on which a greater diversity of activities was taking place. Rowe (1946) describes Inca mining as taking place 4 months out of the year, probably following rotational tax obligations and the agricultural cycle (see also Murra 1982). It is possible that pre-Inca mining in the Ica valley followed similar rhythms and that consequently a relatively small amount of refuse was discarded. Finally, one remaining possibility is that miners did not make much use of ceramic containers at all, especially diagnostic and decorated ceramics. Most likely the present state of the archaeological remains is due to a combination of these factors.

The sites we encountered during the survey were recorded following the site-types put forward by Eerkens et al. (2009). These are *Prospecting sites, Extraction sites* (mines and mining camps), and *Production sites*.

[2] This observation is somewhat subjective and rests on assumptions about what would constitute a "normal" artifactual surface assemblage on a habitation settlement.

The identification of prospecting sites inevitably rests on some degree of speculation because of their ephemeral nature, but Eerkens et al. (2008, 2009) nonetheless outline some expectations regarding this site-type. According to them, prospecting sites are generally small and constructed by lone individuals or small groups. They are strategically located relative to local geography, often on *quebrada* bottoms, and away from habitually settled areas. They serve as base camps where water, food, and tools can be stored while miners are exploring the immediate vicinity. This site-type generally yields small quantities of artifacts.

Extraction sites are mines and surrounding infrastructure, including small specialized settlements that temporarily housed miners. They vary widely in size, but have in common the presence of mining shafts or pits as well as tailings, and may be accompanied by habitation debris, tools, and structures used in initial assaying and preliminary processing of ores. Although Eerkens et al. (2008, 2009) make no mention of ritual components accompanying mines, our survey results indicate that the presence of special ceremonial infrastructure could be a diagnostic trait of extraction sites.

Production sites are metallurgical installations where the ores are processed; for instance, where smelting takes place. The discovery of production sites was not among the primary objectives of this survey. In fact, we expect smelting sites to have been located further down the valley, on plains that are not sheltered by the steep Andean foothills. Lechtman (1976) reports a colonial-period smelting site on the windward part of the middle Ica valley, where it opens onto a wide plain.

Survey Results

In this section we describe the sites discovered during the 2010 survey according to the categories outlined above. Site characteristics are summarized in Table 13.1.

Prospecting Sites

Here we describe the only site that we hypothesize to be a prospecting site based on the characteristics that Eerkens et al. (2009) describe for these sites in the Prehispanic periods and using modern analogs. This site, Tiojate 1 (Fig. 13.2), is composed of a small number of agglutinated rooms located at the bottom of a deep, narrow canyon with a location that maximizes control of movement. A single non-diagnostic sherd was found here, thus its date is unknown, but the sherd's fabric is more consistent with a relatively later rather than earlier ceramic vessel. The canyon in which the site is found is the only access into or out of an extensive network of deep gullies and canyons that are bounded by high and steep crevasses and

Table 13.1 Summary of mining sites discovered in the upper Ica valley survey

Site	Site type	Epoch	Architecture	Size (m^2)	No. of rooms	Other observations
Tiojate 1	Prospection	LIP (?)	Agglutinated rooms	~36	~8	
Coquimbana 1	Extraction	EH, LIP	Cleared space	~750	n/a	
Coquimbana 4	Extraction	LIP	Agglutinated rooms	~300	~14	Storage bins, grinding stones
Coquimbana 5	Extraction	LIP	Agglutinated rooms	~300	~12	Storage bins, grinding stones, stone chisels, terraces with decorated ceramics overlook Coquimbana 4 and 5
Mina Azurita	Extraction	EH, LIP	Elevated platform, ramps, cleared spaces, agglutinated rooms	~3400	~6	Storerooms, elaborate plaza with niches, ramps, and opposing stairs, ancient road
Tortolita	Extraction	MH, LIP	Cleared space	~650	n/a	Cleared space, overlooks small-scale mines. Tenuous association with MH ceramic remains, based on the absence of other settlements
Minas Alfredo	Extraction	LIP	Hilltop cleared space, terraces	~800	n/a	
La Zurita	Production	EH (Nasca 1)	n/a	~600	n/a	15 boulders, well worn with deep channels

Fig. 13.2 The main platform at Coquimbana 1

cliffs. We failed to clearly identify any mines in this area, but some parts of cliffs and hillsides appear to have been partially hacked and burned to spall stone, creating cavities from 1 m to several meters across. The lack of mines in the area beyond Tiojate 1 and the fact that these canyons and gullies are not transit routes incite us to suggest that it is a prospecting site rather than a mining camp or a station meant to either facilitate or control movement. Farther up the canyon LIP sherds and sling-stones were found in association with small natural terraces at strategic passes, but settlements or structures were not.

Extraction Sites

The survey identified six locations that we interpret to be extractions sites, either mines themselves or mines associated with small settlements. Two of these small mining settlements, Coquimbana 4 and 5 (Fig. 13.3), were discovered less than 200 m from each other, on either side of a low ridge. Both are located a small distance from the valley bottom and are directly associated with copper mine shafts that have been exploited until recently. Diagnostic surface ceramics indicate a Late Intermediate Period occupation. In addition, we found at both sites a few grinding stones, the wear on which differs from grinding stones destined for vegetable material processing and is consistent with traces left by crushing mineral material. At Coquimbana 5, a small storage bin was littered with fragments of narrow stone wedges or chisels analogous

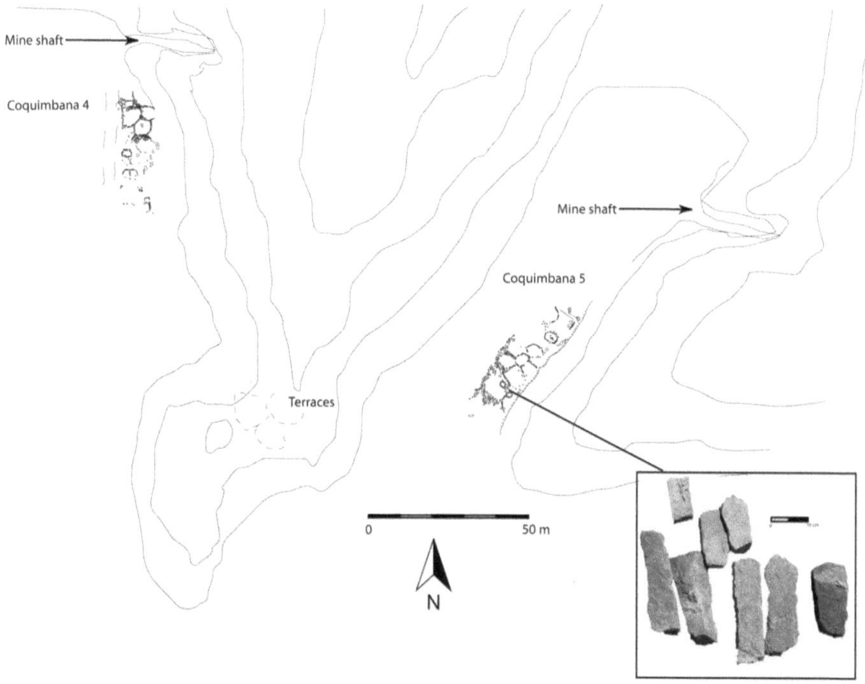

Fig. 13.3 Coquimbana 4 and 5, with a deposit of stone wedges. Many of the LIP decorated ceramics associated with these sites were found on the terraces constructed on the spur between the mining areas

to ones we have found in another ancient mining context (Chap. 8), and to ones described by Petersen (Petersen (2010) [1970]: 36; see also Chap. 7). Toward the edge of the ridge overlooking both sites was a series of small platforms with LIP ceramics that may have formed a lookout, a control station, or a small ceremonial space on which much of this area's rare decorated ceramics were found.

Four additional ancient mines that we encountered share a number of characteristics. The infrastructure varies from case to case, but all involve large platforms either constructed or made of modified and leveled spaces on high crests or spurs overlooking extraction zones. All of these spaces have been reused by miners until recent times. All four of these mining sites show ceramic evidence for an LIP occupation. Two of them, Mina Azurita and Coquimbana 1 (Fig. 13.4), in addition, yielded fragments dating to the Early Horizon (EH ca. BC 800–AD 1). Another mining site, Tortolita, had scattered Middle Horizon (MH ca. AD 600–1000) sherds nearby. All these sites offer extensive vistas of the exploitation zone and of the valley below. Their locations seem to have been meant to maximize visibility and control of the exploitation zone.

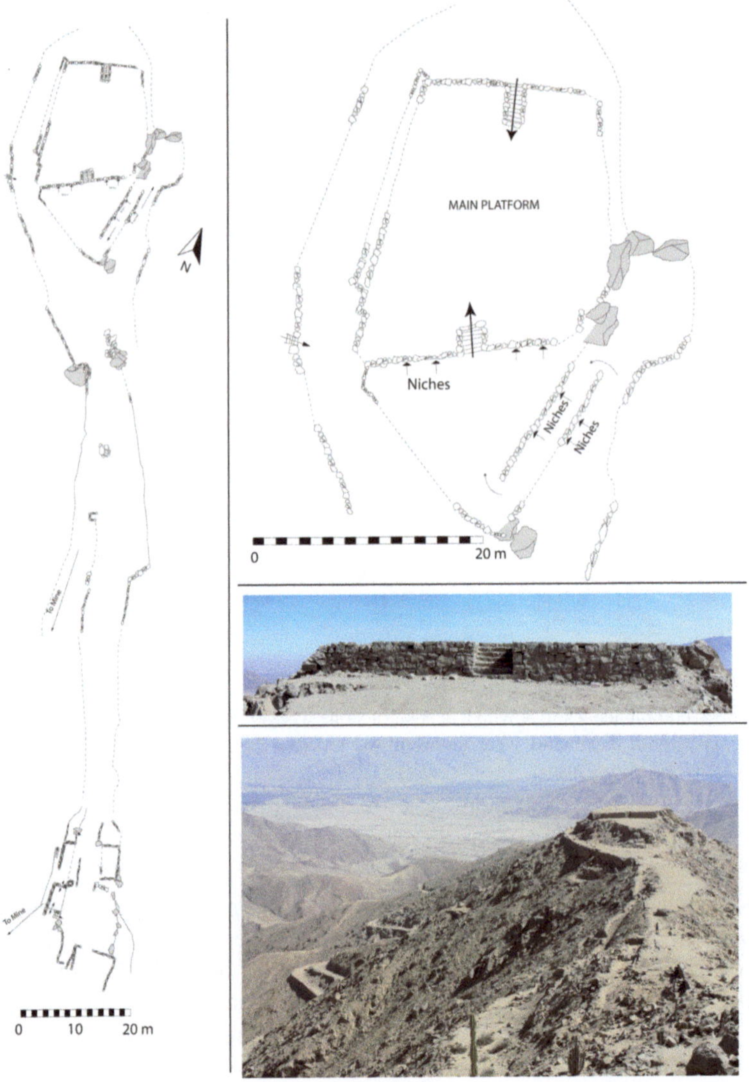

Fig. 13.4 Mina Azurita. Clockwise from left, a general map of the site, a detail of the main platform, a panoramic photo of the main platform's southern side, and a general view of the site looking north-west

Mina Azurita (Fig. 13.5), in particular, is interesting for the sophistication and elaboration of its architecture, composed of platforms, ramps, staircases, and walls of relatively elaborate masonry with niches. Series of storerooms are also found toward the southern part of the site. A main trapezoidal platform has dual sets of stairs on the north and south sides, and is accessed from the east by a series of ramps and the west by stairs descending toward the main extraction zone as it is today.

Fig. 13.5 Details of the carved boulders at La Zurita, a presumed Proto-Nasca production site

Its southern face has several niches, and niches are also found on the double series of ramps leading up to the platform area to the southeast. The platform's superstructure, while undoubtedly Prehispanic in design, shows marks left by steel tools on some of its stone blocks,[3] providing what may be a compelling case of native architectural design employing European technology in the early colonial period for construction or repair. This hypothesis remains to be tested by excavation to verify the presence of early colonial material, which we have not encountered yet. The masonry that composes the series of niched ramps leading up to the platform shows no such distinctive marks, indicating that it is likely to be Prehispanic.

The main platform at Mina Azurita offers a breathtaking view of the quebrada over which it towers and of the Ica valley further west. It also dominates the copper extraction zone as it stands today. Since it has been re-used periodically, it is hard to definitively conclude as to its function. However, its trapezoidal shape, the ramps with niches that lead up to it, as well as its opposing sets of stairs oriented along an east–west axis suggest a ceremonial component linked with procession, performance, and mountain worship in an architectural context that would have required considerable effort. This structure reflects an architectural program that is consistent

[3] This observation was confirmed by Jean-Pierre Protzen (personal communication, 2011) using high-resolution photographs.

with the symbolic use of multiple elevations in ceremonies that involve movement through and along different vertical planes that symbolically parallels movement across the landscape. Its presence indicates that mining at Mina Azurita has at some point been accompanied by a strong ritual component that warranted investment in such a large, sophisticated, and permanent architectural structure, and that this component was carried into the historical period, mandating new investments in architecture following an ancient pattern.

The other mining sites are far more modest than Mina Azurita, but all have large cleared spaces that would have been the only obvious locus of activity, and all are located well above steep slopes and the copper extraction zones. Some of them, such as Minas Alfredo and Coquimbana 4 and 5, have yielded a fair number of decorated LIP sherds in association with the elevated terraces.

Production Sites

A single production site was identified during our survey. It consists of a group of large boulders with deep networks of channels carved in them, called Mina Zurita by Massey (1986: 257, 371). To avoid confusion with Mina Azurita, it is labeled as "La Zurita" on the map in Fig. 13.1. The site is located at the mouth of a gully up which is a copper mine with a high gold content (Mina Monterosa). Massey's collaborators informed her that these boulders were used to crush and wash the native gold found in quartz veins, to separate the heavier gold from other minerals. Our informants corroborate Massey's original observations. The ceramics we found here confirm Sarah Massey's chronological placement of the site to the first century AD (Nasca 1/Proto-Nasca), which would make this one of the earliest mined gold processing sites in the Andes (versus alluvial gold panning, the assumed method of exploitation for the earlier prehistory). Alluvial gold, however, is not known in south coastal geologic formations (Stöllner 2009: 405). This situates the site as contemporaneous with the Necropolis burials of the Paracas peninsula, as well as Rubini's Ocucaje 10 burials from Ocucaje, which contained a fair number of gold objects (Silverman 2002). The location of this site may have been meant to take advantage of the strong winds that blow through the lower part of the upper valley, which traditionally played a part in the process of separating gold from impurities (Petersen (2010) [1970]: 26). We did not observe evidence for smelting, and the mined gold may have simply been cold-hammered.

Summary and Discussion

A number of observations emerge from this survey of mining sites in the upper Ica valley. First, the chronological profile of mining in upper Ica is fairly narrow, at least based on surface ceramic material. Two early mines, Coquimbana 1 and Azurita, yielded a handful of Early Horizon plainware sherds in what are otherwise LIP surface contexts (in addition to later, post-colonial periods). This early presence may reflect an episodic and opportunistic harvest of the various minerals used in pigments, lapidary, and, on occasion, native copper, materials present on Paracas ceramics and in Early Horizon gravelots. The La Zurita processing site constitutes an exception because of the relatively early and elaborate infrastructure presumably destined to process gold quartz (Fig. 13.6).

Other than La Zurita, all mining locations exhibit an LIP period of occupation and use. We note the complete lack of Nasca presence in mining contexts. Nasca miners and metalsmiths may have exploited other locations on the south coast, such as those in the Nasca drainage (Stöllner 2009; Chaps. 8 and 14), or else their mining practices simply did not yield any perceptible chronologically sensitive remains, or those remains have not withstood the passage of centuries if the scale and intensity of mining were small. We also note the relative absence of Wari and Middle Horizon mining activities, although one mining location, Tortolita, had scattered MH sherds nearby that may be associated with a mining platform given the absence of nearby settlements. Nor did the Inca leave any perceptible signs of mining activities at any of these locations. This is perhaps due to the fact that, based on settlement patterns (Williams and Pazos 1974), all sites with Late Horizon material are reoccupied LIP settlements. This continuity in occupation indicates that the Inca presence was not particularly intrusive or disruptive in this part of the valley. This does not necessarily mean that the Inca did not exploit the Ica valleys mineral resources, only that if they did, they may have done so through already-established LIP structures and agents (e.g., Shimada 1985; Ramirez 1994; Chap. 9), thus making it archaeologically virtually undetectable from survey alone. It may also indicate that the upper Ica valley was not seen as economically significant under Inca rule.

This leaves us with a picture in which exploitation of the surveyed mining locations is at its most intensive during the Late Intermediate Period, a time during which a fairly standardized form of mining infrastructure is established. This observation in the field is consistent with a perceived increase in the number of metal objects in various collections that appear to date to this period (Stöllner 2009: 397–398; Root 1949). These LIP sites consist of high-perched platforms, terraces, or, minimally, cleared and leveled surfaces offering broad vistas over the exploitation zones and the valley below. While this may be an artifact of a relatively broad valley, five of the six mining locations identified in the survey are mutually visible. The sixth, Minas Alfredo, which we were able to survey for some 10 minutes before being evicted, also has a high-perched series of hilltop terraces maximizing visibility of both the exploitation zone and the valley below. This emphasis on visibility and infrastructure we interpret as a newfound concern with control, administration, and defense of sources of minerals starting during the LIP, and a new level of elite investment in mineral exploitation and associated ritual practices.

Although we emphasize the control and visibility of the mining settlement pattern (Van Gijseghem et al. 2011), we also suggest that the platforms and cleared

Fig. 13.6 Some of the ceramics recovered in our survey

spaces had, at least partially, a ritual component, especially given the elaboration and care given to Azurita's architecture, and the decorated sherds found almost exclusively on these elevated spaces. Considering Andean populations' multifaceted relationships with mountains, ancestor worship, the earth, in general, and the substances it yields for humans to use, it is expected for the perceived prejudice caused to the mountain's supernatural essence to deserve some form of ritual repayment, as elaborated in the first part of this paper.

The introduction of this pattern does not indicate that mining started during the LIP. Archeological data generally and the Early Horizon sherds we encountered at three sites suggest that these resources were mined in earlier epochs. But the LIP pattern is more elaborate and standardized, suggesting changes in patterns of acquisition of the minerals, as well as changes in the scale and agency of ritual events associated with mining. It is easy to imagine an epoch in which mining was small scale and opportunistic, and that the associated ritual practices did not leave much perceptible archaeological remains prior to the LIP (see Vaughn et al. 2010; Chap. 8; Eerkens et al. 2009) . In fact, we would not expect many of the contemporary mining rituals as observed by ethnographers to leave behind much that would withstand the passage of centuries. However, the collected data reveal that during the LIP the scale and agency of rituals may have changed and been appropriated by political elites, at once monopolizing the extraction and production of precious metals and enjoying the privileges, prestige, and power that accompany their status as ritual officiants.

It is unsurprising that this intensification occurred, as far as archaeological remains indicate, during this period. The LIP on the south coast of Peru was a time during which the region reached its demographic peak (Reindel 2009), and during which hierarchical structures became increasingly complex (Conlee 2003, 2004). Local elites linked by elaborate networks occupied a diversity of social and political positions that were justified and reinforced through various means. Among the strategies they adopted was the control of the production and distribution of both utilitarian and prestige goods. In this perspective, few materials would have acquired as much value in the production of craft goods (i.e., in the creation of new sources of prestige and wealth) as precious metals and stones. The new LIP geopolitical landscape of the upper Ica valley developed to prioritize these new resources, or old resources that gained new socioeconomic importance. This triggered increased investment by prominent political agents in the defense of mining locations, control of access, extraction, associated symbolic practices, and production of metals. It may also have promoted further prospection into hard to reach, sheltered, and seldom-visited canyon networks, as testified by the prospection site at Tiojate 1.

The control of mining locations also very fundamentally positioned individuals and groups, presumably at times miners themselves, but increasingly mining sponsors or administrators, as privileged partners in a reciprocal relationship with the underworld and nature's supernatural forces. The prestige that comes with managing such sensitive affairs is not a trivial one and may have been an important axis of power acquisition, maintenance, but also an instrument of negotiation and contest, depending on the context and the agents involved. We argue that from a relatively free-form, independent, and opportunistic harvest of oxides and other minerals accompanied by personal, intimate ritual practices, the LIP saw an increase in the active control of the minerals and the mythical/symbolic relationships that accompany their harvest, thus prompting a closer control of the mountainous areas in the hinterlands of the upper

Ica valley, and promoting changes in attitudes about the landscape and occupation of that landscape.[4]

Conclusion

During much of prehistory the visible outcrops of the upper Ica valley may have been the sites of independent, episodic, and opportunistic gathering of carbonates and oxides—for lapidary needs and to produce pigments on ceramics and textiles. This type of activity, as Shimada (1994) suggests for the north coast, would have gradually led to the discovery of metal-bearing ores, including native gold at such sites as Monterosa, which can be worked by hammering once impurities have been removed. Usable copper-bearing ores would likewise have been discovered in the other localities we surveyed, prompting a closer control of production as the perceived demand for metals grew. As this process went on, the LIP upper valley populations underwent fundamental changes in their landscape perception, corresponding to new sociopolitical realities and necessities. In a similar way, some locations would have acquired

[4] A few observations about toponymy may shed further light on ancient mines and attitudes on mining and how they are reflected on landscape perceptions. Place-names in the upper Ica valley reflect the ambiguity surrounding mining in traditional Andean cognitive systems. Geographical locations where modern or historical mines have been identified—and presumably where ancient mining would at least have been possible—tend to have names that bear negative or ambiguous connotations, regardless if they are Spanish or Quechua—although we are conscious that we must be careful about assigning such qualitative judgments. Examples of such mountains and *quebradas* where copper and gold mining has been identified bear such Spanish or Quechua names that translate as "serpent," "misstep," "stitches," or "fate/fortune," among others. Furthermore, not a single one of those negative or ambiguous connotations was detected in places where there is no mining, further up the valley and in most of the valley's eastern side, where names are all associated with landforms ("round hill"), ecology ("small branches"), economic activities ("sugar mill"), or names of individuals. Incidentally, it may be warranted to mention that the large *quebrada* over which Mina Azurita is located is called *llancay* (where *llank'ay*=to work, to toil; or alternatively, *llankhay*=to bother, to devastate). Finally, and this last point is very speculative, a supposedly old mine on the northeastern side of the valley, which we could not survey, is called San Miguel Rescate ("rescue/ransom/redemption/recovery"). This specific Saint or aspect of St. Michael (the rescuer?) is not a generally recognized icon in Latin America or elsewhere. Yet, we feel it is worth noting that in one of the main mining festivals in Oruro, Bolivia, it is the figure of St. Michael that leads the dancing devils in processions as they are publicly ridiculed and cast out of the mines (Harris 2000: 63). Roman Catholic hagiography describes St. Michael alternatively as the archangel that leads God's armies against Satan's forces, or as the angel of death who wrestles the souls of the deceased from the devil and carries them to heaven (Holweck 1911). It is tempting to associate this character with the symbolic aspect of mines as extraordinary liminal spaces, gates to the underworld, and the lair of Andean "demons". It is impossible now to say whether this observed pattern is an attempt at social control, that is to say a way to permanently mark these areas as hazardous in order to keep itinerant miners away from the resource, or if it reflects the mythologically dangerous aspects of mines and mining carried over from the prehispanic era into historic toponymy (Van Gijseghem and Whalen 2012).

new symbolic and strategic importance, some of them crystallized by the assignment of *huaca* status by local populations by virtue of their visual, material, and mythological characteristics.

The extraction of precious raw materials from the womb of a mountain is a symbolically charged act. The ritual imperatives that are involved have the effect of creating especially intimate relationships between political agents taking charge of the mining work and the mythological members of sacred genealogies embedded within certain points in the landscape. In that sense, the appropriation and control over mining may not only have yielded economic and political dividends but cosmological dividends as well, imprinting or mapping out sociopolitical agency onto an active, living, and temperamental landscape. Thus, this form of ceremonial obligation constitutes a powerful commodity. Our survey indicates that these responsibilities may have been appropriated by political agents during the Late Intermediate Period, in parallel to the appropriation of mineral exploitation and metal production.

The advantages that control over mining offers certain groups are therefore varied. It legitimizes and reinforces relationships between privileged members of society as well as relationships between these elite groups and the mythological components of the landscape through the intimate act of mining and the requisite responsibilities. It is clear, however, that these responsibilities create a context in which the prestige associated with ritual requirements is not the exclusive domain of the elite and that for much of prehistory these devotional acts could be appropriated by different groups, classes, and factions, providing them with a powerful voice in the negotiation of supernatural affairs. Mines therefore constitute important arenas for reproducing or contesting sociopolitical conditions through the ritual burdens incurred on the miners and on those who position themselves at once as sponsors or administrators of the mining and as curators of the landscape's supernatural forces. Our survey results indicate that by the LIP, sociopolitical elites in Ica increasingly attempted to strengthen their position in the momentous relationships between humans and the underworld as they are expressed in this powerful vector of earthly forces.

Acknowledgments We extend gratitude to the National Geographic Society for funding this phase of a wider program initiated by K.J. Vaughn. We are also grateful of the various forms of assistance and support we received from Ruben Gracia Soto and Susana Arce Torres from Ica's regional Museum and Peru's Ministry of Culture, and logistic help from Henry Vladimir Falcon. We also thank the anonymous reviewer who took the time to carefully review this early draft of this chapter and offer helpful and valuable suggestions.

References

Aldenderfer, M., Craig, N. M., Speakman, R. J., & Popelca-Filcoff, R. (2008). Four-thousand-year-old gold artifacts from the Lake Titicaca basin, southern Peru. *PNAS, 105*(13), 5002–5005.

Barth, F. (1975). *Ritual and Knowledge Among the Baktaman of New Guinea*. Press, New Haven: Yale University.

Berthelot, J. (1986). The extraction of precious metals at the time of the Inca. In J. V. Murra, N. Wachtel, & J. Revel (Eds.), *Anthropological history of Andean polities* (pp. 69–88). Cambridge: Cambridge University Press.

Bouysse-Cassagne, T. (2005). Las Minas del centro-sur andino, los cultos prehispanicos y los cultos cristianos. *Bulletin de l'Institut Français d'Études Andines, 34*(3), 443–462.

Brewer, M. J. (2003). *From myth to reality: performing the devil and Pachamama in the Carnival of Humahuaca* (Master's thesis)., Department of Anthropology, University of Chicago.

Brooks, W. E., Piminchumo, V., Suarez, H., Jackson, J. C., & McGeehin, J. P. (2008). Mineral pigments at Huaca Tacaynamo (Chan Chan, Peru). *Bulletin Français d'Études Andines, 37*(3), 441–450.

Ceruti, C. (2004). Human bodies as objects of dedication at Inca Mountain Shrines (Northwest Argentina). *World Archaeology, 36*(1), 103–122.

Cobo, B. ([1653] 1890–1895). *Historia del Nuevo Mundo*. Sevilla: E. Rasco.

Cobo, B. (1990). *Inca religion and customs*. Austin: University of Texas Press.

Conlee, C. A. (2003). Local elites and the reformation of late intermediate period sociopolitical and economic organization in Nasca, Peru. *Latin American Antiquity, 14*(1), 47–65.

Conlee, C. A. (2004). The expansion, diversification, and segmentation of power in Late Prehispanic Nasca. In K. J. Vaughn, D. Ogburn, & C. A. Conlee (Eds.), *Foundations of power in the prehispanic Andes* (Archaeological papers of the American Anthropological Association, Vol. 14). Arlington, VA: American Anthropological Association.

Costin, C. L. (1996). Craft production and mobilization strategies in the Inca Empire. In W. Bernard (Ed.), *Craft specialization and social evolution: In Memory of V. Gordon Childe* (pp. 211–225). Philadelphia: University of Pennsylvania Museum.

Costin, C. L., & Earle, T. (1989). Status distinction and legitimation of power as reflected in changing patterns of consumption. *American Antiquity, 54*(4), 691–714.

Donnan, C. B. (1992). *Ceramics of Ancient Peru*. Los Angeles: Fowler Museum of Cultural History, UCLA.

Duviols, P. (1977). La capacocha: Mecanismo y función del sacrificio humano, su proyección geométrica, su papel en la política integracionista y en la economía redistributiva del Tawantinsuyu. *Allpanchis, 9*, 11–57. Cuzco.

Earls, J. (1970). The structure of modern Andean social categories. *Journal of the Steward Anthropological Society, 2*, 69–106.

Eerkens, J. W., Vaughn, K. J., & Linares Grados, M. (2009). Pre-Inca mining in the Southern Nasca Region, Peru. *Antiquity, 83*(321), 738–750.

Eerkens, J. W., Vaughn, K. J., Linares Grados, M., & Edwards, M. J. (2008). La Ballena: a mining base camp in the southern Nasca region, Peru. *Antiquity Project Gallery, 82*(315).

Farrington, I. S. (1992). Ritual geography, settlement patterns and the characterization of the provinces of the Inca Heartland. *World Archaeology, 23*, 368–385.

Gil García, F. M., & Fernández Juárez, G. (2008). El culto a los cerros en el mundo andino: estudios de caso. *Revista Española de Antropología Americana, 38*, 105–113.

Glowacki, M. (2005). Food of the Gods or mere mortals? Hallucinogenic Spondylus and its interpretive implications for early Andean society. *Antiquity, 79*, 257–268.

Gose, P. (1986). Sacrifice and the commodity form in the Andes. *Man, 21*(2), 296–310.

Harris, O. (2000). *To make the earth bear fruit: Essays on fertility, work and gender in highland Bolivia*. London: Institute of Latin American Studies.

Holweck, F. (1911). St. Michael the Archangel. In *The Catholic Encyclopedia*. New York: Robert Appleton Company. Retrieved 1 June 1 2011 from New Advent. http//www.newadvent.org/cathen/10275b.htm.

Lechtman, H. (1976). A metallurgical site survey in the Peruvian Andes. *Journal of field Archaeology, 3*(1), 1–42.

Lechtman, H. (1993). Technologies of power: The Andean Case. In J. S. Henderson & P. J. Netherly (Eds.), *Configurations of power: Holistic anthropology in theory and practice*. Ithaca and London: Cornell University Press.

Lechtman, H. (1996a). The Andean world. In E. H. Boone (Ed.), *Andean art at Dumbarton Oaks* (pp. 15–32). Washington DC: Dumbarton Oaks.
Lechtman, H. (1996b). Cloth and metal: The culture of technology. In E. H. Boone (Ed.), *Andean art at Dumbarton Oaks* (pp. 33–43). Washington DC: Dumbarton Oaks.
Lechtman, H. (2007). The Inca, and Andean metallurgical tradition. In R. Burger, C. Morris, & R. Matos Mendieta (Eds.), *Variations in the expression of Inca power* (pp. 313–355). Washington DC: Dumbarton Oaks.
MacCormack, S. (1984). From the sun of the Incas to the virgin of Copacabana. *Representations.*, *8*(Fall), 30–60.
Massey, S. A. (1986). *Sociopolitical change in the upper Ica valley, B.C. 400 to 400 A.D.: Regional states on the south coast of Peru* (Unpublished PhD dissertation). Department of Anthropology, University of California, Los Angeles.
Moore, S. F. (1958). *Power and property in Inca Peru*. New York: Colombia University Press.
Muelle, J. C., & Wells, R. (1939). Las Pinturas del Templo de Pachacamac. *Revista del Museo Nacional, 8*(2), 265.
Murra, J. V. (1982). The Mit'a obligations of ethnic groups to the Inca state. In G. Collier, R. Rosaldo, & J. Wirth (Eds.), *The Inca and Aztec states 1400–1800: Anthropology and history* (pp. 237–262). New York: Academic Press.
Nash, J. (1972). Devil in Bolivia's nationalized tin mines. *Science and Society, 36*, 221–233.
Owen, B. (2001). The economy of metal and shell wealth goods. In T. D'Altroy & C. Hastorf (Eds.), *Empire and domestic economy* (pp. 265–293). New York: Kluwer Academic/Plenum Press.
Paulsen, A. C. (1974). The thorny oyster and the voice of God: *Spondylus* and *Strombus* in Andean prehistory. *American Antiquity, 39*, 597–607.
Peregrine, P. M., & Ember, M. (2002). *Encyclopedia of prehistory* (South America, Vol. 7). New York: Kluwer Academic/Plenum Publishers.
Petersen, G. G. ((2010) [1970]). *Mining and metallurgy in Ancient Peru [W. E. Brooks, Trans.]*. Boulder, CO: The Geological Society of America.
Pozorski, S., & Pozorski, T. (2002). Late coastal Andean Formative. In P. N. Peregrine & M. Ember (Eds.), *Encyclopedia of prehistory* (South America, Vol. 7, pp. 217–227). New York: Kluwer Academic.
Ramirez, S. E. (1994). Ethnohistorical dimensions of mining and metallurgy in sixteenth-century Northern Peru. In A. C. Craig & R. C. West (Eds.), *In quest of mineral wealth: Aboriginal and colonial mining and metallurgy in Spanish America* (Geoscience and man, Vol. 33, pp. 93–108). Baton Rouge: Louisiana State University.
Reindel, M. (2009). Life at the edge of the desert: Archaeological reconstruction of the settlement history in the valleys of Palpa, Peru. In M. Reindel & G. A. Wagner (Eds.), *New technologies for archaeology* (pp. 439–461). Berlin: Springer.
Reinhard, J. (1985). Sacred mountains: an ethnoarchaeological study of high Andean ruins. *Mountain Research and Development, 5*(4), 299–317.
Root, W. (1949). The metallurgy of the Southern Coast of Peru. *American Antiquity., 15*(1), 10–37.
Rostworowski de D.C, M. (1983). *Estructuras Andinas del poder*. Lima: Instituto de Estudios Peruanos.
Rostworowski de D.C., M. (1970). Mercaderes del valle de Chincha en la época prehispánica: Un documento y unos comentarios. *Revista Española de Antropología Americana, 5*, 135–178.
Rostworowski de D.C., M., & Morris, C. (1999). The fourfold domain: Inca power and its social foundation. In F. Salomon & S. B. Schwartz (Eds.), *The Cambridge history of the native peoples of the Americas: South America* (Vol. 3, pp. 769–847). Cambridge: Cambridge University Press.
Rowe, J. H. (1946). Inca culture at the Time of the Spanish Conquest. In J. H. Steward (Ed.), *Handbook of South American Indians* (Vol. 2, pp. 183–330). Washington DC: Smithsonian Institution.
Sallnow, M. J. (1987). *Pilgrims of the Andes: Regional cults in Cusco*. Washington DC: Smithsonian Institution Press.
Sawyer, A. (1966). *Ancient Peruvian ceramics*. New York: Metropolitan Museum of Art.

Schoenberger, E. (2011). Why is gold valuable? Nature, social power, and the value of things. *Cultural Geographies., 18*(1), 3–24.

Schultze, C. A., Stanish, C., Scott, D. A., Rehren, T., Kuehner, S., & Feathers, J. K. (2009). Direct evidence of 1,900 years of indigenous silver production in the Lake Titicaca Basin of Southern Peru. *Proceedings of the National Academy of Sciences of the United States of America, 106*(41), 17280–17283.

Shimada, I. (1985). Perception, procurement and management of resources: Archaeological perspective. In M. Shozo, S. Izumi, & M. Craig (Eds.), *Andean ecology and civilization* (pp. 357–399). Tokyo: University of Tokyo Press.

Shimada, I. (1994). Pre-hispanic metallurgy and mining in the Andes: Recent advances and future tasks. In A. C. Craig & R. C. West (Eds.), *In quest of mineral wealth: Aboriginal and colonial mining and metallurgy in Spanish America* (Geoscience and man, Vol. 33, pp. 37–74). Baton Rouge: Louisiana State University.

Shimada, I., & Merkel, J. F. (1991). Copper-alloy metallurgy in Ancient Peru. *Scientific American, 265*(1), 80–86.

Silverman, H. (2002). Nasca settlement and society on the hundreth anniversary of Uhle's discovery of the Nasca style. In W. H. Isbell & H. Silverman (Eds.), *Andean archaeology I: Variations in sociopolitical organization* (pp. 121–160). New York: Kluwer.

Stöllner, T. (2009). Gold in Southern Peru?: Perspectives of research into mining archaeology. In M. Reindel & G. A. Wagner (Eds.), *New technologies for archaeology: Multidisciplinary investigations in Palpa and Nasca, Peru* (pp. 393–407). Berlin: Springer.

Stone-Miller, R., & McEwan, G. F. (1990/1991). The representation of the Wari State in stone and thread: a comparison of architecture and tapestry tunics. *RES: Anthropology and Aesthetics, 19/20*, 53–80.

Swenson, E. 2004. *Ritual power in the urban hinterland: Religious pluralism and political decentralization in Late Moche Jequetepeque, Peru* (Unpublished PhD. Dissertation). University of Chicago

Taussig, M. T. (1980). *The devil and commodity fetishism in South America*. Chapel Hill: University of North Carolina Press.

Taylor, G. (1980). Supay. *Amerindia Paris, 5*(3–6), 47–63.

van de Guchte, M. (1999). The Inca cognition of landscape: Archaeology, ethnohistory, and the aesthetic of alterity. In W. Ashmore & A. B. Knapp (Eds.), *Archaeologies of landscape* (pp. 149–168). Malden, MA: Blackwell.

Van Gijseghem, H., Vaughn, K. J., Whalen, V. H., Linares G. M., & Olano C. J. (2011). Prehispanic Mining in South America: new data from the upper Ica valley, Peru. *Antiquity 85*(328). Project Gallery.

Van Gijseghem, H., & Whalen, V. H. (2012). *Mining, ritual, and social memory: can place-names reveal ancient attitudes toward landscape?* 77th Meeting of the Society for American Archaeology, Memphis, TN.

Vaughn, K. J., Van Gijseghem, H., Linares, G.. M., & Eerkens. J. (i.p.) Hematita en la Costa Sur del Perú: Investigaciones Arqueológicas en Mina Primavera. *Chungará, Chilean Journal of Anthropology.*

Vaughn, K. J., Van Gijseghem, H., Eerkens, J., & Linares G.. M. (2010). *Two millennia of Prehispanic Mining in Nasca, Peru.* Paper presented at the *First International Meeting on Prehispanic Mining in the Americas*, Taltal-San Pedro de Atacama. Chile, 22–26 Nov 2010.

Whitehouse, R. D. (1992). *Underground religion: Cult and culture in prehistoric Italy* (Specialist studies on Italy 1). London: Accordia Research Centre, University of London.

Williams, C. A., & Pazos, M. (1974). *Inventario, Catastro, y Delimitacion del Patrimonio Arqueologico del Valle de Ica*. Ica, Peru: Unpublished report submitted to the Instituto Nacional de Cultura.

Chapter 14
Mining Archaeology in the Nasca and Palpa Region, South Coast of Peru

Markus Reindel, Thomas R. Stöllner, and Benedikt Gräfingholt

Introduction

Since 1997 the German Archaeological Institute has engaged in archaeological investigations in the northern Nasca region, on the south coast of Peru (Reindel and Wagner 2009). In addition to the documentation and archaeological investigations of the Nasca geoglyphs, settlement patterns have been the main focus of the project. During our surveys and investigations the discovery of metal artifacts, mineral products, mines and quarries, as well as places for the processing of mining products indicated the importance of mining activities in the region. Still today, gold and copper mining is a major income source in the region, which is characterized by a geological belt of rich ore and mineral deposits.

In order to investigate further the occurrences of minerals and ores in the research area and to understand the mining and processing of these raw materials, as well as the socioeconomic importance of the distribution of raw materials and artifacts, in 2006 the German Archaeological Institute began to cooperate closely with the

M. Reindel (✉)
German Archaeological Institute, Commission for Archaeology of Non-European Cultures, Dürenstr. 35-37, 53173 Bonn, Germany
e-mail: Reindel@kaak.dainst.de

T.R. Stöllner
Deutsches Bergbau-Museum Bochum, Forschungsstelle Archäologie und Materialwissenschaften, Fachbereich Montanarchäologie, Herner Straße 45, 44787 Bochum, Germany

Institut für Archäologische Wissenschaften, Lehrstuhl Ur- und Frühgeschichte, Universitätsstr. 150, 44780 Bochum, Germany
e-mail: thomas.stoellner@bergbaumuseum.de; thomas.stoellner@ruhr-uni-bochum.de

B. Gräfingholt
Institut für Archäologische Wissenschaften, Fach Ur- und Frühgeschichte, Am Bergbaumuseum 31, 44791 Bochum, Germany
e-mail: Benedikt.Graefingholt@ruhr-uni-bochum.de

N. Tripcevich and K.J. Vaughn (eds.), *Mining and Quarrying in the Ancient Andes*, Interdisciplinary Contributions to Archaeology, DOI 10.1007/978-1-4614-5200-3_14,

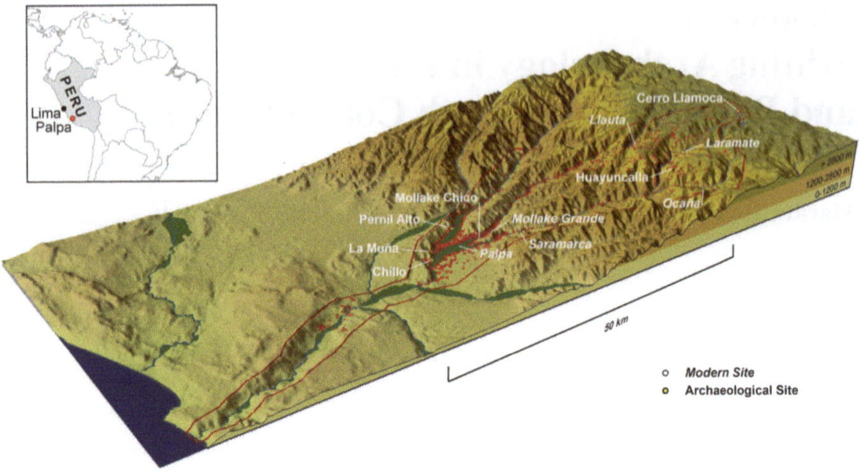

Fig. 14.1 Research area of the Andean Transect project indicating Prehispanic settlements (*red dots*) and important archaeological sites mentioned in the text (*yellow dots*) (Design: V. Sossna)

German Mining Museum in Bochum that specializes in the investigation of mining activities of ancient cultures worldwide (Stöllner and Reindel 2007; Stöllner 2009, 2011). The German Mining Museum has world-wide expertise in the study of mining and mining products and is therefore the ideal partner to address the numerous questions related to quarrying, mining, and processing of metals and minerals in our research area on the south coast of Peru (Stöllner et al. 2003, 2008).

In this paper we present the preliminary results of the first surveys focused on mining archaeology in the Nasca and Palpa region. Following the history of our research project, in the first section of the paper we present archaeological evidence of the use of artifacts produced from mineral or metallic raw materials, as well as evidence of mining activities discovered during our archaeologically oriented settlement surveys. In the second section, we focus specifically on mining archaeology, reviewing the available literature about mining archaeology on the south coast of Peru and presenting the basic geology of the region. Finally, we present the results of our own surveys of raw materials and mining sites, as well as the preliminary results of some representative laboratory analysis of raw materials, which allow us to formulate hypotheses about the importance of mining activities and products in the cultural process of the region.

The research area comprises the valleys of Palpa, in the northern Nasca region (what Vaughn and colleagues refer to as the "Central Nasca region," see Chap. 8; Fig. 14.1). During our settlement surveys between 1997 and 2007 we concentrated on the coastal region of the foothills of the Andes. Beginning in 2008 we expanded our research area to the highlands and to the Pacific and defined a transect line through the different ecological zones between the desert coast and the highest peaks of the western slope of Andean highlands, which we call the "Andean Transect". Along this Andean Transect we have registered about 1,600 archaeological sites so far, most of them settlements. These sites span duration from the first occupation of the region in the Palaeoindian period until the end of the Prehispanic period.

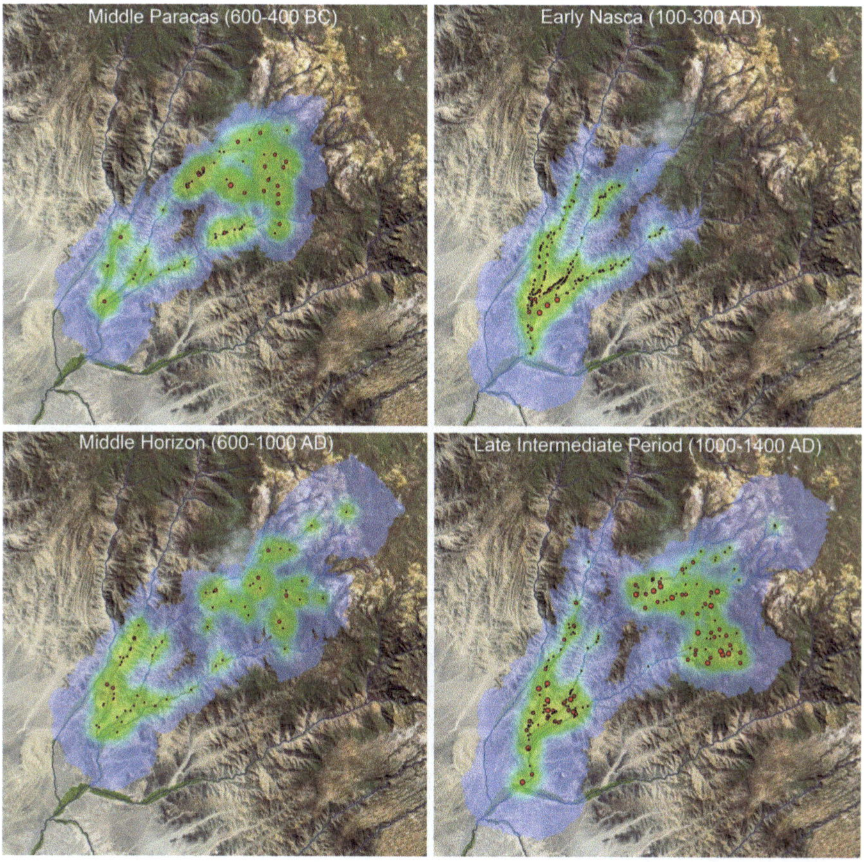

Fig. 14.2 Settlement patterns within the research area in different periods (Design: V. Sossna)

The aim of our project is to reconstruct settlement shifts as influenced by climatic variations in this representative region of the Andes. Archaeologically we focus on the changing patterns of movement of people and goods over time. Following the model of verticality, in this context the movement and exchange of mining products is of crucial importance. In this paper we will present the preliminary results of our surveys and investigations of mining activities.

We know from our geoarchaeological studies that over time, settlement centers shifted considerably between the highlands and the coast due to changes in climate and landscape conditions (Fig. 14.2; Eitel et al. 2005; Eitel and Mächtle 2009). Our settlement distribution maps show that the coastal region experienced an important increase in population from the highlands in the Paracas period, a time which was characterized by generally more humid conditions than today. In the early and middle Nasca periods, settlements concentrated on the coast. It was only in the late Nasca period, when climatic conditions became extremely dry, that this tendency reversed and the settlement movements shifted towards the middle reaches of the valleys and to the highlands. In the Middle Horizon the coastal region was nearly completely abandoned and people moved to the highlands. Only in the Late Intermediate period, when climatic conditions became more humid again and therefore favorable for settlement and agriculture, did people return from the highlands in order to establish new settlements on the coast and on the western slopes of the Andes. This sequence shows that constant movement between the highlands and the coast took place over the millennia and that presumably people had constant contact and moved or exchanged goods between these regions to obtain the raw materials for their economic activities.

Stone, Mineral, and Metal Artifacts Through the Prehispanic History of Palpa and Nasca

In this section of the paper we review the occurrences of artifacts made from different mineral and metallic raw materials documented during our archaeological surveys and excavations in the Palpa region. Furthermore, we present some evidence of quarries and extraction places of different raw materials discovered during our settlement surveys. The latter were revisited during our specific mining surveys and served as a starting point for further studies in mining archaeology. The review of contexts related to the production of raw materials or artifacts from mineral or metal sources is presented in a chronological order.

The earliest evidence documented so far in the Andean Transect date to the Early Archaic period (8000 cal. BC). The Middle Archaic period is represented by the site Pernil Alto (3800–3000 BC). Very few sites have been documented for the Initial period (1500–800 BC), while numerous settlements are known from the Paracas period (800–200 BC). The following Nasca period is divided into Initial Nasca (200 BC–100 AD), Early Nasca (100–300 AD), Middle Nasca (300–450 AD), and Late Nasca (450–650 AD). The end of the Middle Horizon is not well defined in the Palpa area, but according to other regions it lasted until about 1000 AD. It is followed by the Late Intermediate period (hereafter LIP), which came under Inca influence about 1400 AD. In the research area, however, the Inca presence was limited to very few administrative installations, suggesting that the Inca administration in Palpa did not alter substantially the cultural and economic activities in the region. As the Inca presence is hardly perceivable in the archaeological record, in our presentation of the cultural process of the Palpa valleys we do not differentiate between the LIP and a specific Inca period (Reindel 2009).

One of the most common products that were brought from the highlands to the coast from the earliest to the latest periods was obsidian. We found obsidian flakes and artifacts, for example, in all layers of a test excavation in a rock shelter at an altitude of 4,300 m, near the Cerro Llamoca, where the lower levels dated to 8000 BC. In the oldest levels we found the obsidian artifacts mixed with other lithic materials like chert and other fine crystalline minerals (Fig. 14.3). On the coast we recovered obsidian projectile points in the Archaic layers of Pernil Alto, which date to the fourth millennium BC (Fig. 14.4; Reindel and Isla 2006; Reindel 2009, 2010; Isla Cuadrado 2009). Pernil Alto is a site which represents the stage of initial sedentism on the south

Fig. 14.3 Artifacts recovered in the lowest layers of a test excavation at the Llamoca rock shelter (eighth millenium BC), millimeter scale at the *left*. Large object is about 5 cm in diameter (Photo: M. Reindel)

Fig. 14.4 Obsidian projectile points recovered at the excavation of the archaic settlement Pernil Alto (fourth millenium BC) (Photo: B. Gräfingholt)

coast of Peru. The settlement is composed of round or oval pit houses, similar to the ones excavated at Paloma and Chilca (Benfer 1999; Engel 1966, 1980, 1988).

The earliest use of metals in Palpa, in this case a gold ring, dates to the early Paracas period (Fig. 14.5). This gold ring was found in a burial chamber associated with typical Early Paracas ceramics with some influence of the Cupisnique culture (Isla Cuadrado and Reindel 2006a; Reindel and Isla 2006). The Chavín horizon saw the first widespread use of gold objects that were produced with fairly simple metallurgical techniques (Lothrop 1951).

Gold objects from the Nasca period are known from the elite burials of La Muña (Fig. 14.6; Isla Cuadrado and Reindel 2006b; Reindel and Isla Cuadrado 2001). The use of gold objects in the Nasca period was more common, although it never reached the level of the cultures of the north coast of Peru, like the Moche or Sicán. The finds of La Muña show also that in the Nasca period copper came into use, as evidenced by the copper beads recovered from the tombs of La Muña. Several

Fig. 14.5 Obsidian projectile point, gold ring and stone beads recovered at the excavation of an Early Paracas tomb at Mollake Chico (eighth century BC) (Photo: J. Isla)

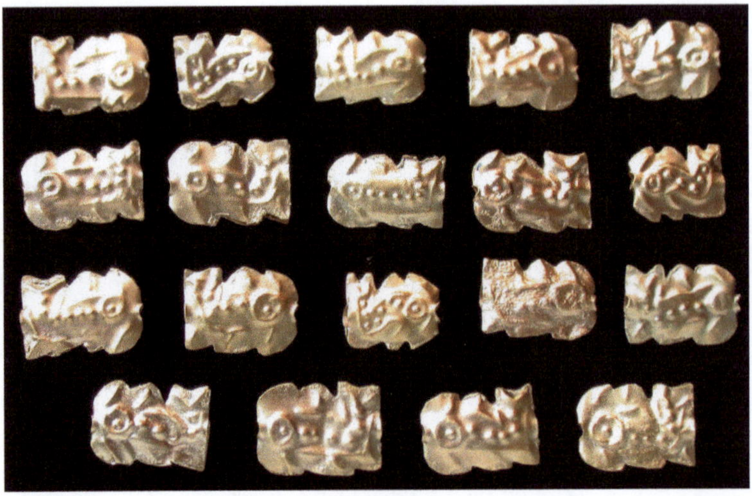

Fig. 14.6 Gold beads representing fishes recovered at the excavation of a Middle Nasca tomb at La Muña (400 AD). Beads are approximately 1.5 cm in length (Photo: M. Reindel)

secondary copper minerals like malachite, chrysocolla, and turquoise were also common. These semi-precious minerals were used also as pigments during the Nasca period. Another pigment which is often found in burials is hematite.

In the Middle Horizon the use and the number of copper objects rose significantly. In 2010 we excavated several burials at the highland Huari site Huayuncalla. Huayuncalla is located at an altitude of 3,200 m at the upper reaches of the Viscas river (Fig. 14.1). Two rectangular burial chambers were excavated that were

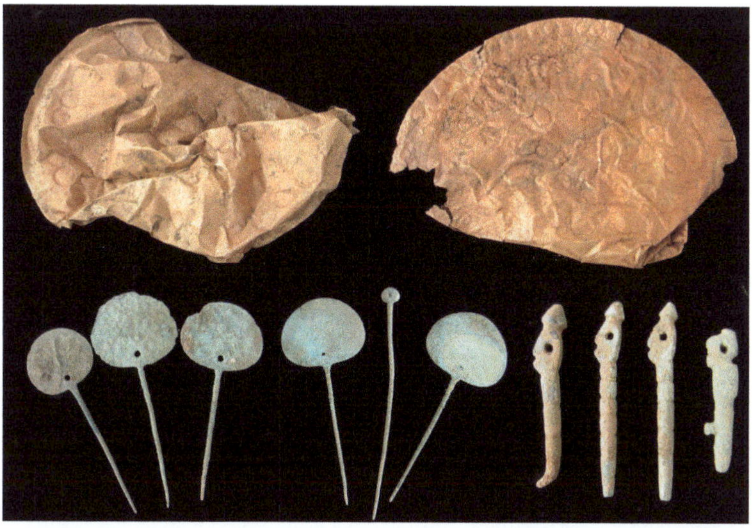

Fig. 14.7 Gold and copper objects recovered at the excavation of Huari tombs at Huayuncalla (700 AD). Gold disks are about 25 cm in diameter, tupus at *lower left* are 15–20 cm, knobs at *lower right* are 4.5–6 cm (Photo: M. Reindel)

enclosed by circular walls. In each burial chamber we found secondary burials with 20 individuals, accompanied by Huari ceramics and metal objects. The copper objects included tupus, different adornments, spear throwers, and other tools (Fig. 14.7). At the bottom of each burial chamber were found large discs of gold foil. The sheer amount of copper objects shows that a real boom in the use of copper occurred during the Middle Horizon.

This picture continues into the LIP. Although our excavations of sites of this period were limited, the presence of copper and gold objects even on the surface of sites of the LIP is obvious. This general picture resembles the state of the knowledge about metallurgy on the south coast of Peru during the LIP presented by Root (1949) and Lechtman (1976).

Geology and Research History of the Palpa and Nasca Region

Considering the metal resources in southern Peru, the main formations with gold or copper-bearing mineralization are the "Batolito de la Costa" and the "Complejo Bella Union". The metal deposits consist of epithermal or hydrothermal veins oriented parallel to the western cordillera in most cases. Host rocks are either Jurassic or Cretaceous volcanic or sedimentary–volcanic rocks. The copper deposits regularly also contain a smaller amount of noble metals, lead and zinc. Also native gold–silver fine disseminated in hydrothermal quartz veins are found. The whole metal ore district is designated as the Nasca–Ocoña belt (Petersen 1979, 1989; Fig. 14.8).

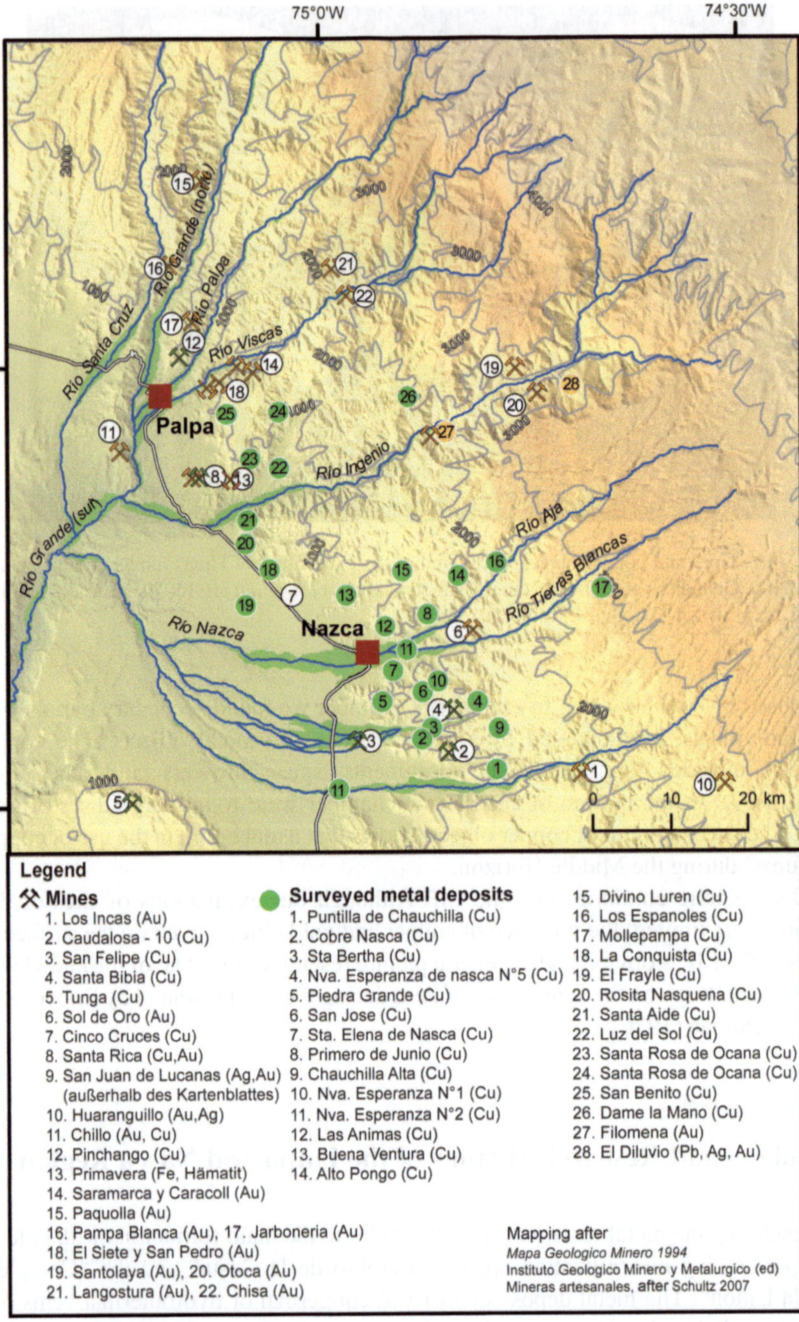

Fig. 14.8 The copper and gold deposits in the surroundings of Palpa and Nasca and their prehistoric and recent use (after Stöllner 2009)

Some metal bearing veins appear near the surface and constitute a rich zone for mining today and probably also in Prehispanic times. Some of the ores are extremely rich in gold and copper, sometimes iron oxide is present. Root (1949) and Lechtman (1976) have reviewed this situation and see it as a basis for the extensive use of copper objects mainly in the late periods of the south coast of Peru.

In contrast to the northern coast of Peru and also to the central Andean Altiplano, southern Peru never was a focal area of metallurgical inventions or innovations. At least this is the general picture obtained by the older literature about the topic (Lechtman 1976; Lothrop 1937; Root 1949; Stöllner 2011). Metals came into use in the context of the Paracas culture (Tello 1959, Tello and Mejía Xesspe 1979; Uhle 1913). And also during the Nasca period metallurgy principally remained on a basic level of cold working of gold as well as simple copper working techniques. Consequently Root (1949) concluded that the Paracas and Nasca had developed a comparatively simple and low level gold-metallurgy (see also Lothrop 1937).

Natural gold alloys are dominant in regional metal assemblages. The metallurgy especially of the Nasca period remains enigmatic up to now: while some slag-sites and remnants of copper metallurgy are known (e.g., Lechtman 1976), the smelting of complex sulfides or polymetal copper ores is not to be expected even before the end of the Nasca-culture (around AD 600)—but this certainly needs more detailed field-work for confirmation.

According to the available literature, a considerable increase of metallurgy is notable only from the Middle Horizon (600–1000 AD) onward. In this phase not only is there an increase of metal artifacts, but also the variation of metal compositions indicate external influences on the basis of an intensified trade especially with the Altiplano. On the other hand, after studying and analyzing artifacts from museums and excavations, Ruppert (1982, 1983) postulated that turquoise deposits existed in southern Peru (see also Petersen 1970). So far no major deposits have been located. Many sources of turquoise, however, have been located in northern Chile and their pre-colonial exploitation recently has been investigated (González and Westfall 2005, 2008; see also Chap. 12).

Finally, obsidian also needs careful discussion: In southern Peru and northern Chile and Bolivia many obsidian sources have been investigated and geochemically characterized. Of these, however, there were major sources that have played an important role in the long-distance trade networks such as Quispisisa and Jampatilla in the highlands of Ayacucho (Burger and Asaro 1979; Burger et al. 2000; Tripcevich 2011; Chap. 2).

Mining Archaeology in Palpa and Nasca—The Field-Surveys

Mines and quarries were not easy to identify because in most cases, mining activities in colonial and in modern times had altered the evidence of Prehispanic activities. Our initial survey comprised the whole Nasca area, with visits to the sites previously described by Eerkens et al. (2009) (Fig. 14.9). Also our field procedure

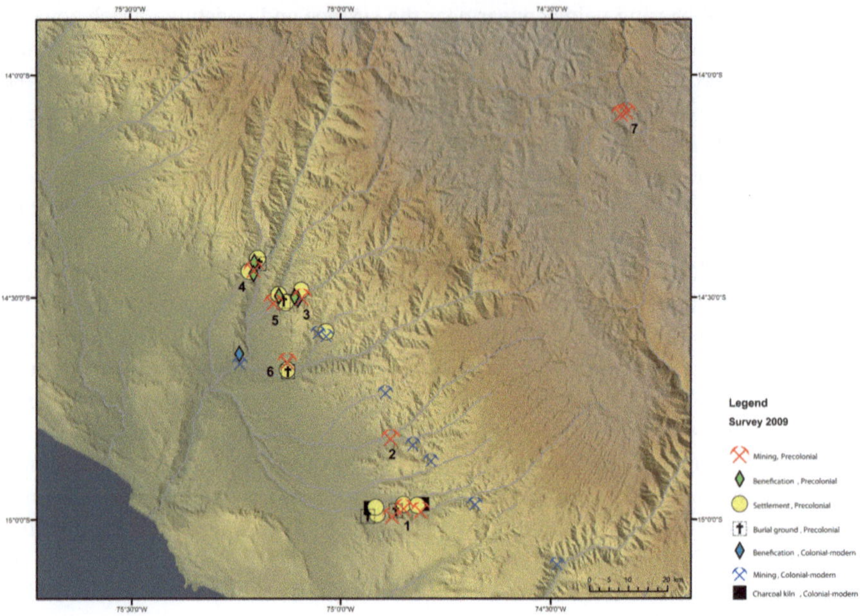

Fig. 14.9 Research areas in South Peru: Visited sites between 2006 and 2009 (Design: DBM, A. Hornschuch)

was very similar to the one described by Eerkens and colleagues. With the background of our excavation finds, our interests were focused on mines and quarries of different materials. Regarding lithic raw materials, we were interested in obsidian, chert, and fine crystalline rocks. Furthermore, we looked for gold and copper ores. Regarding the minerals that were used for ornaments and for pigments, we concentrated on malachite, turquoise and other copper sulfates, hematite, and ochre. Many of the places with mining activities were identified by locations where ores and minerals were processed. At these sites we documented mining tools and cultural finds. Finally we took samples for geochemical analysis in the laboratory.

Important new discoveries were achieved in the highlands: chert and obsidian quarries still are investigated only to a small extent. The quarry shown in Fig. 14.10 is close to the rock shelter at Cerro Llamoca mentioned before, where artifacts of the same material were recovered in excavations of layers that dated back to the eighth millennium BC. The waste dump in front of the quarry was clearly visible, where crudely worked preforms and typical artifacts like unifacial artifacts were found.

During our survey we visited the obsidian source of Jichja Parco which is located on the way to Huancasancos, near the upper reaches of the Rio Grande river [Fig. 14.9(7)]. Large amounts of obsidian debris lie inside deep pits, from where the obsidian was extracted from deposits that were close to the surface. The deposits are located close to an ancient pathway that connected this site with the Palpa region. According to recent research the Jichja Parco quarries can be allotted to a larger area

Fig. 14.10 Chert quarry with refuse dump at the foot of Cerro Llamoca (4,200 masl) (Photo: M. Reindel)

of obsidian outcrops: geochemically all of them can be assigned to the site called Quispisisa by Burger (Burger and Glascock 2000; Tripcevich and Contreras 2011; see our analysis below).

In the coastal areas field-work was carried out either in a northern part around Palpa as well as in the surrounding of the Nasca-valley in the south (Figs. 14.9, 14.11, and 14.13). South of the Nasca-Valley new insights were gained in the Las Trancas valley where settlements and grave-yards of different periods are known [e.g., Chauchilla; Figs. 14.9(1) and 14.11]. Flanks of mineral deposits are exposed northwards of the flanks of this fertile valley. Eerkens et al. (2009: 744–745) have mentioned possible smelting sites at the site of Media Luna located at the mouth of the valley. Further investigation however made clear that the charcoal heaps scattered in the plains in front of the hillock of Media Luna may be related to colonial charcoal burning, but not to smelting from the sixteenth and seventeenth century. Some of the very old radiocarbon dates published by Eerkens et al. (2009: 744, Table 2) may belong to charcoal burning of fossilized wood found in the sediment layers of the valley. More interesting were several small scale mining sites: they lined up especially along the northern flanks of the valley near various settlements of the late Middle Horizon and the LIP. Most of them can be described as polymetallic copper-deposits and occurrences. On some sites the surface is composed of small tailings and platforms with diagnostic sherd scatters (Fig. 14.11: Pc007, Pc014). Most likely these platforms have served as ore-beneficiation sites while settlement areas were situated near the fertile grounds of the valley.

Similar traces also were reported for the Aja river (Eerkens et al. 2009) where our team was able to locate another pre-colonial site (Pg002). The site is located at the flanks of the Aja valley at a mountain ridge that is oriented south to the Tierras Blancas valley. Quartzite veins may indicate the usage of gold-bearing ores. At site

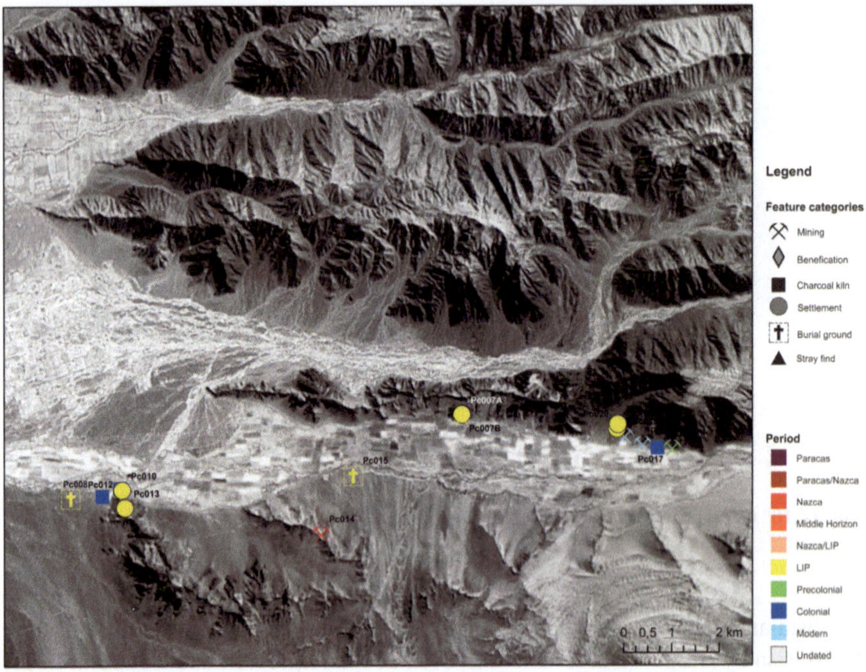

Fig. 14.11 Peru, Valle Las Trancas, sites visited during the survey 2009 (Design: DBM, A. Hornschuch)

Pg002 three small near-surface exploitations were visible [Figs. 14.9(2) and 14.12]. One of these was a typical exploitation pocket, another one followed a quartz-vein along a natural fracture. A smaller amount of stone-tools was found at dumps beneath these extractions. Some of those certainly were rather related to ore-beneficiation than to the actual extraction work.

After this general survey of the Nasca region we concentrated on the valleys of Palpa. Today gold mining is a major activity in the middle reaches of the valleys of Palpa. With simple tools and dynamite, informal miners exploit the gold deposits of the fairly rich veins of the Nasca-Ocoña belt. A major location for gold processing is Saramarca, in the Viscas valley, where nowadays the minerals are crushed in simple stone mills, washed with water and extracted with mercury (Schulz 2007; Stöllner 2011; Stöllner and Reindel 2007).

Interesting results were achieved nearby settlements of the Paracas and Nasca periods near Saramarca, an area that still is in operation and famous for its gold hosting ores [Fig. 14.9(3)]. At Saramarca it was difficult to find undisturbed mines, because most of the recent mining activities have redeveloped and destroyed older traces. In some cases (e.g., site Pe 029) it was possible to distinguish near-surface extractions with hammering traces from underground galleries worked with iron tools. In most cases evidence could be collected by ore-beneficiation sites that were located above the valley. These sites are well-dated by ceramic assemblages and are

Fig. 14.12 Peru, Aja Valley, mine 2 at site Pg002 (Photo: DBM, T. Stöllner)

contemporary to the settlements of the valley ground (Paracas and Nasca). This certainly means that most of the mining activity was organized from these settlements. There was no need to have special mining camps, contrary to the current situation where temporal camps reflect the volatile situation of a gold rush driven by market economy and high gold prices (Kuramoto 2001; Schulz 2007). In contrast to the assumption of Eerkens et al. (2009) we could not yield any evidence of precolonial temporal mining camps. This is true also for other sites that we visited in the surrounding valleys of Palpa. In all cases permanent settlements were situated nearby or even in short, easy reachable distances.

This is especially true for the ore-processing and the mining sites that have been located at a mountain ridge that divides the Rio Grande and the Santa Cruz-valley [Fig. 14.9(4)]. The small mining sites hosted in polymetallic ores (Au/Cu?) are situated near a larger series of settlements from the Nasca and the LIP. At these settlements we detected a stone quarry and more important ore-beneficiation consisting of crushing plates and mallets or hammers. The site of Locari is therefore another good example for extraction and organizational pattern mentioned before.

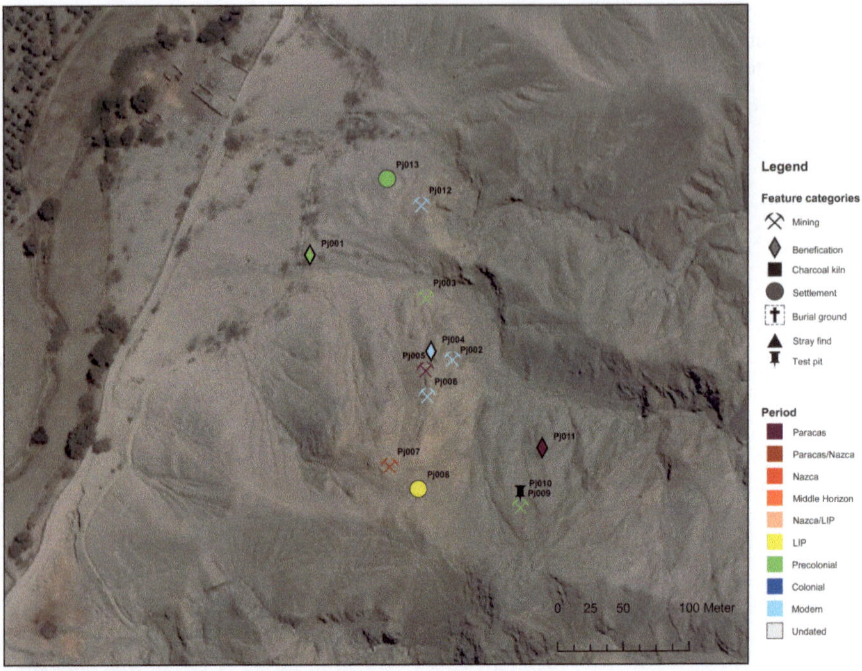

Fig. 14.13 Peru, Viscas-Valley, area of Mollaque Grande, sites visited and discovered during the survey 2009 (Design: DBM, A. Hornschuch)

Also at Mollake Grande we located many modern test pits, but also several Prehispanic mines and places for ore processing that were not modified by modern activities [Figs. 14.9(5), 14.13, and 14.14]. The mill stones, mortars, hammer stones, and waste dumps were associated mainly with Nasca and Paracas, and occasionally with LIP surface remains. In some cases a clear stratigraphic sequence was observed: modern, coarse debris dumps are superimposing older dumps consisting of finer gravel and debris as well as stone tools and pre-colonial pottery (Fig. 14.15). It is interesting that most of the mining traces were re-mined in younger periods and it would not be surprising if older *chungo* and *batanes*-sets also would have been reused. The ancient mining activities certainly belonged to the late Paracas and early Nasca-period while settlement terraces found above are much younger and delivered ceramic sherds from the LIP. It is not by mere chance that especially the LIP is represented at nearly all mountainous locations in the surrounding Palpa valley. Investigations of the settlement patterns have provided the insight that especially during that period such hilltop locations were favored places for settlements, either to avoid the settling of the fertile valley bottoms or to search for better defendable locations. If the close relation to mineral resources was another reason is debatable but not proven at the moment. The results of our settlement surveys however suggest that this region in all time periods was settled or even used for the exploitation of the rich gold deposits.

Fig. 14.14 Mollake Grande. Superposition of refuse of modern mining activities over ancient mining refuse characterized by Prehispanic ceramic sherds. The rounded stone at the *right* is about 50 cm in height (Photo: Th. Stöllner)

Fig. 14.15 Mollake Grande. Mortars, hammer stones and pestles used for mining activities (Photos/Design: DBM, B. Gräfingholt)

The mines in Palpa were quite small. No large mine like the Mina Primavera in the Ingenio valley [see Chap. 8; Fig. 14.9(6)] has been detected so far. Many smaller mines from different time periods, however, suggest a long-lasting tradition of mining in the region. The different types of evidence for mining activities include horizontal and vertical pits, tailings of mining activities, mining tools, and local processing areas of the ores.

It is not always clear if ancient mines and quarries in the Palpa area date from Prehispanic times or from colonial times. Modern quarries, however, can easily be recognized by the traces of fresh cuts or metal tools. Ancient mines have associated remains of ceramic artifacts, mining tools like simple stone hammers and a set for beneficiating the ores (Fig. 14.15). The ores and minerals extracted from the different deposits generally were processed in a preliminary way at places close to the mines and quarries. In some cases these processing places were directly associated with settlements. What is surprising was the fact that fire-setting could not be observed in one single case. This may be the reason why larger quantities of stone hammers—which are so typical for exploitation methods of this kind—cannot be found. We stress that no case of fire-setting before the colonial period has been documented so far in the Andean regions (Shimada pers. comm. I; Stöllner 2011).

Geochemical Analyses of Ores, Metals, and Minerals

The first results of the geochemical analysis of the obsidian samples allow us to trace the sources of the obsidian that was used in Palpa. During a field survey in 2009 we visited the open pit obsidian quarry in the Altiplano region near Huanca Sancos called Jichja Parco (Stöllner 2011; see above). Tripcevich and Contreras (2011; Chap. 2) who had visited and mapped this site as well, demonstrated that Jichja Parco is part of the greater flow of obsidian that has been known as Quispisisa, and that it had to be viewed as the main supplier for obsidian in the South of Peru.

Obsidian samples from 18 securely dated archaeological sites of the Andean Transect research area (highland and coastal sites), as well as 4 samples from the obsidian quarry site Jichja Parco were taken for chemical analysis, in order to determine the provenance and raw material sources of the people living in the Andean Transect area in different time periods. Neutron Activation Analysis (NAA), which is commonly used to determine the provenance of obsidian especially in South America and Peru (Glascock et al. 2007), enabled us to characterize the trace elements of the Obsidian samples and to compare our results with the published sources. For our analysis we concentrated on the elements Na; K; Sc; Cr; Fe; Co; Zn; As; Rb; Zr; Sb; Cs; Ba; La; Ce; Nd; Sm; Eu; Tb; Yb; Lu; Hf; Ta; Th and U. The analyses were undertaken by the Curt-Engelhorn Zentrum Archäometrie in Mannheim, Germany, with a Coaxial HPGe. The TRIGA-research reactor at the University of Mainz was used to irradiate the specimens.

To determine the origin of our samples we plotted the concentrations of Hf (ppm), Cs (ppm), Th (ppm), La (ppm), Fe (%), Ba (ppm), Fe (%), Rb (ppm) and Zr (ppm),

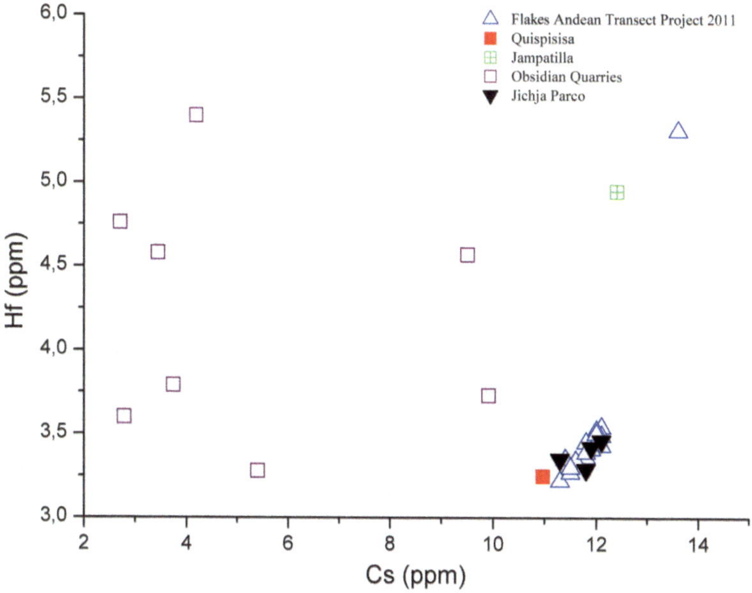

Fig. 14.16 Bivariate plot of Cs versus Hf for obsidian flakes from the Andean Transect Project and Jichja Parco source. Element concentration of Cs and Hf for the Obsidian Quarries (Alca 1, Alca 2, Alca 3, Chivay, Puzolana, Potreropampa, Lisahuacho, Aconcahua), Quispisisa and Jampatilla are taken from Glascock et al (2007)

Zn (ppm) in combination with the data for the obsidian quarries provided by Glascock et al. (2007). The results of the Hf and Cs relations presented in Fig. 14.16 indicate that all but one flake from the LIP site Chillo fall within the chemical group defined as Quispisisa and more precisely Jichja Parco. Our results confirm the hypothesis of Tripcevich and Contreras (2011) that Jichja Parco is the main quarry used in the central Andean region. We proved that the Palpa region used the Jichja Parco source as main supplier for obsidian throughout the Archaic up to the Middle Horizon. The LIP seemed to alter the trade routes and used other sources of obsidian which were farther away than Jichja Parco (Tripcevich and Contreras 2011), which certainly is the background for the provenance results of the LIP-flake of Chillo. A similar conclusion was reached for the Chivay obsidian source where obsidian procurement declined during the LIP (Tripcevich 2009).

First results of compositional analysis of samples extracted from modern and ancient mines in the Palpa region show surface near oxides and carbonides with high enrichments in copper and gold (up to 100 g/t; Fig. 14.17). There existed small but highly enriched "bonanzas" which typically have been exploited in earlier stages of metallurgy (Kuramoto 2001; Montoya et al. 1994). This resembles the general picture that we obtained from the excavation finds. Metal objects mainly are made of gold and copper or alloys of both materials.

Proben	SiO2 in %	Al2O3 in %	Fe2O3 in %	Cu in %	Ag in %	Au in %
PA 12	80,2	1,34	8,15	3,97	0,0002	0,0101
PC 01	67,9	2,77	12,97	3,5	0,0002	0,0149
PC 02	94,2	0,64	1,23	0,0063	0,0006	0,0013
PC 07	90,0	2,27	2,34	0,2365	0,0012	0,0071
PC 14	10,6	3,06	24,16	2,96	0,00004	0,0038
PD 01	32,3	0,20	26,35	18,37	0,0002	0,0418
PD 04	38,7	13,15	29,57	2,03	0,0005	0,0057
PD 17	9,9	2,23	75,45	1,54	0,00006	0,0148
PE 01	28,0	1,72	62,88	1,2	0,0009	0,0304
PE 06	98,1	0,04	1,60	0,0096	0,0082	0,1051
PE 25	91,9	0,39	4,25	0,1372	0,0002	0,0011
PH 02	89,2	4,59	2,59	0,9730	0,0003	0,0013
PJ 01	68,2	6,30	10,80	4,68	0,00013	0,0034
PE 29	75,7	1,00	21,78	0,0919	0,0002	0,0024

Fig. 14.17 Results of ICP-OES compositional analysis of copper and gold concentrations of ore samples recollected from mining places in the research area

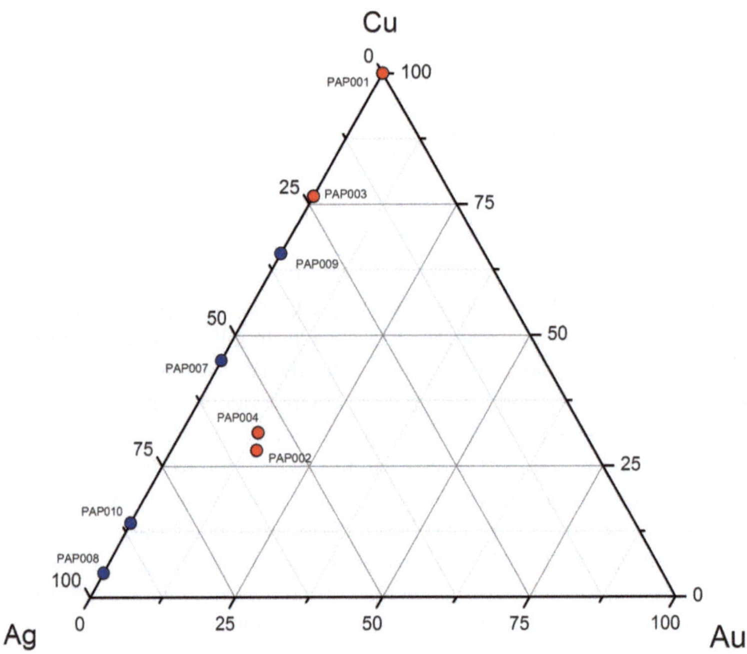

Fig. 14.18 Results of XRF compositional analysis of some archaeological metal objects recovered in the research area

First results of X-ray fluorescence analysis of metal artifacts of the Nasca period (the red dots in the diagram Fig. 14.18) show that the copper and natural gold alloys are typical for the metal ores of the Nasca-Ocoña formation of the coast. One pure copper artifact from Botiqiriayoq may derive from primary native copper deposits and could therefore have been transported from the coastal areas to the Altiplano.

The Middle Horizon artifacts, however (the blue dots in the diagram of Fig. 14.18), are copper arsenide alloys with silver from deposits which seem typical for the polymetallic ores from Northern Peru or in the Lake Titicaca region. The silver of one sample of Lucriche may derive from silver rich ores (cerussites). It is interesting that the only tin bronze analyzed so far comes from the Montegrande site at the river-banks of the Rio Grande within the coastal cordillera. Based on the abundance of surface finds of precious materials like metal objects and Spondylus shells and artifacts, as well as the amount of llama bones, it is assumed that Montegrande was an important trading post between the coastal and the highland regions from the Paracas through the LIP and Inca Period. The tin bronze therefore fits into such a pattern (generally since the late Middle Horizon: Lechtman 2007; Stöllner 2011: 194). To understand the origin of the ores that have been used for fabrication of many of those metal artifacts, more delicate methods of analysis are necessary.

Most of the ore samples have been analyzed in laboratories of the German Mining Museum in Bochum with ICP-OES (inductively coupled plasma optical emission spectrometry). The method allows to detect up to 70 chemical elements on the level of trace elements and to determine their quantities. The metal ores from the Nasca-Ocoña geological formation show surface near oxides and carbonides with high enrichments in copper (e.g., PAP 12: 20 % of copper) and gold (up to 105 g/t; Fig. 14.17). There are small but high enriched "bonanzas" which typically have been exploited in earlier stages of metallurgy.

A selection of some of the ores was investigated also with ICP-MS to determine the lead-isotopes (Klein et al. 2009). The latter investigations are currently carried out in collaboration with the University of Frankfurt. Both the lead-isotopes and the trace elements are indispensable requisites if metal artifact provenances may answered undoubtedly and by crosscheck. As far as the small sample-series allows a first statement, there is a characteristic isotopic pattern of the southern Peruvian coastal area that even overlaps also with some of the silver–copper artifacts found during the excavations and surveys in the highlands. The isotopic variation separates clearly Northern Coastal and Central Peruvian ore samples (Lechtman 1991), but also colonial silver objects and coins produced from Potosí silver bearing ores (Desaulty et al. 2011; Fig. 14.19). A rough evaluation indicates a further south Peruvian ore field not yet sufficiently investigated (Lechtman and Macfarlane 2005). It should be mentioned that some of the ores investigated from the Nasca-Ocoña formation nicely matches with silver–copper alloys from the Palpa-Altiplano survey. As the data processing of the Palpa samples is not completed yet, no further evaluation is possible at the moment. However, there is undoubtedly still a high potential for a continuation in analyzing metals by these methods in southern Peru.

Secondary copper minerals like malachite, chrysocolla, and turquoise were used in Palpa from Archaic times on. The use of turquoise objects in Palpa is especially interesting. According to previous studies, no turquoise deposit has been identified so far in the Central Andes (see above). It is believed that turquoise was imported from outside the Central Andean area.

In a study based on geochemical analysis of museum and excavation materials, however, the German geologist Hans Ruppert postulated in 1982 the presence of a hitherto unrecorded turquoise deposit in southern Peru. In the surrounding of Palpa

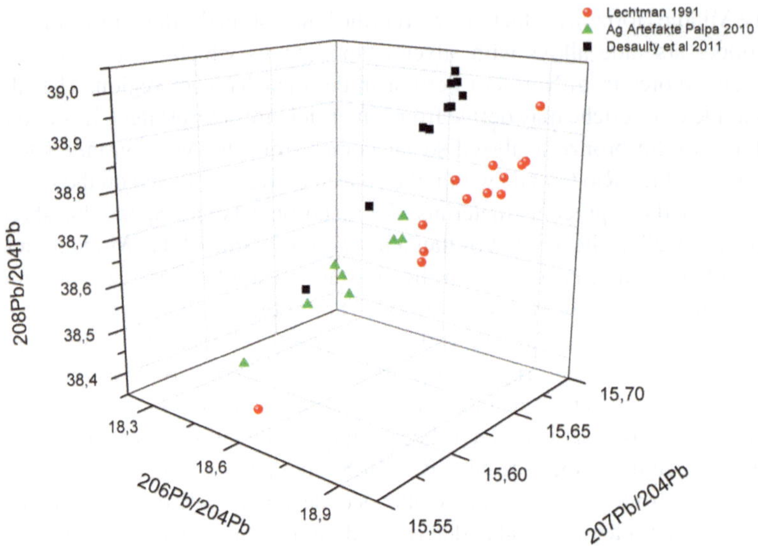

Fig. 14.19 Lead isotope data from some metal objects from the Palpa highland Survey in comparison to data from central and northern Peruvian ore sources (data from Lechtman 1991) and from colonial silver objects and coins produced from ores of northern Bolivia (Potosí) (data from Desaulty et al. 2011)

several small turquoise occurrences were identified by our collaborator, the geologist Ulrich Glasmacher from Heidelberg University. They are found in context with the various copper sources there [e.g., Fig. 14.9(3–5)]. It is therefore possible that Palpa or the Nasca region in general was one of the sources of Peruvian turquoise, but are much too small to compete against the major sources of northern Chile (e.g., recently discussed by Stöllner 2011: 190–191).

Work for the Future

In the future we plan to continue our systematic investigation of lithic, mineral, and metallic raw materials. We need to further investigate the development of techniques of mining and processing of the raw materials. We will continue our search for the evidence of smelting of the ores and we will use geochemical methods to identify the provenance of the raw materials of the objects that we recovered from our excavations.

Finally, we plan to continue with systematic studies of the geochemical and typological characteristics of lithic materials. At the moment B. Gräfingholt is processing more than 400 obsidian projectile points recovered from surveys and excavations at sites of all time periods. The results of these studies will be analyzed in the context of exchange systems between the cultures of the coast and the highlands in the study area of the Andean Transect project.

Summary and Conclusions

Summing up the preliminary results of our investigation about mining and quarrying activities in the valleys of Palpa, we can state that we detected several places related to the extraction and the processing of lithic material, ores, and minerals in Prehispanic times. All places detected so far are locations with evidence of small scale mining. It is also important to point out that so far we have not found any smelting place or any furnaces where the ores may have been smelted.

The available evidence demonstrates, however, that lithic materials, minerals, and ores were produced and moved along the Andean Transect since the earliest periods of Prehispanic occupation of the region. Obsidian was transported to the coast since the archaic period and can be considered as evidence of the mobility of the earliest settlers. Gold is present in the region at least since the Early Paracas period. In part, the gold was produced locally, but some of the gold was also imported from the north (Schlosser et al. 2009). The small number of known Paracas gold objects shows that these were rare and precious items, just as in other regions of the Central Andean Area in the Early Horizon.

This situation changed obviously in Nasca times. Nasca gold objects are present in many museum and private collections. The objects found in the elite tombs of La Muña show that gold played an integral role in the upper class of Nasca society. In the Nasca period the use of copper also begins. The associated Nasca ceramic sherds at nearly every Prehispanic mining place that we identified during our surveys may indicate increased mining activities in the Nasca period.

While in Nasca times clearly gold was the favored ore to be mined, in the Middle Horizon mining and metallurgy of copper and silver rich ores dominated. Copper objects were found particularly in the highlands. Due to the presence of rich deposits of copper in the Nasca-Ocoña belt, however, it is possible that the mines in this region constituted part of the sources from where the raw material was transported to the highlands.

Copper and gold production continued into the LIP, a period when the whole research area of the Andean Transect was repopulated after a time of nearly complete abandonment.

Acknowledgments Funding for this research has been provided by the Federal German Ministry for Education and Research (BMBF) and the German Research Foundation (DFG). The mining surveys took place in March and April 2009 and were accompanied by the authors and Dr. Ch. Dekant (Munich), P. Segura (Palpa) and Prof. Dr. G.A. Wagner (Heidelberg). Thanks to Johny Isla Cuadrado the ceramic finds were dated according to the detailed chronology of the Palpa-project; additional thanks to Annette Hornschuch and Stefanie Menic for their assistance in preparing the maps. Thanks to the laboratory of the DBM, especially Dr. Michael Bode, Dipl. Min. Dirk Kirchner, and Dr. Michael Prange as well as PD Dr. Sabine Klein at the University of Frankfurt and the Laboratories of the CEZ Mannheim (Prof. Dr. E. Pernicka, Dr. J. Lutz).

References

Benfer, R. (1999). Proyecto de excavaciones en Paloma. Valle de Chilca, Perú. *Boletín de Arqueología PUCP, 3*, 213–237.

Burger, R. L., & Asaro, F. (1979). Análisis de los rasgos significativos en la obsidiana de los Andes. *Revista del Museo Nacional, 43*, 281–326.

Burger, R. L., & Glascock, M. D. (2000). Locating the Quispisisa Obsidian Source in the Department of Ayacucho, Peru. *Latin American Antiquity, 11*, 258–268.

Burger, R. L., Mohr Chavez, K. L., & Chavez, S. J. (2000). Through the glass darkly: Prehispanic obsidian procurement and exchange in Southern Peru and Northern Bolivia. *Journal of World Prehistory, 14*, 267–362.

Desaulty, A.-M., Telouk, P., Albalat, E., & Albarède, F. (2011). Isotopic Ag-Cu–Pb record of silver circulation through 16th-18th century Spain. *Proceedings of the National Academy of Sciences, 108*(22), 9002–9007.

Eerkens, J. W., Vaughn, K. J., & Linares Grados, M. (2009). Pre-Inca mining in the Southern Nasca Region, Peru. *Antiquity, 83*, 738–750.

Eitel, B., Hecht, S., Mächtle, B., Schukraft, G., Kadereit, A., Wagner, G. A., Kromer, B., Unkel, U., & Reindel, M. (2005). Geoarchaeological evidence from desert loess in the Nazca-Palpa region, southern Peru: Palaeoenvironmental changes and their impact on Pre-Columbian cultures. *Archaeometry, 47*(1), 137–158.

Eitel, B., & Mächtle, B. (2009). Man and environment in the Eastern Atacama Desert (Southern Peru): Holocene climate changes and their impact on pre-columbian cultures. In M. Reindel & G. A. Wagner (Eds.), *New technologies in archaeology. Multidisciplinary investigations in Nasca and Palpa, Peru* (pp. 17–37). Berlin/Heidelberg: Springer.

Engel, F. A. (1966). *Geografía humana prehistórica y agricultura precolombina de la quebrada de Chilca*. Lima.

Engel, F. A. (1980). Paloma village 613: A 6000 year old "Fog Oasis" village in the Lower Central Andes of Peru. In F. A. Engel (Ed.), *Prehistoric andean ecology. Man, settlement and environment in the Andes* (pp. 103–135). New York: Humanities Press.

Engel, F. A. (1988). *Ecología prehistórica andina. El hombre, su establecimiento y el ambiente de los Andes. La vida en tierras áridas y semiáridas. Chilca, pueblo 1. Implementos de hueso*. Lima.

Glascock, M. D., Speakman, R. J., & Burger, R. L. (2007). Sources of archaeological obsidian in Peru. Descriptions and geochemistry. In M. Glascock (Ed.), *Archaeological chemistry. Analytical techniques and archaeological interpretation* (pp. 522–552). Washington, DC: ACS symposium series 968.

González, C., & Westfall, C. (2005). Consideraciones sobre la prehistoria de Atacama: El Salvador y sus aportes locales e interregionales. *Boletín de la Sociedad Chilena de Arqueología, 38*, 53–70.

González, C., & Westfall, C. (2008). Atacameños en El Salvador: Nuevas apreciaciones sobre un fardo funerario del Cementerio Las Turquesas. *Estudios Atacameños, 35*, 49–73.

Isla Cuadrado, J. (2009). From hunters to regional lords: Funerary practices in Palpa, Peru. In M. Reindel & G. A. Wagner (Eds.), *New technologies for archaeology. Multidisciplinary investigations in Palpa and Nasca, Peru* (pp. 119–139). Berlin/Heidelberg: Springer.

Isla Cuadrado, J., & Reindel, M. (2006a). Una tumba Paracas Temprano en Mollake Chico, valle de Palpa, costa sur del Perú/Ein Grab der frühen Paracas-Zeit in Mollake Chico, Palpa-Tel, Südküste Perus. *Zeitschrift für Archäologie Außereuropäischer Kulturen, 1*, 153–181.

Isla Cuadrado, J., & Reindel, M. (2006b). Burial patterns and sociopolitical organization in Nasca 5 society. In W. Isbell & H. Silverman (Eds.), *Andean archaeology III: North and south* (pp. 374–400). New York: Springer.

Klein, S., Domergue, C., Lahaye, Y., Brey, G. P., & von Kaenel, H.-M. (2009). The lead and copper isotopic composition of copper ores from the Sierra Morena (Spain). *Journal of Iberian Geology, 35*, 59–68.

Kuramoto, J. (2001). La Minéria Artesanal e Informal en el Perú. *Mining, Minerals and Sustainable Development, 82*, 1–47.
Lechtman, H. (1976). A metallurgical site survey in the Peruvian Andes. *Journal of Field Archaeology, 3*(1), 1–42.
Lechtman, H. (1991). The production of copper-arsenic alloys in the Central Andes: Highland ores and coastal smelters. *Journal of Field Archaeology, 18*(1), 43–76.
Lechtman, H. (2007). The Inca , and Andean metallurgical tradition. In R. Burger, R. Morris, & R. Matos Mendieta (Eds.), *Variations in the expression of Inca power. A symposium in Dumbarton Oaks, 1997* (pp. 313–355). Washington, DC: Dumbarton Oaks Research Library.
Lechtman, H., & Macfarlane, A. W. (2005). La metalurgia del bronce en los Andes Sur Centrales: Tiwanaku y San Pedro de Atacama. *Estudios atacameños, 30*, 7–27.
Lothrop, S. K. (1937). Gold and silver from Southern Peru and Bolivia. *Journal of the Royal Anthropological Institute of Great Britain and Ireland, 67*, 305–325.
Lothrop, S. K. (1951). Gold artifacts of Chavin Style. *American Antiquity, 3*, 226–240.
Montoya, M., Garcia, W., Caldas, J. (1994). *Geologia de los cuadrangulos de Lomitas, Palpa, Nasca y Puquio. Carta Geológica Nacional 53*. Lima.
Petersen, G. (1970). *Minería y Metalurgia en el Antiguo Perú*. Lima.
Petersen, U. (1979). Metallogenesis in South America: Progress and problems. *Episodes 4*, 3–11.
Petersen, U. (1989). Geological framework of Andean Mineral Resources. In G. E. Erickson, M. T. Cañas Pinochet, J. A. Reinemund (Eds.), *Geology of the Andes and its relation to hydrocarbon and mineral resources*. Circum-Pacific Council for Energy and Mineral Resources, v. 11, Houston, Texas, 213–232.
Reindel, M. (2009). Life at the edge of the desert. Archaeological reconstruction of the settlement history in the valleys of Palpa, Peru. In M. Reindel & G. A. Wagner (Eds.), *New technologies for archaeology. Multidisciplinary investigations in Palpa and Nasca, Peru* (pp. 439–462). Berlin/Heidelberg: Springer.
Reindel, M. (2010). Archäologische Forschungen der Jahre 2007 und 2008 im Anden-Transekt, Süd-Peru. *Zeitschrift für Archäologie Außereuropäischer Kulturen, 3*, 207–224.
Reindel, M., & Isla, J. (2006). Culturas tempranas de la costa sur del Perú: sus evidencias en los valles de Palpa. *Boletín de Arqueología PUCP, 10*, 237–284.
Reindel, M., & Isla Cuadrado, J. (2001). Los Molinos und La Muña. Zwei Siedlungszentren der Nasca-Kultur in Palpa, Südperu. Los Molinos y La Muña. Dos centros administrativos de la cultura Nasca en Palpa, costa sur del Peru. *Beiträge zur Allgemeinen und Vergleichenden Archäologie, 21*, 241–319.
Reindel, M., & Wagner, G. A. (2009). Introduction: New Methods and Technologies of Natural Sciences for Archaeological Investigations in Nasca and Palpa, Peru. In M. Reindel & G. A. Wagner (Eds.), *New technologies for archaeology. Multidisciplinary investigations in Palpa and Nasca, Peru* (pp. 1–13). Berlin/Heidelberg: Springer.
Root, W. (1949). The metallurgy of the Southern Coast of Peru. *American Antiquity, 15*(1), 10–37.
Ruppert, H. (1982). Zur Verbreitung und Herkunft von Türkis und Sodalith in präkolumbischen Kulturen der Kordilleren. *Baessler Archiv (NF), 30*, 69–124.
Ruppert, H. (1983). Geochemische Untersuchungen an Türkis und Sodalith aus Lagerstätten und präkolumbischen Kulturen der Kordilleren. *Berliner Beiträge zur Archäometrie, 8*, 101–210.
Schlosser, S., Kovacs, R., Pernicka, E., Günther, D., & Tellenbach, M. (2009). Fingerprints in gold. In M. Reindel & G. A. Wagner (Eds.), *New technologies for archaeology. Multidisciplinary investigations in Palpa and Nasca, Peru* (pp. 409–436). Berlin/Heidelberg: Springer. Schulz, G.G. (2007). Kleinstbergbau auf Gold im Gebiet von Palpa/Peru. World of Mining 1: 41-47.
Schulz, G. G. (2007). Kleinstbergbau auf Gold im Gebiet von Palpa/Peru. *World of Mining, 59*(1), 42–47.
Stöllner, T. (2009). Gold in Southern Peru? Perspectives of Research into Mining Archaeology. In M. Reindel & G. A. Wagner (Eds.), *New technologies for archaeology. Multidisciplinary investigations in Palpa and Nasca, Peru* (pp. 393–407). Berlin/Heidelberg: Springer.
Stöllner, T. (2011). Präkolumbischer Bergbau in den Anden. In R. Slotta (Ed.), *Schätze der Anden—Chiles Kupfer für die Welt*. Exhibition catalogue, Bochum, 181–212.

Stöllner, T., Gambaschidze, I., & Hauptmann, A. (2008). The earliest gold mining of the ancient world? Research on an Early Bronze Age Gold Mine in Georgia. In Ü. Yalcin, H. Özbal, A. G. Paamehmetolu (Eds.), *Ancient Mining in Turkey and the Eastern Mediterranean*. International conference AMITEM 2008, Ankara, pp. 271–288.

Stöllner, T., Körlin, G., Steffens, G., Cierny, J. (2003). (Eds.), *Man and mining. Studies in honour of Gerd Weisgerber*. Der Anschnitt, Beiheft 16, Bochum.

Stöllner, T., & Reindel, M. (2007). Vorspanische Goldgewinnung im Bereich von Palpa-Nasca? Bericht über eine einwöchige Prospektionskampagne an der Südküste Perus, *Der Anschnitt 1, 59. Jahrgang*, 2–12.

Tello J.C. (1959). *Paracas—primera parte*. Lima.

Tello J. C., Mejía Xesspe T. (1979). *Paracas, segunda parte: Cavernas y Necrópolis*. Lima.

Tripcevich, N. (2009). Exotic goods, chivay obsidian, and sociopolitical change in the South-Central Andes. In C. D. Dillian & C. L. White (Eds.), *Trade and exchange. Archaeological studies from history and prehistory* (pp. 59–74). New York: Springer.

Tripcevich, N., & Contreras, D. A. (2011). Quarrying evidence at the Quispisisa Obsidian source, Ayacucho, Peru. *Latin American Antiquity, 22*, 121–136.

Uhle, M. (1913). Zur Chronologie der alten Kulturen von Ica. *Journal de la Société des Américanistes de Paris N.S., 10*, 341–367.

Part IV
Discussion

Chapter 15
Some Thoughts on Mining and Quarrying in the Ancient Andes

Richard L. Burger

In the following discussion, I offer some thoughts stimulated by the papers brought together in this volume. At the suggestion of the editors, I emphasize the themes in the initial portion of the volume dealing with the extraction of non-metallic deposits.

In my opinion, *Mining and Quarrying in the Ancient Andes*: Sociopolitical, Economic and Symbolic Dimensions represents a milestone in Andean archaeology. It demonstrates that for the first time there is a critical mass of high-quality research on Prehispanic mining and quarrying. One of its strengths is that it is not limited to Peru; on the contrary, it is also drawn from field investigations in Ecuador, Chile, and Bolivia. This new wave of research on Prehispanic mining and quarrying has been complemented by fresh ethnographic work on contemporary extraction practices among traditional indigenous groups, as well as a critical appreciation of relevant historical documents. This advance in research on mining and quarrying has been a long time coming, considering that its foundations were laid down decades ago by a relatively small number of archaeologists and geologists.

Perhaps most notable among these pioneering scholars was Georg Petersen (1898–1985), a geologist of German birth and Peruvian nationality who spent most of his life in Peru, first working in the oil and mining industries, and then as a professor of geology in Lima at the Universidad Nacional de Ingeniería and the Pontificia Universidad Católica del Perú (C.R. Petersen 2010). Unlike many Peruvian geologists, Petersen's interests extended well beyond economic concerns and much of his scholarship focused on the interface between archaeology and geology. A personal friend of Julio C. Tello, Jorge Muelle, Junius Bird, Duccio Bonavia, and other archaeologists, Petersen pioneered the study of Prehispanic mining and metallurgy, with interests that reached beyond metals to the full range of

R.L. Burger (✉)
Anthropology, Yale University, P. O. Box 208277, New Haven, CT, USA
e-mail: richard.burger@yale.edu

N. Tripcevich and K.J. Vaughn (eds.), *Mining and Quarrying in the Ancient Andes*, Interdisciplinary Contributions to Archaeology, DOI 10.1007/978-1-4614-5200-3_15,

rocks and minerals utilized by the people of the ancient Andes. And he was as interested in the procurement of the raw materials via quarrying and mining, as he was in the subsequent production of tools and other objects. As far back as the 1953 he worked with Junius Bird at the American Museum of Natural History in New York City on the investigation of the Prehispanic Chilean miner known as the Copper Man (Petersen 2010: 39)

Petersen's decades of research culminated with the publication of *Minería y Metalurgia en el Antiguo Perú* by the Museo Nacional de Antropología y Arqueología in Pueblo Libre in 1970. Forty years later, an English edition of Petersen's book was published by the Geological Society of America as a result of the efforts of geologist William Brooks (Petersen 2010), and the volume still has much to offer.

Although an earth scientist by training, Petersen developed into a true Peruvianist, who was almost as comfortable with the Spanish chronicles and archaeological evidence as with geological stratigraphy. He traveled throughout the Peruvian highlands and coast and had an encyclopedic knowledge of Peru's metal and mineral deposits, as well as an extensive professional network that gave him access to a large body of unpublished information. All this is summarized in his remarkable 1970 volume (Petersen 1970). When I began my research on obsidian, I naturally sought Petersen out in Lima and received valuable advice as well as encouragement and support. As a testimonial by Duccio Bonavia (2010) demonstrates, I was just one in a long line of scholars to benefit from Petersen' s generosity.

Considering Petersen's scientific background, it is noteworthy that he wrote the following passage in the Prologue of his monograph:

> Mining during ancient times was mystical and religious. The ancient Andeans felt intimately connected with nature and the mines that provided the metals were worshipped and given offerings. Padre Calancha (1638) described how the ancient Peruvians worshipped their gold mines (*chuqui*), their silver mines (*coya*), and mines that yielded other metals (*corpa*). Pyrite, cinnabar (*llimpi*), and the furnaces (*huayras*) used for smelting were also revered. Padre Bernabe Cobo (1653, Libro XIII, c.XI, p.166) observed that those who worked the mines also worshipped the hills and the mines, which were called *coya*, and were asked to yield their mineral wealth. The miners danced, drank and left candles in order to glorify their prayers
>
> (Petersen 2010: xix)

It should be emphasized that despite the technical orientation of Petersen's work, he did not lose sight of the human experience reflected in the archaeological record. For example, in his account of a mummified miner discovered in 1899 at Mina Chuquicamata in the province of Antofagasta, Chile, Petersen (2010: 40) wrote "Copper Man" is a mute testament to human tragedy and a symbol of uncounted generations and sacrifices during millennia of ancient mining."

The numerous contributions gathered together in this volume follow in the tradition of Petersen, as well as that of Junius Bird, Heather Lechtman, and other pioneering scholars. Lechtman's work has been particularly influential in drawing attention to the cultural construction of technology, and she has used Andean metallurgy to illustrate the powerful ideological forces at work in the choices that are made in all technological processes (Lechtman 1980). Lechtman also followed Petersen's lead in searching for and investigating ancient mining areas using insights drawn from both archaeology and materials science (e.g., Lechtman 1976). After reading Lechtman's work, Prehispanic mining and quarrying cannot be understood without detailed reference to the society and cosmology of ancient Andean peoples. Bird's special interest in mining technique and the tools that were used was also a valuable contribution (Bird 1979).

While the work of these early investigators has been a source of inspiration, the sudden surge of interest in mining and quarrying may also be interpreted as a function of the changing political and economic realities in Peru and its Andean neighbors over the last decade. Bruce Trigger (1994) has argued that attempts to understand the history of archaeology can be divided into internalist and externalist approaches. Internalist approaches concentrate on the changing understanding of a problem within archaeology, whereas externalist approaches focus on the relationship between archaeological understanding and the sociocultural context in which archaeology is practiced. While an internalist approach would emphasize the pioneering research of early scholars and their accomplishments, as I have done above, an externalist approach would focus on other factors outside of archaeology (Trigger 1994: 118). An externalist approach would note how the end of nationalistic military regimes and civil strife in Chile and Peru eventually led to the widespread adoption of neo-liberal economic policies encouraging large foreign investments in mining and the extraction of oil and gas. In many Andean nations, most notably Peru, this policy has fueled an economic boom. As William Brooks recounts (2010: ix), by 2009 Peru led the world in silver production, was the world's second leading producer of copper, and was the leading producer of gold in Latin America. From an externalist perspective (e.g., Patterson 1995), the quantum leap of research on mining and quarrying seems to be a result of this general phenomenon. Trigger ultimately concluded that the internalist and externalist approaches are complementary rather than antithetical, and that a rounded explanation requires both (Trigger 1994: 118).

The current boom in mining and quarrying is of relevance not only to understanding the timing of intensification of archaeological interest in the subject, but also because of the threat these economic developments pose to surviving Prehispanic mines and quarries. The enormous scale of investment and landscape modification by foreign mining companies in Peru and elsewhere in the Andes cannot be exaggerated, and the search for new zones to exploit is ongoing, often impacting mining areas abandoned since colonial times. The opportunity to study mining and quarrying is diminishing each year, as the activities of mining companies become more pervasive.

On the other hand, many of these enterprises sponsor archaeological surveys and excavation as part of their environmental and cultural impact statements and subsequent mitigation efforts, so it might be argued that the affect on archaeological research is not altogether problematic. Unfortunately, the results of these sponsored projects are little in evidence in this volume. This may reflect the emergence of an archaeological "gray literature" consisting of technical reports prepared to comply with government regulations but never adequately disseminated or incorporated into academic discussion. This is particularly tragic in the case of the work funded by mining enterprises, since these are often located in remote high-altitude locations rarely studied in depth by archaeologists. In at least one case, that of the Yanacocha gold mine in the highlands of Cajamarca, contract archaeologists have encountered abundant evidence of religious rituals carried out at or near the mines, but to date, this information has remained unpublished (Jose Pinilla Blenke, personal communication).

The economic prosperity in Peru and several of its neighbors has stimulated an upsurge in local demand for a multitude of consumer goods, such as stone kitchen counters and wall veneers that require the quarrying of materials from what were once ancient quarries. For example, a visit in 2012 by the author to the well-known andesite deposit of Huaccoto near the modern city of Cuzco encountered workmen with electric saws and generators rapidly obliterating evidence of this Inca quarry (see Chap. 3), despite a commitment from the Ministry of Culture to protect it. As new four-star hotels and luxury homes are built in many Andean cities as a result of increased tourism and local prosperity, the impact on a range of mines and quarries is sure to rise. In reality, the intact sites relevant to the study of Prehispanic mining and quarrying have survived until today due to the underdevelopment of many Andean regions rather than the government policies protecting the cultural patrimony.

A corollary of this observation is that the mines or quarries of those rocks and minerals of least interest to contemporary populations are the ones most likely to have survived undisturbed. The studies of the Quispisisa obsidian source (Chap. 2, Tripcevich and Contreras 2011) and the Mina Primavera hematite quarry (Chap. 8) are good illustrations of this phenomenon. Since colonial times obsidian has had little if any economic value, since its main function as a raw material used for cutting implements was made almost obsolete by the European introduction of inexpensive metal tools. Most Peruvian geologists do not even bother to mention the presence of obsidian in their survey reports because of its irrelevance to the modern economy. But even in a case such as obsidian, the remote but vast quarry area has come under threat from a proposed government hydroelectric project (Chap. 1), in response to the demand for power touched off by rapid economic expansion.

Despite all of the current forces threatening mining and quarrying sites, it is difficult to read the contributions to this volume without being impressed by the richness and potential of the numerous sites that have been investigated and the many more that remain to be studied. A good illustration of this was the chapter devoted to hematite mining at the Mina Primavera mine in Nasca (Chap. 8). Initially, the quarry site appeared to be unpromising from an archaeological perspective since the mouth of the mine was closed by a modern masonry wall within a concrete frame. Yet excavations behind the entryway revealed not only evidence of Prehispanic mining, but also the hammerstones used in this quarrying and ground stone mortars and other artifacts used to produce red pigment from the hematite. In this, and several of the other chapters, excavation has proved essential for uncovering intact evidence of extractive activities hidden by mining debris and later deposits.

One of the distinctive contributions of this volume is to demonstrate the power of archaeology to document changing patterns of mining and quarrying. In many cases, what emerges is the unique life history of a mine or quarry, with its trajectory of initial use, its period of intensive exploitation, followed by its eventual decline and abandonment. In some cases, a single mine might experience more than one such cycle over the *longue dureé*. In the cases documented in this volume, what accounts for the timing of these cycles often seems to be a function of local socioeconomic forces, rather than broad pan-regional cultural or evolutionary trends. In the case of the hematite mine in Nasca, for example, there is meager evidence for initial use in late Early Horizon times, followed by intensification during the first few centuries of the first millennium; the latter corresponds to the apogee of the Nasca ceremonial center of Cahuachi, a site known for the abundance of fine polychrome ceramics bearing images of temple's religious cosmology. As Vaughn and his colleagues hypothesize, it is likely that the demand for a high-quality hematite used in the fine clay slips in Early Nasca ceramics at Cahuachi and other sites probably goes a long way to explaining the timing of the fluorescence of the Mina Primavera mine. It is intriguing and significant, however, that the exploitation of the mine decreased precipitously in Late Nasca times with the decline of Cahuachi and only rebounded weakly during the Middle Horizon, despite evidence at the mine of pottery styles often associated with Huari. Even though the Huari state had the power to mobilize large amounts of labor for public ends, as exemplified by sites such as Pikillaqta, this did not have any direct impact on the exploitation of the Mina Primavera hematite mine.

The general observation of the relevance of local economic forces in interpreting the life histories of mines and quarries raises the frequently mentioned issue of the relation of quarries and mines to Prehispanic elites and government rulers. In the Andes, the image of Tahuantinsuyu as a totalitarian state controlling a wide range of resources and activities, including the mining of precious metals, has had a profound impact on archaeological interpretation of pre-Inca archaeological remains. While this image of Inca power was initially based on Spanish historical accounts, it has received some archaeological support.

Judging from the chapters in this volume, the Incas did have a direct and deep involvement in mining activities, and not just of precious metals. As the chapters on the exploitation of copper ores at sites in the Atacama Desert of northern Chile (Chap. 12) and of chryscolla and opaline silica at Los Infieles Mining Complex in north-central Chile (Chap. 9) vividly demonstrate, Inca impact was powerful and unprecedented at these mineral deposits. At Los Infieles the Inca presence involved the imposition of buildings planned according to Inca architectural canons, the supply of food and provisions to those involved in the mining, the introduction of Inca pottery forms and designs to symbolize the role of the Inca state as the provider of food and drink in the contexts of public feasting associated with the mines, and the promotion of Inca state religious practices and paraphernalia at the mines. Salazar and his colleagues made similar observations about the Inca presence at late prehistoric mining sites in Atacama. Significantly, they also note Inca construction of small posts in strategic locations in order to control goods and people coming in and out of the mining complex. These findings not only strengthen confidence in the descriptions of Inca state involvement chronicled by Sancho de la Hoz and other early Spanish writers (Petersen 2010: 41–43) but also provide an enriched appreciation of the strategies and tactics implemented by Inca administrators as they expanded their political and military power into the mineral-rich southern Andes.

Yet it is dangerous to apply the Inca model to earlier times. The empire of the Incas appears to be an anomalous phenomenon in the Prehispanic history of the Andes. Few if any earlier states appear to have exercised coercive power to the degree practiced by the Incas. It is noteworthy that none of the examples of quarrying and mining discussed in this volume show evidence of direct state involvement or control prior to the Incas. Nor, with the exception of the account of Tiwanaku stone quarries (Chap. 4), do they seem to require positing the involvement of the large numbers of workers that we associate with Inca mining and quarrying. On the contrary, most of the mines and quarries described here seem to indicate the participation of small numbers of workers using relatively simple tools and simple methods that changed little over time.

In fact, as more time is spent exploring quarries and mining areas, the model of "elite control" comes increasingly hard to accept in most cases. For example, the obsidian mining pits documented at Quispisisa are spread over a 90-hectare area, and the zone in which obsidian can be quarried is much larger (Chap. 2; see also Burger and Glascock 2002; Tripcevich and Contreras 2011). The other major obsidian sources in Peru, Chivay in the Colca Valley and Alca in the Cotahuasi canyon, are also massive (Tripcevich and Mackey 2011; Rademaker 2012). Kurt Rademaker has estimated that the Alca source extends over an area that is roughly 320 km^2. It is probably significant that none of the recent surface surveys carried out at Peru's three major obsidian sources has encountered state buildings or a government administrative presence of any kind, despite the excellent preservation of these obsidian deposits areas and the exploitation of these deposits for over 10,000 years.

The same observations could be made about the Prehispanic exploitation of salt deposits. The famous salt source at Cerro de Sal in the Tarma area discussed in passing by Jennings and his colleagues (Chap. 6) ran for 16.5 km. Given the scale of these deposits, it is hard to imagine how a local *curaca* or state administrator could "restrict access" to these deposits. Many of the stone, salt, copper, and other deposits utilized in prehistoric times were similarly too large to control, except, perhaps, with the kind of hegemonic power possessed by an imperial state such as Tahuantinsuyu. Even at salt deposits covering a more limited area, such as the salt production from brine at San Blas in Junin or the quarrying of salt blocks along central Huallaga in San Martin, archaeological surveys and excavations have failed to find any evidence of state involvement in salt extraction during Prehispanic times (DeBoer 1984; Morales 1977).

Given my doubts about the applicability of the Inca model to pre-Inca mining and quarrying, it is interesting that so many ethnographic observations in this volume document the role of communal organization of extractive activities. As the chapter describing the extraction of salt from the Huarhua Salt Mine vividly illustrates, salt was traditionally considered an open resource and salt production was organized and carried out effectively by the salt makers themselves (Chap. 6); a similar pattern seems to have characterized salt production at Maras in Cuzco.

Indeed, judging from the contributions to this volume, the assumption that large numbers of workers or a high degree of complexity needs to be posited for Prehispanic quarries where considerable material has been removed is simply unwarranted. For example, it is clear from the case of the red hematite mine in northern Chile exploited between 11,200 and 11,500 BP that the extraction of large quantities of minerals does not require a complex sociopolitical structure or a well-developed economic system. As documented at the San Ramón 15 Mine (Chap. 7), Preceramic hunters, gatherers, and fisherfolk were responsible for extracting some 3,000 metric tons of rock during the Early Holocene. After processing the hematite and goethite, this mass of rocks would have ultimately produced about 300 tons of red and yellow pigment. Given this accomplishment by the very ancient small-scale groups of northern Chile, it not unreasonable to attribute many of the later mining activities in the Andes to the communal or individual efforts linked to modestly sized and relatively unstratified rural communities.

Since the seminal work of John V. Murra, Andean archaeologists and ethnographers have had a special fascination with the possibility of multi-ethnic involvement in the extraction of scarce raw materials, and this theme appears repeatedly in this volume (e.g., Chap. 6). Murra (1972) concluded that the interviews in the 1562 Visita de la Provincia de León de Huánuco attested to multi-ethnic salt production. Based on archaeological evidence, Daniel Morales (1977) likewise argued that his excavations at San Blas salt deposit in Junin confirmed that this pattern of multi-ethnic exploitation of minerals dated to long before Inca times. Nonetheless, despite these assertions, this model has never been rigorously tested. A convincing demonstration of this model will require larger-scale excavations and more thorough analysis of remains. The studies summarized here demonstrate that such a critical evaluation is now feasible with the appropriate field methodology.

I have already commented that one of the strengths of this volume is the focus on diachronic change in mining. Many of the mining and quarrying sites reported here show change over time in the intensity of extraction, the activities carried out at the mine or quarry, and the range of other activities associated with extractive process, such as the playing of music or the offering of exotics to the supernatural forces associated with the deposits. While direct evidence of patterns of long-term change is especially valuable, complimentary approaches have also begun to make a contribution. One promising new technique is the study of the elemental compositions in lake cores (Chap. 10). This approach is illustrated by Colin Cooke's study of mercury isotopes in the lakes close to the Huancavelica cinnabar deposits. In this study he is able not only to identify the beginning of cinnabar mining, but he is also able to document the increased intensity of extractive activities at the mines during Chavín and Inca times, followed by an even greater spike in the Colonial period (Cooke et al. 2009). The methodology used in this study is broadly applicable and will serve as a compliment to archaeological work at the mines themselves, and in some cases as the sole source of information when archaeological evidence at the mines has been badly damaged or totally obliterated.

It is also worth noting that there are many materials mined in the Prehispanic Andean world have yet to be researched. Anthracite coal, lapis lazuli, quartz crystal, and cinnabar, for example, were all valued commodities in the ancient Andes and were widely traded, but the mines and quarries from which they were extracted have yet to be located and researched. In the case of cinnabar, prior to the Spaniards, due to its strong vermillion color, it (mercury sulfide) was used for body and face painting as well as a pigment in textiles, and it appears to have also been used in the amalgamation of precious metals (Chap. 10). There is an archaeological and historic evidence that the Santa Barbara mine above the town of Huancavelica was used before the Spaniards (Burger et al. 2002), but no field research has been conducted. While the Santa Barbara deposits were heavily exploited until 1825, it is likely that evidence of the Prehispanic mining activity remains somewhere in the vast mine complex.

Georg Petersen observed over 40 years ago, "The themes of mining and metallurgy in ancient Peru were aesthetic, utilitarian and religious," and one of the richest themes in this volume is the symbolic and religious importance of mines and mining. This realization resonates with contemporary insights in Andean ethnohistory and ethnography. Carolyn Dean (2010), for example, recently has written an entire book exploring the Inca conception of stone as potentially animate, sentient, and sacred. The notion that stones or other minerals were imbued with supernatural forces appears to go back long before the Incas, judging from the stones dressed and treated as sacred at Late Preceramic sites such as Bandurria and El Paraiso (Fung 1988; Engel 1957).

In his contribution to this volume Dennis Ogburn (Chap. 3) observes that in Father Bernabe Cobo's list of the sacred places (*wakas*) in Cuzco, three of the *wakas* were stone quarries. According to Cobo, these quarries were worshipped and offerings were made to them. Archaeological documentation of offerings and remains suggesting other ritual behavior at Prehispanic mines and quarries are described in several of this volume's chapters; evidence for this has been recovered from Inca mines (Chaps. 3 and 12), but it is also found at quarries used a millennium or more before the Incas (Chap. 8). The symbolic significance of particular kinds of stone may even be crucial for understanding the selection of particular quarries or the reasons for why activity was shifted from one quarry to another, as appears to be the case for Tiahuanaco (Chap. 4). The discovery of three ancestral tombs (*chullpas*) at the Chivay source adjacent to obsidian quarries in the puna is evidence of the symbolic importance of these deposits of natural glass (Burger and Glascock 2002, 2011: 290). The discovery by Hiram Bingham III of an offering of small obsidian nodules brought to Machu Picchu from the Chivay source, hundreds of kilometers away as the condor flies, is another dramatic expression of the religious significance of materials frequently treated as "utilitarian" (Burger 2004). The documentation of *chullpas* at the Chivay source is paralleled by reports of *chullpas* at the Inca granite quarry near Ollantaytambo and a stone platform with religious offerings of Spondylus shell found overlooking the Inca turquoise mines at the San José del Abra complex in Atacama. The religious significance of mines and quarries is by no means limited to the Andes. The ancient Egyptians considered quarries to be sacred and built shrines there for worship, as in the case of turquoise quarries in the Sinai Peninsula (e.g., Valbelle and Bonnet 1996). Yet the special sensitivity that Andean archaeologists have shown to this theme is unusual and contrasts sharply with more narrow economic perspective of many Mesoamerican archaeologists (e.g., Cobean 2002).

The themes mentioned here and many others make the reading of this volume an exhilarating experience. It is my hope that these chapters will encourage a new generation of archaeologists to investigate an even broader range of quarries of mines and to delve still deeper into the technological, political, economic, and religious forces that shaped the complex life histories of Prehispanic Andean mines and quarries.

Acknowledgments I would like to thank Nico Tripcevich and Kevin Vaughn for inviting me to participate in the original session and, when I was unable to attend, to still extend an invitation to me to submit this discussion for the published volume.

References

Bird, J. (1979). The copper man, a prehistoric miner and his tools from northern Chile. In E. P. Benson (Ed.), *Pre-Columbian metallurgy in South America* (pp. 105–131). Washington, D.C.: Dumbarton Oaks Research Library and Collection.

Bonavia, D. (2010). Letter from Duccio Bonavia. In W. E. Brooks (Ed.), *Mining and metallurgy in Ancient Perú [W. E. Brooks, Trans.]* (Special Paper 467, p. xv). Boulder, Co: The Geological Society of America.

Brooks, W. E. (2010). Introduction to the English translation of mining and metallurgy in ancient Perú. In W. E. Brooks (Ed.), *Mining and metallurgy in Ancient Perú [W. E. Brooks, Trans.]* (Special Paper 467, pp. vii–ix). Boulder, Co: The Geological Society of America.

Burger, Richard L. and Ramiro M. M. (2002). Atalla: A Center on the Periphery of the Chavín Horizon. *Latin American Antiquity, 13*(2), 153–177.

Burger, R. (2004). Scientific insights into daily life at Machu Picchu. In R. L. Burger & L. Salazar (Eds.), *Machu Picchu: Unveiling the mystery of the Incas* (pp. 85–106). New Haven: Yale University Press.

Burger, R. L., & Glascock, M. (2002). Tracking the source of Quispisisa type Obsidian from Huancavelica to Ayacucho. In W. Isbell & H. Silverman (Eds.), *Andean archaeology I* (pp. 341–368). New York: Kluwer Academic/Plenum Publishers.

Calancha, Antonia de 1638 (1981). *Corónica moralizada del Orden de San Augustín en el Perú.* In Ignacio Prado Pastor (Ed.). Lima: Imprenta de la Universidad Nacional Mayor de San Marcos.

Cobean, R. H. (2002). *Un Mundo de Obsidiana. Minería y Comercio de un Vidrio Volcánico en el México Antiguo.* Mexico, D.F.: Instituto Nacional de Antropología e Historia.

Cobo, P. B. 1653 (1956). *Historia del Nuevo Mundo*, 2 vol. Madrid: Biblioteca de los Autores Españoles.

Cooke, C. A., Balcom, P. H., Biestar, H., & Wolfe, A. P. (2009). Over three millennia of mercury pollution in the Peruvian Andes. *Proceedings of the National Academy of Sciences of United States of America, 106*(22), 8830–8834.

Dean, C. (2010). *A culture of stone. Inka Perspectives on Rock.* Durham: Duke University Press.

DeBoer, W. R. (1984). *Archaeological reconnaisance in the Central Huallaga, Department of San Martín, Northeastern Peru.* New York: Laboratory of Archaeology, CUNY, Queens College.

Engel, F. (1957). Sites et établissements sans céramique de la Côte péruvienne. *Journal de la Société des américanistes NS, 46*, 67–155.

Fung, R. (1988). The late Preceramic and Initial Period. In R. Keatinge (Ed.), *Peruvian prehistory* (pp. 67–96). NY: Cambridge University Press.

Lechtman, H. (1976). A metallurgical site survey in the Peruvian Andes. *Journal of Field Archaeology, 3*(1), 1–42.

Lechtman, H. (1980). The Central Andes: Metallurgy without iron. In T. A. Wertime & J. D. Muhly (Eds.), *The coming of the age of iron* (pp. 267–334). New Haven: Yale University Press.

Morales, D. (1977). *Excavaciones en Las Salinas de San Blas. Seminario Arqueológico (pp. 27–48). Seminario de Historia Rural Andina*. Lima: Universidad Nacional Mayor de San Marcos.

Murra, J. V. (1972). El "control vertical" de un máximo número de pisos ecológicos en la economía de las sociedades andinas. In J. V. Murra (Ed.), *Visita de la Provincia de León de Huánuco (1562), Iñigo Ortiz de Zuñiga,Visitador* (Vol. I, pp. 429–476). Huánuco, Peru: Universidad Hermilio Valdizán.

Patterson, T. C. (1995). *Towards a Social History of Archaeology in the United States*. Ft. Worth, Texas: Harcourt, Brace and Company.

Petersen, C. R. (2010). Prologue: Georg Petersen Gaulke, 1898–1985. In G. Petersen (Ed.), *Mining and metallurgy in Ancient Perú [W. E. Brooks, Trans.]* (Special Paper 467, pp. xiii–xiv). Boulder, CO: The Geological Society of America.

Petersen, G. (1970). *Minería y Metalurgia en el Antiguo Perú* (Arqueológicas 12). Pueblo Libre, Lima, Perú: Museo Nacional de Antropología y Arqueología.

Petersen, G. (2010). *Mining and metallurgy in Ancient Peru [W. E. Brooks, Trans.]* (Special Paper 467). Boulder, CO: The Geological Society of America.

Rademaker, K. (2012). *Early human settlement of the high-altitude Pucuncho basin, Peruvian Andes* (Unpublished Ph.D. dissertation). Department of Anthropology and Quaternary Studies, University of Maine, Orono

Trigger, B. (1994). The coming of age of the history of archaeology. *Journal of Archaeological Research, 2*(1), 113–136.

Tripcevich, N., & Contreras, D. (2011). Quarrying evidence at the Quispisisa obsidian source, Ayacucho, Peru. *Latin American Antiquity, 22*, 121–136.

Tripcevich, N., & Mackey, A. (2011). Procurement at the Chivay obsidian source, Arequipa, Peru. *World Archaeology, 43*(2), 271–297.

Valbelle, D., & Bonnet, C. (1996). *Le sanctuaire d'Hathor, maîtresse de la turquoise: Sérabit el-Khadim au Moyen Empire*. Paris: Picard and Aoste.

Chapter 16
Discussion: Mineral Resources and Prehispanic Mining

Izumi Shimada

My discussion primarily addresses the second half of this book (Chaps. 7–14) that presents case studies of Prehispanic mining, the process of extracting a wide range of minerals from the earth, commonly metallic ores for smelting and decorative purposes. Richard Burger comments on the first half that relates more closely to quarrying, a type of mining that in archaeology commonly refers to the extraction of dimension, or architectural, stones for construction and decorative purposes and non-crystalline and microcrytalline minerals and rocks for tool making. My comments are thematically organized and written from the vantage of personal knowledge and experience garnered over 3 decades on diverse aspects of the integrated process of Prehispanic mining and metallurgy in the Andes, particularly on the north coast of Peru.

Preliminaries: Why Has not Prehispanic Mining Received the Attention It Deserves?

This book effectively reminds us that Andean civilization past and present has been inseparable from mining and quarrying. These activities and their products have played integral parts during much of the human existence in the Andes up to this day. The early Holocene mining of hematite and goethite (reddish pigments) at San Ramón 15, a site dated to cal. 10,200 and 11,500 BP that Salazar et al. (Chap. 7) have documented, is a striking case in point. Dynamic and complex geological processes have given birth to impressive mineral diversity and wealth throughout much of the Andes, particularly in the highlands of the modern nations of Bolivia, Chile, and Peru. Today Chile and Peru rank among the largest global producers of economically

I. Shimada (✉)
Southern Illinois University, Carbondale, IL, USA
e-mail: ishimada@siu.edu

N. Tripcevich and K.J. Vaughn (eds.), *Mining and Quarrying in the Ancient Andes*, Interdisciplinary Contributions to Archaeology, DOI 10.1007/978-1-4614-5200-3_16,

important metals as copper, gold, lead, silver, tin, and zinc (see e.g., Cavanagh 2011; Cavanagh and Glover 1991; Slotta and Schnepel 2011). Contributed chapters in this book discuss additional metallic ores (hematite and mercury). Not surprisingly, there is a close correlation between the availability of appropriate ores and the development of sophisticated, versatile, and innovative metallurgical traditions in the Prehispanic Andes (e.g., Lechtman 1976, 1979, 1988). The geological processes also account for the availability of a multitude of igneous dimension stones (ranging from andesite, diorite, granite, and rhyolite) that were extensively utilized by the Tiwanaku and Inca for their justly famed constructions and stelae (see Chaps. 3 and 4).

Given the long-standing and widespread importance that mining and quarrying and their products have held in the Andes, why have not they received much attention from archaeologists? This is the question I have pondered for many years. Was it the perceived rarity of Prehispanic mines? In her influential work, Lechtman (1976) did not identify conclusively (clearly she was very cautious) any Prehispanic mines during her macro-scale mine survey. Subsequently, Oehm (1984: 27) concluded that the Cerro Blanco copper mine adjacent to the town of Batán Grande in the mid-La Leche Valley on the north coast of Peru (Shimada 1994; Shimada and Craig in press; Shimada et al. 1982) was the only credible Prehispanic mine in Peru identified up to that point in time. Were Prehispanic mines truly as rare as these conclusions implied?

These questions and conclusions are even more baffling given that scientific investigation into Prehispanic Andean mining had an early and propitious start (see below) and Andean archaeology has had a long-standing interest in paleoenvironmental reconstruction and natural resource management (Shimada and Vega-Centeno 2011).

True, there are various practical difficulties that hamper our efforts as discussed in Chap. 13 by Van Gijseghem et al. and others (e..g., Eerkens et al. 2009; Lechtman 1976; Núñez 1999; Shimada 1994) such as disturbances or even obliteration of ancient mines by more recent, often larger scale exploitations. Owners jealously control access to their mines and surrounding areas by fencing and/or armed guards, effectively preventing us from examining them. Mineralized areas where mines are likely to be found are in rugged, mountainous terrains outside of typical archaeological study areas, thus requiring separate surveys. In addition, the widespread political violence during the 1980s and 1990s clearly constrained our ability to conduct surveys in such areas in Peru (Shimada and Vega-Centeno 2011). From an Old World perspective, Bloxam (2011: 149) lists other factors that have contributed to ancient quarries and mines being neglected by archaeologists and the broader public: their material remains are generally mundane, nonmonumental, and often scattered over large areas and "are still rather peripheral to archaeological research as they are often seen in a limited, usually technological, context." The last factor is particularly influential and relates to what I consider to be a basic, conceptual problem that has hindered advances in archaeological investigation of ancient mines. I consider it below by means of a summary discussion of earlier pioneering studies.

Antonio Raimondi (1826–1890), a natural historian/scientist from Milan, Italy, who arrived in Peru in 1850, laid a sound empirical foundation by publishing valuable detailed information and observations from his many field trips throughout Peru (accompanied by numerous meticulously prepared maps and other illustrations) in the first three volumes of the encyclopedic series on the geology (especially mineralogy and mining) of Peru that he conceived and named *El Perú* (Raimondi 1874, 1876, 1880; Villacorta 2004). His 1878 publication (Raimondi 1878) offers a rather complete inventory of mineral sources and resources of Peru. His effort to properly contextualize mines by discussing their social and technological dimensions as well as challenges they face (e.g., Raimondi 2004 [1887]) still serves as a standard against which to measure our modern anthropology of mining and other technologies.

A generation later, the fortuitous discovery of a remarkably well-preserved Prehispanic miner, the so-called Copper Man, inside a collapsed copper mining shaft at Chuquicamata in what is today the Far North Coast of Chile in 1899 (Bird 1979) and at least one other well-preserved, mummified corpse, presumably of miner, at the same mine (Mead 1921; see detailed discussion by Figueroa et al. in press) provided valuable information and insights into labor intensive, manual mining techniques using simple equipment that included hafted stone hammers, hoes or shovel-like hafted tools, and tightly woven baskets.

These important developments, however, did not spawn follow-up studies until the 1970 publication of *Minería y Metalurgia en el Antiguo Perú* (2010 [1970]) by the German-born Peruvian geologist, Georg Petersen. Although many archaeologists did not properly appreciate it, its importance to the subject at hand lies in the fact that he embraced a very broad vision of mining—from the identity and locations of numerous minerals utilized to mining methods and tools. The temporal–spatial and topical coverage of his book closely matches that of this book, except for the social and symbolic dimensions. He attempted to integrate (1) mining and metallurgy, (2) field and laboratory works, and (3) mineralogical, archaeological, historical, and ethnographic data. His conceptual holism (a nascent form of which was already seen in the works of A. Raimondi) in treating mining and metallurgy as phases of a single productive process is noteworthy.

Subsequent pan-Central Andean survey mostly of then-known and accessible ancient mines and metallurgical sites and systematic and detailed archaeometric analyses of collected samples by Heather N. Lechtman (1976) represented an important expansion and refinement of the line of investigation that Raimondi (1876, 1878) initiated and Petersen (1970) refined. Lechtman (1976, 1988) provided much empirical support of the logical connection between locally available ores and smelted metals thereby effectively reminding us of the importance of examining mining and metallurgy (or more broadly, different facets and stages of a given production system) as an integrated whole. The same systemic conception underlies what I have called a "holistic approach to craft production" (Shimada 1985, 1994; Shimada and Craig in press: Shimada and Wagner 2007) that entails a balanced, interdisciplinary investigation of the raw materials (including their acquisition), production technology, organization and personnel, and utilitarian and symbolic use and significance of products. In essence, it is an attempt to broadly contextualize production systems.

The above, integrated conception seems logical enough, but in actuality, I believe many archaeologists treat raw material acquisition and the subsequent stages of the production (including products) as analytically distinct and independent, *prioritizing* attention on the *finished (more familiar) products*. While Andean archaeological literature is replete with studies dealing with crafting, its products and their roles in the political economy, studies that give due attention to raw material acquisition (i.e., mining-quarrying) as *an* integral *part of a production process* are quite rare. I argue that it is this conceptual problem, the disarticulated treatment of the production process and system, together with other factors mentioned above, that largely explain why mining and quarrying have not received the attention they deserved from archaeologists.

The preceding discussion was intended to provide an intellectual context and basis for assessing the significance of this and one other new publication that focus much deserved attention to Prehispanic mining and quarrying. The latter is an upcoming special issue of the Chilean journal, *Chungará*, presenting the proceedings of the First International Meeting on Prehispanic Mining in the Americas held in Chile in November, 2010. These publications effectively showcase a new generation of archaeological investigations of varying scale and intensity exploring diverse aspects of mining. They reinforce the point that the presumed rarity of Prehispanic mines reflects not true paucity but insufficient archaeological attention (Shimada 1994), particularly on the south coast of Peru and in the adjacent South-Central Andes (the Titicaca Basin and the adjoining *altiplano* and coastal valleys) that have long been known for their mineral wealth, innovative and advanced metallurgy, and/or broader cultural developments. With multiple, concurrent projects working in these areas, we anticipate important synergetic advances and testing and refinement of existing views (e.g., Lechtman and Macfarlane 2005). Encouragingly, many of the projects in this volume are keenly aware that mining is only the first stage and facet of what is in reality a multistage, multifaceted, and multisite production system and have accordingly adopted a long-term, multi or interdisciplinary and regional approach (see below).

Methodological Challenges and Solutions

This volume focuses on "archaeological research at primary deposits of raw materials extracted through mining or quarrying in the Andean region." Although it is not one of its stated goals and archaeological study of Prehispanic mining is still young in terms of formalization, sophistication, and intensity, we need to better resolve various basic methodological challenges.

Locating Prehispanic Mines

How do you locate and date ancient mines? These are the challenges faced by all contributors. As for the first question, some contributors are explicit about the methods they employed. Vaughn et al. (Chap. 8) and Van Gijseghem et al. (Chap. 13), for example, both relied heavily on information culled from modern local informal and itinerant miners. The turquoise and chrysocolla mines discussed by Salazar et al. (Chap. 12) and Cantaturri (Chap. 9), respectively, were discovered prior to their investigations and thus the pertinent discovery methods not mentioned. For Reindel et al. (Chap. 14), mine surveys and investigations were an integral part of their ambitious coast-highland transect of Nasca and Palpa drainages aimed "to reconstruct settlement shifts as influenced by climatic variations … . " Mine location was not an end in and of itself; mining and their products were instead conceived as a means of documenting the "changing patterns of movement of people and goods over time." Discoveries of mines of different periods described in other chapters, on the other hand, were the result of more focused archaeomineralogical surveys.

Given that archaeologists rarely have a comprehensive, in-depth knowledge of the natural resources of their study areas (Shimada 1998), local informants such as miners, hunters, and herders are the most accessible sources of pertinent information (see Shimada and Craig in press). At the same time, as with any informants, we should be mindful that all are not equally knowledgeable, reliable or open-minded, or willing to share what they know. It is prudent to interview a range of informants and show them good specimens of what you seek.

It is surprising that local information sources are not complemented by independent and/or larger, regional scale approaches such as the search for mineralization sources. On coastal Peru, eroded Cretaceous igneous intrusions that mineralized the surrounding areas can be easily recognized by their conical form (often called *panes de azúcar*) either on remote-sensing images and/or in the field. Accordingly, specific areas can be targeted for intensive mine surveys. "Floats" or ore fragments in dry streambeds that indicate minerals of interest originating somewhere upstream can be used to narrow the survey area (Shimada and Craig in press). While many authors of this volume have adopted a broad landscape approach to Prehispanic mining particularly its symbolic dimension, we should consider the possibility that Prehispanic miners recognized the critical connection between readily recognizable mineralization sources (e.g., igneous intrusions) and the occurrence of desired minerals. Such sources could well have been considered *huacas* for their form and association with valued minerals.

For locating mines, elsewhere I have emphasized the value of identifying and tracing roads that head to rugged, uninhabited mountainous terrain (Shimada and Craig in press). This approach offers the important advantage of allowing us to identify mines, campsites, and production sites at termini and along the way that operated synchronously. Although questions surrounding the transport of dimension stones concern Ogburn (Chap. 3) and Janusek et al. (Chap. 4), no one discusses them in regard to extracted ores, presumably because of the small scale of mining. These issues, however, may be pertinent for the period of intensified and larger scale mining in the Ica valley mentioned by Van Gijseghem et al. (Chap. 13). Vacant enclosures along the roads leading to Prehispanic mines in the Batán Grande area have been interpreted as corrals for llamas used in ore transport and as an indication of intensity and organization involved in mining (Shimada and Craig in press).

As all the mining chapters note, *preserved and securely dated* Prehispanic mines—regardless of the mineral extracted—were generally small in scale and shallow in depth with no winzes (vertical or steeply inclined shafts between different levels for access and/or ventilation) or stopes (step-wise excavations to extract ore from steeply inclined veins; Shimada and Craig in press). Miners also carefully and selectively traced high-grade veins resulting often in irregular trenches and pits and relatively small accumulations of gangues and tailings. The small size of gangues and their relatively limited weathering distinguish them from the surrounding landscapes. This point, in combination with sinuous access roads, irregular prospection and mining pits and trenches, and associated artifacts and constructions, can help us identify ancient mines. There is no single effective means of discovering Prehispanic mines and, in essence, I recommend a multipronged approach that utilizes combinations of the aforementioned methods.

The small size of identified prehistoric mines does not mean, however, that there were no large Prehispanic mines. It is more likely that mines with large, high quality ore deposits did exist but were subsequently reworked to the point we cannot no longer ascertain their Prehispanic origins. Consider the case of the impressive Chuquicamata copper mine (until recently, the world's largest open-pit copper mine) in the Atacama region of north Chile and the fortuitous preservation and discovery of Prehispanic miners and their tools discussed earlier. We should consider the distinct possibility that mines that have been securely dated to the Prehispanic era only present us with *a truncated vision* of the true variability and that the economically, politically, and symbolically most important mines have been lost to our study.

Speaking of the scalar variability of Prehispanic mines, Van Gijseghem et al. (Chap. 13) adopt a typology proposed by Eerkens et al. (2009) that consists of "prospecting sites," "extraction sites" (mines and mining camps), and "production sites." As with any typology, it imposes some sense of the existing variability and order out of the data amassed, but also raises some issues. For example, a relatively small pit that in terms of modern mining standards might represent a mere prospecting pit may well have been an extraction site that was initiated but soon abandoned for some reason. In addition, an extraction site close to a production site could well have been worked on a daily commuting basis and may not been associated with any "mining camps." Reindel et al. (Chap. 14) in their coast-highland transect of the Nasca-Palpa drainage point to the rarity of mining camps. The distinction between the first two categories can be nebulous and should not be based on size or the presence/absence of associate structures. The quality, extent, and ease of extraction of the ores should be considered, together with the correspondence between the extracted minerals and what were actually utilized in the associated production site. New data from the Inca site of El Abra adjoining a turquoise mine (Salazar et al. [Chap. 12])—what might be classified as a mining camp in this typology—suggest that at least some lapidary work using gold sheets and mined turquoise may have been conducted there, blurring the mining camp-production site distinction. Use of this typology requires caution. Our understanding of ancient mines, regardless of their scale and intensity and the identity of minerals extracted, *cannot be adequately achieved without basic knowledge of associated production site(s)*. In the case of Sicán copper-arsenic production, the placement of smelting sites appears to have been primarily dictated by their proximity to mines supplying the necessary ores and secondarily, to a labor force and fuels (Shimada and Merkel 1991; Shimada and Craig in press). Importantly, well-defined roads connected mines directly with some of the documented smelting sites. While logistical and financial constraints make it difficult to concurrently search for both mines and production sites, the contributors to this volume must not lose sight of the integrated nature of mines and production sites.

Dating Mines

Reliable dating of Prehispanic mines is a persistent and challenging task due to: (1) the limited range and number of diagnostic artifacts (probably reflecting the basic simplicity of the mining tool kit and nature and status of personnel involved) and datable organic remains; (2) disturbance from later human exploitation; (3) the effects of taphonomic processes including downslope movement and animal activity (e.g., bats). Fire-setting or -quenching to aid in ore extraction would leave datable charcoal and soot, but the technique appears not to have been employed in Prehispanic mining (see comments by Reindel et al. [Chap. 14]; also Shimada and Craig in press).

As with locating Prehispanic mines, their dating is also best approached in a multipronged manner (Shimada and Craig in press). The presence or absence of abandoned or broken stone hammers (as detailed by Salazar et al. [Chap. 7] in their discussion of the Archaic hematite mine of San Ramón 15), straight audits (horizontal entrances to mines), stopes and winzes, as well as the distinct marks that steel tools leave on walls are ways of distinguishing Prehispanic from Hispanic/modern mining. For more precise dating, we have to rely on associated artifactual and architectural remains. An extension of this method is to examine features found along associated roads. A dating method that may be productively employed in dating ancient mines that do not have associated artifactual or architectural remains is optically stimulated luminescence, or OSL, dating (Aitken 1998; Murray and Olley 2002), that can date unexposed sand that may have accumulated inside or at the mouths of mines. The method measures the amount of ionizing radiation that sand grains emit to estimate the last time they were exposed to light.

Research Methods and Organizations

Analytical Methods

This volume, with a few exceptions, does not concern itself much with analytical methods used in mineralogical or chemical compositional determination of extracted minerals. Janusek and his colleagues (Chap. 4) summarize their use of the ever more popular portable X-ray florescence spectrometer complemented by X-ray diffraction analysis. Schultze (Chap. 11) passingly speaks of scanning electron microscopy (energy-dispersive X-ray spectroscopy?) in her discussion of evidence for silver extraction in the northern Lake Titicaca Basin. Reindel et al. (Chap. 14) mention their use of lead isotope analysis for provenience and their distribution study of metal objects in the Nasca-Palpa drainage. The same technique would be very much applicable to Schultze's (Chap. 11) study of silver mining and the political economy of its products as its purification requires the mixing of silver-bearing with lead containing ores.

Chapter 10 deserves careful attention as it is built upon the results of inductively coupled plasma (ICP) analysis. Brooks and Schwörbel (Chap. 10) summarize known recovery and concentration methods in the Andes past and present, including panning of placer gold and mercury amalgamation-*refogado* processing (burning to volatilize mercury) of vein gold that is often finely disseminated in quartz. It is for the recovery of the latter type that amalgamation is widely employed throughout the world. In support of their claim for pre-Incaic mercury amalgamation, the authors adduce varied lines of evidence including ethnohistorical accounts. Pre-Incaic use of cinnabar, reddish mercuric sulfide (HgS), as a pigment has long been known and recent isotopic analysis of sediments from a lake near the famed Huancavelica mercury mine (Cooke et al. 2009) suggests cinnabar mining started around 1400 BC R. Larco (2001: 135) emphatically argues that the Mochica (aka Moche; ca. AD 100–750/800) already knew and employed this method, but his evidential basis is quite weak.

Their case, however, rests primarily on ICP results of pre-Incaic gold alloy sheet metal objects from Peru and Colombia. They argue that the significantly lower mercury contents detected by ICP on these samples (<100 ppm with an average of ca. 10 ppm) in comparison to naturally occurring mercury levels on modern placer gold samples (i.e., 1,000 to 10,000 ppm) "can only be explained by heating and volatilization of the mercury" presumably used for amalgamation.

Their explanation is quite plausible, but not entirely convincing yet. As the authors acknowledge, the placer gold samples they used as the comparative baseline may have been contaminated with mercury from modern gold mining and processing upstream. It would be useful to have a comparison of surface and interior mercury contents by means of a microprobe analysis of cross-sections of both gold and sheet metal samples. Additionally, we cannot ignore the potential effects of alloying and sheetmetal preparation on the mercury contents of the analyzed sheet metal samples, which are gold–silver–copper–arsenic alloys (see Table 10.3 in Chap. 10 by Brooks and Schwörbel). The extent of volatilization and lowering of mercury concentrations from numerous cycles of heating and hammering conducted with furnaces readily capable of achieving temperatures over 1,100 °C (Shimada et al. 2007) should be properly evaluated. Experimental testing is urged.

This volume also offers additional new data on *pre-Inca* gold mining that has received far less attention than the resultant gold *objects*. Reindel et al. (Chap. 14) briefly address gold mining on the south coast, noting that in Nasca times gold was clearly the desired ore to be mined and that placer gold was not available on the south coast (Stöllner 2009). These views are reinforced by the findings by Van Gijseghem et al. (Chap. 13) at Mina Zurita in the Ica valley that seems to date to Nasca 1/Proto Nasca. The technological and organizational details of gold extraction at these sites, however, are not yet available.

Productive Research: Multi- Versus Inter-Disciplinary Approach

"The archaeologist who works alone is a species in danger of extinction," so declared Killick (2008: 58) some year ago to impress upon us that the knowledge and expertise required to conduct cutting edge archaeological research had expanded beyond the capacity of a single individual archaeologist and that a well-integrated collaboration with relevant specialists was essential. While Andean archaeology appears to be prospering, this prophetic statement cannot be readily dismissed particularly for those interested in the exploitation and management of natural resources and their broader significance. In fact, we should ask whether our research on various aspects of Prehispanic mining is productively organized and conducted with pertinent expertise and knowledge. The authorship and the contents of constituent chapters indicate that there is a good deal of variation in the manner and extent to which specialists of diverse fields have participated in attaining results reported.

Most chapters are the result of some degree of *multi*disciplinary collaboration with one or more specialists in some facets and stages of the research, often postfieldwork examination or analysis of samples collected by archaeologists who may not have been properly trained to undertake the task alone. What we should strive for, however, is maximum integration of specialists, ideally from the conception and design of research to field implementation and onto the final publication of results. Bringing to bear different but complementary perspectives, knowledge and expertise to the widest range of research activities, particularly in fieldwork, is what is meant by an *inter*-disciplinary approach (also called trans- or cross-disciplinary) that offers valuable synergy and a stimulating in situ question-and-answer and self-corrective process (Shimada 2011). These are far from original recommendations (see e.g., Buikstra 1991), but deserve to be repeated as they are often not heeded.

In this regard, two projects stand out: those of the "Andean transect" by Reindel et al. (Chap. 14) and Mina Primavera by Vaughn et al. (Chap. 8). The former is unique in having the unparalleled long-term and wide-ranging (including "archaeo-scientific") support of the German Archaeological Institute (DAI), a federal research institution dedicated to comparative studies of ancient civilizations worldwide. The Nasca-Palpa project forms part of a global study of the transformation from sedentism to complex society and focuses on "trans-disciplinary investigation" of the interrelationship among settlement, economy, and environment (Reindel and Wagner 2009). Thus, their investigations of mines and quarries are nested within two broader sets (regional and international) of research issues and aims and conducted by archaeologists working closely in and out of the field with appropriate specialists (two co-authors, Stöllner and Gräfingholt are members of other research institutes, the German Mining Museum and the Institute of Archaeologic Sciences [Ruhr University] both in Bochum). Although their chapter has a preliminary character, with their trans-disciplinary character, we can anticipate in the near future a much more comprehensive picture of regional mines and quarries, their products and significance. In essence, they should be able to properly *contextualize* mining and quarrying through their investigation into the broader environmental, sociopolitical, and technological conditions and processes. They have already shown a significant correlation between the prevailing climatic conditions and settlement pattern; i.e., coastal occupation is emphasized during periods of relative humidity while the occupation shifts toward higher elevations during periods of relative dryness. Whether mining and quarrying activities mirror this pattern is not clear.

The complementary Mina Primavera project directed by Vaughn in the adjacent Ingenio valley is an outgrowth of his "Early Nasca Craft Economy" project (2002–2007). These two projects have a clear logical thread as the mine was discovered in the process of locating sources of raw materials used in the manufacture of well-known polychrome Nasca ceramics. The Mina Primavera project is thus clearly focused on a comprehensive understanding of the technology, organization, and significance of the Nasca crafts. It is similar to the Nasca-Palpa project in applying an interdisciplinary approach and aiming to contextualize the hematite mine that was primarily exploited during the Early Nasca phase. Vaughn and his colleagues (Chap. 8) make a strong case of interconnection between mining, on the one hand, and the ebb and flow of Cahuachi as the primary Nasca ceremonial-civic center together with the concurrent shift of emphasis from polychrome textiles to polychromic slip-painted ceramics that extensively utilized hematite, on the other. Hematite exploitation at the mine declined significantly after this period. The authors should test the quality of the hematite that remains in the mine to dispel the possibility that its disuse owes to the qualitative decline or exhaustion of hematite.

As expected, contributing chapters of the book present a wide range of analytical methods and approaches that, in general, await refinement and formalization.

Interpretive Models and Interpretations

Chaîne opératoire (Operational Sequence) and Technological Choice

The editors of this volume justly observe that, "Regardless of the materials that originate in mines and quarries, many of the technological considerations that an ancient miner must face are shared." Solutions to the shared tasks and challenges have been described in terms of the two related concepts of *chaîne opératoire* and technological choice formalized by Leroi-Gourhan (1993; also see Dobres and Hoffman 1994; Edmonds 1990; Schlanger 1994; Sellet 1993) and Lemmonier (1992, 1993; see also Sillar and Tite 2000), respectively. As used in this volume, the concepts typically refer to specific technical steps and sequences and the associated decision-making processes by which raw materials such as clays, ores, and microcrystalline minerals are chosen and physically and/or chemically shaped and transformed into desired products (artifacts) and used. The documented steps and sequence, in turn, form the basis to examine the factors that shaped or guided specific decisions and steps (e.g., inherent limitations and potential of the raw materials selected, and the expectations of users and social value of finished products) as well as interrelationships between steps, pertinent equipment, knowledge and skill, and other issues. They can chart the directions of our inquiries at different levels of abstraction and map specific technical steps and cognitive processes of their social actors, including the symbolic significance of their own acts and products.

In this book, for example, Tripcevich and Contreras (Chap. 2) discuss how obsidian knapping choices relate to intensity of use, while Roddick and Klarich (Chap. 5), from an ethnoarchaeological perspective, examine factors shaping the selection of raw materials for pottery production. Salazar et al. (Chap. 7) discuss the criteria for selecting appropriate rocks for making stone hammers for Archaic hematite mining. Ogburn (Chap. 3) and Janusek et al. (Chap. 4) describe the inherent material and symbolic factors that may have influenced the selection of dimension stones. Schultze (Chap. 11) combines excavation, settlement survey, archaeometric, ethnohistoric, and other lines of evidence to propose a partial *chaîne opératoire* for silver extraction.

These and other discussions of the *chaîne opératoire* and technological choices in this volume are in general preliminary or partial in character and based only on one or a few case studies. An exception is the similar operational sequences (five stages) of chrysocolla mining activities that Cantarutti (Chap. 9) documented at five mining areas in the Los Infieles area of central Chile. They are, however, based on limited or no excavations of primary extraction and/or production sites that are situated at different points of the *chaîne opératoire.* Clearly, the nature and plausibility of inferences at different levels of abstraction as well as the interconnections between technical and social factors and acts hinge on how securely and thoroughly the specific technical steps and sequences of the *chaîne opératoire* were established. I find many of the behavioral and organizational inferences found in this book necessitate *logical leap-frogging.* Various chapters speak of the significance of copper, silver, and chrysocolla to the political economy and prestige economy of associated polities such as the Tiwanaku, Ica, and Inca without adequately establishing such critical factors as the organization and management of their production sites, productive outputs, and the contexts and manners in which products were utilized. The social and symbolic values of raw materials and products cannot be assumed based on their presumed rarity, labor cost, and other variables (see e.g., Helms 1993).

The establishment of a *chaîne opératoire* and discussion of technological choices need to be based more on pertinent empirical data from primary context excavations and archaeometric and experimental testing of excavated evidence (Shimada and Wagner 2007; Shimada and Craig in press) than on inferences or assumptions built on ethnohistoric, ethnographic, and/or modern material scientific data. A solid empirical basis is essential in coping with interpretive problems stemming out of the principles of multiple solutions (that there are multiple solutions to any given problem), equifinality (that a given end state [product] can be reached by different potential means and/or pathways), and functional equivalence (that two or more items that differ in specific aspects can serve basically the same function or have the same basic meaning). Detailed documentation of specific technical steps and sequences and an in-depth understanding of the limitations and potential of raw materials are prerequisites to understanding why specific solutions or pathways were selected.

Formulation of a well-documented, working model of craft production with its material, behavioral, social, and ideological correlates and ramifications such as the one illustrated here (Fig. 16.1) requires a thorough contextual understanding of the

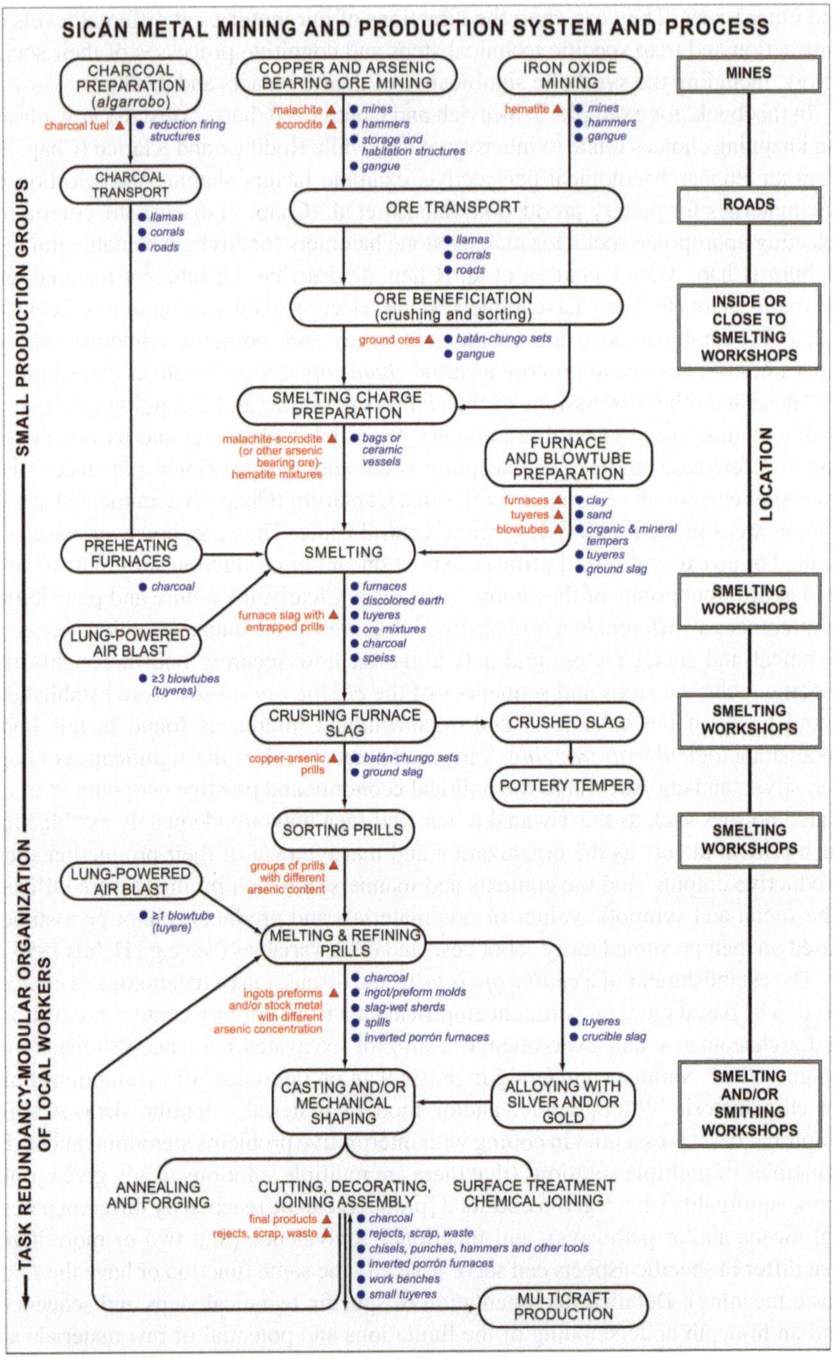

Fig. 16.1 Working model of Sicán metal mining and production system and process. Prepared by Izumi Shimada and Steve Mueller

production. Such understanding, in turn, results from long-term, interdisciplinary, regional research that actively searches and focuses on multiple sites that will illuminate the different facets and stages of a given production process and system (Shimada and Wagner 2007; Shimada and Craig in press).

In spite of the above criticism concerning the preliminary and/or partial understanding of the *chaîne opératoire*, there are various projects that hold much promise for establishing a comprehensive and referential *chaîne opératoire*. Silver smelting and concurrent multimineral extraction in Puno Bay in the northern Titicaca Basin that Schultze (Chap. 11) describes is such a case. The area seems to have already provided varying quantities and qualities of evidence that together suggest the presence of much, if not all of the *chaîne opératoire* for Prehispanic silver smelting. Just as exciting is the distinct possibility that there are sites relatively close to each other (or even within same sites) that together form a case of multicraft production or co-production (Shimada 2007), a type of "synergistic co-evolution … between a variety of mineral industries" and perhaps even copper and silver production. Multicraft production refers to the "concurrent practice of multiple crafts by different individuals or groups, each specialized in one or more craft, in the same or in a series of adjacent spaces," whereas co-production describes a form of multicraft production "in which artisans specializing in different crafts … collaborate in the design and manufacture of products...." (Shimada 2007: 5–6).

To fulfill the potential this area holds, however, Schultze (Chap. 11) needs to improve the dating of mines and other pertinent contexts, and better establish the association of key material evidence through excavations, among other tasks. The specific steps and sequences involved in silver purification should be empirically established rather than simply applying the knowledge and understanding of historical or modern silver extraction methods. In-field collaboration with pertinent specialists should also help identify relevant mineral and metallurgical remains. Only then, will her discussion of another type of synergistic co-evolution, that of co-evolution between ritual practices and mineral craft industries, as well the political economic significance of the products of these crafts be convincing.

Ritual and Symbolic Significance of Mines and Minerals: Lo Andino (Invariable Andean Beliefs and Worldviews)?

Following documentation of ritual offerings of *Spondylus princeps* at Prehispanic mines in the Batán Grande area and an associated smelting site (Shimada 1994: 54; Shimada and Merkel 1991; Shimada and Craig in press), I urged more attention be paid to the ritual and symbolic aspects of mining and associated metallurgical activities (Shimada in press). Salazar et al. (Chap. 12), Cantarutti (Chap. 9), and Vaughn and his colleagues (Chap. 8) show that ritual activities at and the significance of mines cross-cut time, space, and the nature of the mineral extracted (turquoise, chrysocolla, and hematite). In fact, most chapters discuss not only the sacred character and ritual significance of mines, but also how they relate inseparably to the landscape and in process acquire importance.

Most contributors seem to have been inspired by ethnohistoric accounts of the Incaic conception of sacred landscape and, to a lesser degree, by Ingold's (2000) dwelling perspective of the landscape. For example, Salazar et al. (Chap. 12) and Cantarutti (Chap. 9) discuss Inca exploitation of turquoise and chrysocolla in northern and central Chile, respectively. In the process, they argue for the critical role of Inca state religion and rituals in displaying and legitimizing their appropriation and restructuring of the physical and symbolic (sacred geography) resources of each area. Their arguments are based on inferred rituals atop the nearby ridge or nearby elevations and "wide spread Andean beliefs" or ethnohistoric accounts of the Inca's conception of the sacred landscape.

The zeal with which some authors pursued the documentation and interpretation of the ritual significance of mines, particularly by invoking Inca and ethnographic beliefs and ritual practices raises serious methodological and theoretical concerns. While I strongly endorse their exploration into the ritual, symbolic, and sociopolitical dimensions of mines and mining, I find it problematical that Van Gijseghem and his colleagues (Chap. 13) directly apply "general principles on mining and landscape in the Andes" culled from ethnohistoric and ethnographic information from Peruvian and Bolivian highlands to the Early Nasca and later Ica situations in the Ica valley. They point out that the "extraction of a mountain's substance" is conceived as "a profound transgression," a violation of mountain's sanctity, and that "mining is fundamentally dangerous" physically and supernaturally. They cite ethnographic accounts of beliefs that unpredictable and powerful spirits or devils (*supay*) reside in mines and that miners' and mine owners' (or political leaders') must make offerings to minimize accidents and other risks that accompany mining.

To what extent do these beliefs reflect their Prehispanic and pre-Inca counterparts, particularly given the fact that many of the preserved and known Prehispanic mines were relatively small, and nearly always shallow, open pits posing little risk to the miner? Of course, the "Copper Man" described earlier does indicate that Prehispanic mining was by no means risk-free. At the same time, historical mining that went to the unprecedented depths starting early in the colonial era (e.g., Bakewell 1984; Young 1994) brought concomitant risks to miners and constant struggles for improved working conditions. Have beliefs in mines and mining remained unchanged in spite of significant changes in its practices and numerous accidents and deaths over 450 years? The invariability of beliefs and practices or "general principles" needs better substantiation.

Their accompanying discussion of metal symbolism and the argument that "the social and political capital that comes with metal manufacture, ownership, and exchange" (applied to the Late Intermediate Ica) are again based on extrapolation of data and views from different areas and times (heavily based on the Incaic situation; e.g., Lechtman 1993, 2007). Unfortunately, there is a minimum of data for and understanding of *local* Ica metallurgical production and associated symbolism to judge if the Incaic situation is indeed applicable here. To infer socially and historically specific symbolic behaviors by reference to those of other cultures is a problematical approach. Overall, I urge greater caution in applying ethnohistorical and ethnographic information and insights.

Conclusions

Nearly 2 decades have passed since I wrote a critical overview of Prehispanic Andean mining and metallurgy. This book is strong testimony for how many advances have been made since then in archaeological investigation of Prehispanic Andean mining. Various of the recommendations and criticisms I made back then—for example, broadening of our conception of mining and minerals to be studied (Shimada 1994: 67) and attention to the ritual and symbolic aspects of mining (Shimada in press)—are now being addressed. This volume showcases many of the new generation Andean and Andeanist researchers and their emerging results. I can state emphatically that the subject of Prehispanic Andean mining is finally coming to receive attention it deserves. Contrary to the widespread tendency in archaeometallurgy to focus on the technical analysis of finished metal products, this book is definitely broader and anthropological in its aims, the issues it targets, and the analytical and interpretive models it adopts. Particularly notable in regard to the last is its use of the *chaîne opératoire* and technological choice concepts, and attempt to illuminate how mining and minerals were embedded in other aspects of Prehispanic Andean life including the physical and symbolic landscapes and political economy of social elites involved. The last point serves to remind us of the inherent danger of the reductionist/rationalist thinking that seems widespread in regard to the strategies for acquiring minerals (including those used for lithics). At least Inca practices tell us that concern with time and labor costs were often secondary to political and symbolic factors.

At the same time, there remain many unresolved issues and tasks. For example, while it is very welcome that attention is now being paid to a wide range of minerals, we (more likely, our archaeological colleagues) need to better educate ourselves so that minerals are properly analyzed and/or correctly identified (e.g., sodalite misidentified as lapis lazuli and/or turquoise distinguished from chrysocolla). In addition, we need to continue searching for Prehispanic sources of diverse minerals as has been done for obsidian. For cinnabar, there appears to have been at least one important pre-Hispanc source in the south highlands of Ecuador (Loma Gashuin near Azogues; Truhan et al. 2005). Where are the sources for the turquoise and chrysocolla that were extensively utilized for decorative purposes in pre-*Inca* times? In other words, we still face the basic task of improving our knowledge of mineral sources.

Reflecting the recent nature of many of the research projects represented, this book has a definite preliminary character. Likewise, the methods and broader approaches employed can and should be refined and formalized so that research aims can be effectively attained. Although it is not within the purview of the book (or the studies represented within) to deal with the entire production process and system (e.g., mining-metallurgy) including products and their use and significance, its focus on "primary evidence for raw material extraction" has resulted in a series of tenuous and highly speculative visions of how mining and quarrying articulate with other aspects of Prehispanic Andean life, particularly world views and political economy.

The sort of holistic vision and comprehensive *chaîne opératoire* I have sought to define takes many years of focused, interdisciplinary teamwork. What I find exciting about this book is that many of its contributors are ready to pursue, if not already well into pursuing, similar goals and making lasting contributions to Prehispanic Andean craft production studies and beyond. I hope the criticisms and suggestions presented here help in this pursuit.

Acknowledgments I thank the editors Kevin Vaughn and Nicholas Tripcevich for their invitation to serve as a discussant of their symposium at the 2011 annual meeting of the Society for American Archaeology in Sacramento and this volume. I am grateful to Melody J. Shimada for providing me with valuable comments and editorial assistance on a draft of this paper.

References

Aitken, M. J. (1998). *An introduction to optical dating: The dating of quaternary sediments by the use of photon-stimulated luminescence*. Oxford: Oxford University Press.

Bakewell, P. (1984). Mining in colonial Spanish America. In L. Bethell (Ed.), *The Cambridge history of Latin America* (Colonial Latin America, Vol. II, pp. 105–152). Cambridge: Cambridge University Press.

Bird, J. (1979). The "Copper Man": A prehistoric miner and his tools from Northern Chile. In E. P. Benson (Ed.), *Pre-Columbian metallurgy of South America* (pp. 105–132). Washington, DC: Dumbarton Oaks.

Bloxam, E. (2011). Ancient quarries in mind: pathways to a more accessible significance. *World Archaeology, 43*(2), 149–166.

Buikstra, J. E. (1991). Out of the appendix and into the dirt: Comments on ten years of bioarcheological research. In M. L. Powell, A. M. Mires, & P. Bridges (Eds.), *What mean these bones? Studies in Southeastern bioarchaeology* (pp. 172–188). Tuscaloosa: University of Alabama Press.

Cavanagh, J. (2011). *Las Minas del Perú—Top mining companies in Peru 2009–2011*. Lima: Peru Top Publications S.A.C.

Cavanagh, J., & Glover, M. (1991). *Mines and miners in Peru/Las Minas del Perú*. Lima: Peru Top Publications S.A.C.

Cooke, C. A., Balcom, P. H., Biestar, H., & Wolfe, A. P. (2009). Over three millennia of mercury pollution in the Peruvian Andes. *Proceedings of the National Academy of Science of the United States of America, 106*(22), 8830–8834.

Dobres, M.-A., & Hoffman, C. (1994). Social agency and the dynamics of prehistoric technology. *Journal of Archaeological Method and Theory, 3*, 211–258.

Edmonds, M. M. (1990). Description, understanding and the chaîne opératoire. *Archaeological Reviews from Cambridge, 9*(1), 55–70.

Eerkens, J. W., Vaughn, K. J., & Linares Grados, M. (2009). Pre-Inca mining in the Southern Nasca Region, Peru. *Antiquity, 83*(321), 738–750.

Figueroa, V., Manríquez, G., Salazar, D., Salinas, H., Núñez-Regueiro, P., & Aspillaga, E. (in press). *Pre-Hispanic miners from the Atacama Desert (Region of Antofagasta, Chile): Archaeological and bioanthropological evidence*. Special issue of *Chungará*. Selected papers of the first international meeting on Prehispanic mining in the Americas edited by Diego Salazar, Valentina Figueroa, Victoria Castro and Germán Manríquez.

Helms, M. W. (1993). *Craft and the kingly ideal: Art, trade, and power*. Austin: University of Texas Press.

Ingold, T. (2000). *The perception of the environment. Essays on livelihood, dwelling, and skill*. London: Routledge.

Killick, D. (2008). Archaeological science in the USA and in Britain. In A. Sullivan (Ed.), *Archaeological concepts for the study of the Cultural Past* (pp. 40–64). Salt Lake City: University of Utah Press.
Larco, R. (2001). *Los Mochicas* (Tomo II). Lima: Museo Arqueológico Rafael Larco Herrera.
Lechtman, H. N. (1976). A metallurgical site survey in the Peruvian Andes. *Journal of Field Archaeology, 3*, 1–42.
Lechtman, H. N. (1979). Issues in Andean metallurgy. In E. P. Benson (Ed.), *Pre-Columbian metallurgy of South America* (pp. 1–40). Washington, DC: Dumbarton Oaks.
Lechtman, H. N. (1988). Traditions and styles in Central Andean metalworking. In R. Maddin (Ed.), *The beginning of the use of metals and alloys* (pp. 344–378). Cambridge, MA: MIT Press.
Lechtman, H. N. (1993). Technologies of power: The Andean case. In J. S. Henderson & P. J. Netherly (Eds.), *Configurations of power: Holistic anthropology in theory and practice*. Ithaca/London: Cornell University Press.
Lechtman, H. N. (2007). The Inka, and Andean metallurgical tradition. In R. Burger, C. Morris, & R. Matos Mendieta (Eds.), *Variations in the expression of Inka Power* (pp. 313–355). Washington DC: Dumbarton Oaks.
Lechtman, H. N., & Macfarlane, A. (2005). La metalurgia del bronce en los Andes Sur Centrales: Tiwanaku y San Pedro de Atacama. *Estudios Atacameños, 30*, 7–27.
Lemonnier, P. (1992). *Elements for an anthropology of technology* (Anthropological papers, Vol. 88). Ann Arbor: Museum of Anthropology, University of Michigan.
Lemonnier, P. (1993). Introduction. In P. Lemonnier (Ed.), *Technological choices: Transformation in material cultures since the neolithic* (pp. 1–35). London/New York: Routledge.
Leroi-Gourhan, A. (1993). *Gesture and speech [Anna Bostock Berger, Trans., from French]*. Cambridge, MA: The MIT Press.
Mead, C. W. (1921). Prehistoric mining in Western South America. *Natural History, 21*(5), 453–456.
Murray, A. S., & Olley, J. M. (2002). Precision and accuracy in the optically stimulated lumniscence dating of sedimentary quartz: a status review. *Geochronometria: Journal on Methods and Applications of Absolute Chronology, 21*, 1–16.
Núñez, L. (1999). Valoración minero-metalúrgica circumpuneña: menas y mineros para el Inka rey. *Estudios Atacameños, 18*, 177–221.
Oehm, V. P. (1984). *Investigaciones sobre Minería y Metalurgia en el Perú Prehispánico: Una Visión Crítica Actualizada* (Bonner Amerikanistische Studien, Vol. 12). Bonn: Universitat Bonn.
Petersen, G. G. (1970). *Minería y metalurgia en el antiguo Perú* (Arqueológicas, Vol. 12). Lima: Museo Nacional de Antropología y Arqueología.
Petersen, G. G. (2010 (1970)). *Mining and metallurgy in ancient Peru [Brooks, W. E., Trans.]*. Boulder, CO: The Geological Society of America.
Raimondi, A. (1874). *El Perú, Parte Preliminar* (Tomo I). Lima: Imprenta del Estado.
Raimondi, A. (1876). *El Perú, Historia de la Geografía del Perú* (Tomo II). Lima: Imprenta del Estado.
Raimondi, A. (1878). *Minerales del Perú o Catálogo Razonado de una colección que representa los principals tipos de minerals de la República con muestras de huano y restos de aves que lo han producido*. Lima: Imprenta del Estado.
Raimondi, A. (1880). *Perú, Historia de la Geografá del Perú* (Tomo III). Lima: Imprenta del Estado.
Raimondi, A. (2004[1887]). Minas de Oro del Perú. In *Colección Estudios Geológicos y Mineros para la obra "El Perú"*, Vol. I, compiled by Villacorta, L. F. (pp. 93–159). Compañía de Minas Buenaventura, Asociación Educacional Antonio Raimondi, Fondo Editorial de la Universidad Nacional Mayor de San Marcos, Lima.
Reindel, M., & Wagner, G. A. (Eds.). (2009). *New technologies in archaeology. Multidisciplinary investigations in Nasca and Palpa, Peru*. Berlin: Springer.

Schlanger, N. (1994). Mindful technology: Unleashing the *chaîne opératoire* for an archaeology of mind. In C. Renfrew & E. Zubrow (Eds.), *The ancient mind: Elements of cognitive archaeology* (pp. 143–151). Cambridge: Cambridge University Press.

Sellet, F. (1993). Chaîne opératoire: the concept and its applications. *Lithic Technology, 18*, 106–112.

Shimada, I. (1985). Perception, procurement and management of resources: Archaeological perspective. In S. Masuda, I. Shimada, & C. Morris (Eds.), *Andean ecology and civilization* (pp. 357–399). Tokyo: University of Tokyo Press.

Shimada, I. (1994). Pre-hispanic metallurgy and mining in the Andes: Recent advances and future tasks. In A. C. Craig & R. C. West (Eds.), *In quest of mineral wealth: Aboriginal and colonial mining and metallurgy in Spanish America* (Geoscience and man, Vol. 33, pp. 37–74). Baton Rouge: Louisiana State University.

Shimada, I. (1998). Bienes Exóticos y Metodología Arqueológica: ¿Qué Tan Viables Son Nuestros Datos Sobre Interacción Interregional. In F. Cárdenas-Arroyo & T. L. Bray (Eds.), *Intercambio y Comercio entre Costa, Andes y Selva: Arqueología y Etnohistoria de Suramérica* (pp. 223–254). Bogotá: Universidad de los Andes.

Shimada, I. (2007). Introducción. In I. Shimada (Ed.), *Craft production in complex societies: Multi-crafting, sequential production, and producers* (pp. 1–21). Salt Lake City: University of Utah Press.

Shimada, I. (2011). Hacia una mejor integración de la arqueometría y la arqueología andina: El caso de Pachacámac. In Vetter, L., Vega-Centeno, R., Petrick, S., & Olivera, P. (Eds.), *Actas del II Congreso Latinoamericano de Arqueometría. Lima, 19 al 21 Octubre de 2009* (pp. 29–54). Fondo Editorial de la Universidad Nacional de Ingeniería, Lima. In press assessment and prospect. Special issue of *Chungará.* Proceedings of the first international meeting on Prehispanic mining in the Americas, Taltal-San Pedro de Atacama, Chile. 29 Nov to 5 Dec.

Shimada, I. (in press). Assessment and prospect. Special issue of *Chungará.* Selected papers of the first international meeting on Prehispanic mining in the Americas edited by Diego Salazar, Valentina Figueroa, Victoria Castro and Germán Manríquez.

Shimada, I., & Craig, A. K. (in press). *The style, technology and organization of Sicán mining and metallurgy, Northern Peru: insights from holistic study.* Special issue of *Chungará.* Selected papers of the first international meeting on Prehispanic mining in the Americas edited by Diego Salazar, Valentina Figueroa, Victoria Castro and Germán Manríquez.

Shimada, I., Epstein, S. M., & Craig, A. K. (1982). Batán Grande: a prehistoric metallurgical center in Peru. *Science, 216*, 952–959.

Shimada, I., Goldstein, D. J., Wagner, U., & Bezúr, A. (2007). Pre-Hispanic Sicán furnaces and metalworking: Toward a holistic understanding. In R. Lleras Pérez (Ed.), *Metalurgía en la América Antigua: Teoría, Arqueología, Simbología y Tecnología de los Metales Prehispánicos* (Travaux de l'Institut Français d'Études Andines, Vol. 253, pp. 337–361). Lima: Fundación de Investigaciones Arqueológicas Nacionales, Banco de la Republica, Bogotá and Instituto Frances de Estudios Andinos.

Shimada, I., & Merkel, J. F. (1991). Copper alloy metallurgy in ancient Peru. *Scientific American, 265*, 80–86.

Shimada, I., & Vega-Centeno, R. (2011). Peruvian archaeology: Its growth, characteristics, practice and challenges. In L. Lozny (Ed.), *Comparative archaeologies* (pp. 569–612). New York: Springer.

Shimada, I., & Wagner, U. (2007). Craft production on the pre-Hispanic North Coast of Peru: A holistic approach and its results. In J. Skibo, M. Grave, & M. Stark (Eds.), *Archaeology as anthropology: Theoretical and methodological approaches* (pp. 163–197). Tucson: University of Arizona Press.

Sillar, B., & Tite, M. (Eds.). (2000). Technological choices in ceramic production. *Archaeometry, 42*(1).

Slotta, R., & Schnepel, I. (Eds.). (2011). *Schätze del Anden: Chiles Kupfer für die Welt.* Bochum: Deutschen Bergbau-Museums Bochum.

Stöllner, T. (2009). Gold in Southern Peru? Perspectives of research into mining archaeology. In M. Reindel & G. A. Wagner (Eds.), *New technologies for archaeology: Multidisciplinary investigations in Palpa and Nasca, Peru* (pp. 393–407). Berlin: Springer.

Truhan, D., Burton, J. H., & Olsen Bruhns, K. (2005). *El cinabrio en el mundo andino* (Vol. 18, pp. 193–206). Cuenca, Ecuador: Revista de Antropología.

Villacorta, L. F. (2004). Antonio Raimondi, context histórico de su labor geológica y minera. In *Colección Estudios Geológicos y Mineros para la Obra "El Perú"*, Vol. I, compiled by L. F. Villacorta (pp. 19–66). Lima: Compañía de Minas Buenaventura, Asociación Educacional Antonio Raimondi, Fondo Editorial de la Universidad Nacional Mayor de San Marcos.

Young, O. E., Jr. (1994). Black legends and silver mountains: Spanish mining in colonial Latin America reconsidered. In A. C. Craig & R. C. West (Eds.), *In quest of mineral wealth: Aboriginal and colonial mining and metallurgy in Spanish America* (Geoscience and man, Vol. 33, pp. 109–118). Baton Rouge: Louisiana State University.

MIX
Papier aus verantwortungsvollen Quellen
Paper from responsible sources
FSC® C105338

If you have any concerns about our products,
you can contact us on
ProductSafety@springernature.com

In case Publisher is established outside the EU,
the EU authorized representative is:
Springer Nature Customer Service Center GmbH
Europaplatz 3, 69115 Heidelberg, Germany

Printed by Libri Plureos GmbH
in Hamburg, Germany